Civil Engineering and Symmetry

Civil Engineering and Symmetry

Special Issue Editors

Edmundas Kazimieras Zavadskas
Romualdas Bausys
Jurgita Antucheviciene

MDPI • Basel • Beijing • Wuhan • Barcelona • Belgrade

MDPI

Special Issue Editors

Edmundas Kazimieras Zavadskas
Vilnius Gediminas Technical University
Lithuania

Romualdas Bausys
Vilnius Gediminas Technical University
Lithuania

Jurgita Antucheviciene
Vilnius Gediminas Technical University
Lithuania

Editorial Office
MDPI
St. Alban-Anlage 66
4052 Basel, Switzerland

This is a reprint of articles from the Special Issue published online in the open access journal *Symmetry* (ISSN 2073-8994) from 2017 to 2019 (available at: https://www.mdpi.com/journal/symmetry/special_issues/Civil_Engineering_Symmetry).

For citation purposes, cite each article independently as indicated on the article page online and as indicated below:

LastName, A.A.; LastName, B.B.; LastName, C.C. Article Title. *Journal Name* **Year**, *Article Number*, Page Range.

ISBN 978-3-03921-002-2 (Pbk)
ISBN 978-3-03921-003-9 (PDF)

Contents

About the Special Issue Editors

Edmundas Kazimieras Zavadskas, Ph.D., DSc, Professor at the Department of Construction Management and Real Estate, Chief Research Fellow at the Laboratory of Operational Research, Research Institute of Sustainable Construction, Vilnius Gediminas Technical University, Lithuania. He received his Ph.D. in Building Structures (1973), Dr. Sc. (1987) in Building Technology and Management. He is a member of Lithuanian and several foreign Academies of Sciences, Doctore Honoris Causa from the Poznan, Saint Petersburg and Kiev Universities, and Honorary International Chair Professor in the National Taipei University of Technology. Awarded by the International Association of Grey System and Uncertain Analysis (GSUA) for his huge input in the Grey System field, he was elected to Honorary Fellowship of International Association of Grey System and Uncertain Analysis, a part of IEEE (2016), awarded by "Neutrosophic Science—International Association" for distinguished achievement in neutrosophics, conferred an honorary membership (2016), and awarded the Thomson Reuters certificate for access to the list of the most highly cited scientists (2014). A highly Cited Researcher in the field of Cross-Field (2018), recognized for exceptional research performance demonstrated by production of multiple highly cited papers that rank in the top 1% by citations for field and year in Web of Science. Main research interests: multicriteria decision-making, operations research, decision support systems, multiple-criteria optimization in construction technology and management. Over 460 publications in Clarivate Analytic Web of Science, h = 55, a number of monographs in Lithuanian, English, German, and Russian. Editor in Chief of journals *Technological and Economic Development of Economy* and *Journal of Civil Engineering and Management*, as well as Guest Editor of over twenty Special Issues related to decision making in engineering and management.

Romualdas Bausys, Dr. Sc, Professor at the Department of Graphical Systems, Vilnius Gediminas Technical University (VGTU), Faculty of Fundamental Sciences, Lithuania. Education: Civil Engineer at Vilnius Civil Engineering Institute, now Vilnius Gediminas Technical University, Vilnius, Lithuania in 1982; Dr. Sc. Vilnius Gediminas Technical University, Vilnius, Lithuania in 2000. Research interests: multimedia processing, multicriteria decision making, information technologies. Over 50 publications in Clarivate Analytic Web of Science and Scopus.

Jurgita Antucheviciene, Ph.D., Professor at the Department of Construction Management and Real Estate at Vilnius Gediminas Technical University, Lithuania. She received her Ph.D. in Civil Engineering in 2005. Her research interests include multiple-criteria decision-making theory and applications, sustainable development, construction technology and management. Over 90 publications in Clarivate Analytic Web of Science, h = 23. A member of IEEE SMC, Systems Science and Engineering Technical Committee: Grey Systems, and of two EURO Working Groups: Multicriteria Decision Aiding (EWG—MCDA) and Operations Research in Sustainable Development and Civil Engineering (EWG—ORSDCE).

symmetry

MDPI

Editorial

Civil Engineering and Symmetry

Edmundas Kazimieras Zavadskas [1,2] , **Romualdas Bausys** [3] **and Jurgita Antucheviciene** [1,*]

1   Department of Construction Management and Real Estate, Vilnius Gediminas Technical University,
    Sauletekio al. 11, LT-10223 Vilnius, Lithuania; edmundas.zavadskas@vgtu.lt
2   Institute of Sustainable Construction, Vilnius Gediminas Technical University, Sauletekio al. 11, LT-10223
    Vilnius, Lithuania
3   Department of Graphical Systems, Vilnius Gediminas Technical University, Sauletekio al. 11, LT-10223
    Vilnius, Lithuania; romualdas.bausys@vgtu.lt
*   Correspondence: jurgita.antucheviciene@vgtu.lt; Tel.: +370-5-274 -5233

Received: 2 April 2019; Accepted: 2 April 2019; Published: 5 April 2019

Abstract: A topic of utmost importance in civil engineering is finding optimal solutions throughout
the life cycle of buildings and infrastructural objects, including their design, manufacturing, use,
and maintenance. Operational research, management science, and optimisation methods provide a
consistent and applicable groundwork for engineering decision-making. These topics have received
the interest of researchers, and, after a rigorous peer-review process, eight papers have been
published in the current special issue. The articles in this issue demonstrate how solutions in civil
engineering, which bring economic, social and environmental benefits, are obtained through a
variety of methodologies and tools. Usually, decision-makers need to take into account not just a
single criterion, but several different criteria and, therefore, multi-criteria decision-making (MCDM)
approaches have been suggested for application in five of the published papers; the rest of the papers
apply other research methods. The methods and application case studies are shortly described further
in the editorial.

Keywords: multiple-criteria decision-making (MCDM); hybrid MCDM; fuzzy sets; rough sets;
D numbers; 3D modelling; image processing; experimental testing; civil engineering; manufacturing
engineering; transportation; logistics

---

1. Introduction

A topic of utmost importance in civil engineering is finding optimal solutions throughout the
life cycle of buildings and infrastructural objects, including their design, manufacturing, use, and
maintenance. Operational research, management science, and optimisation methods provide a
consistent and applicable groundwork for engineering decision-making. Real-world decision problems
are usually solved by applying a multi-criteria decision making (MCDM) framework, which means that
decisions are constructed by considering multiple criteria or points of view and taking them into account.
Therefore, MCDM has become a universal tool for the solution of real-world problems. The MCDM
method review, performed by Mardani et al. [1], distinguished 15 fields of real-world problems: Energy,
environment and sustainability, supply chain management, material selection, quality management,
geographic information systems, construction and project management, safety and risk management,
manufacturing systems, technology and information management, operations research and soft
computing, strategic management, production management, and tourism management.

The evolution of the MCDM methods has been directed to take into account the uncertainty of the
initial information. Due to the different application areas, modern decision-making solutions quite
frequently include linguistic valuations of the different aspects of the considered alternatives. This
information type is characterised by not-strictly-defined meanings. Symmetry-based techniques play

quite an important role in considering systems involving uncertainty in the information. Considerable research concerning decision-making has been performed by applying neural networks, fuzzy logic, and interval numbers. The success of these approaches relies on the fact that all these methods are derived through utilising the appropriate symmetries. Therefore, different fuzzy approaches have been proposed to model this type of information. For the most popular MCDM methods, such as DEMATEL, PROMETHEE, TOPSIS, AHP, ANP, VIKOR, COPRAS, ARAS, and WASPAS, fuzzy extensions have been proposed [2–4].

Neuro-fuzzy systems have been proposed to cover more complicated formulations of decision-making problems. A comprehensive review, concerning numerous innovation aspects in neuro-fuzzy systems and the applications of these systems in various real-life issues, is presented in [5].

Intensive research has been performed in order to extend their capabilities, concerning the more accurate modelling of the uncertain and vague initial information in decision-making problems. The various "fuzzy" approaches have been proposed to model different aspects of the information uncertainty. Neutrosophic sets, recently introduced by Smarandache, opened up new possibilities for representing the uncertain and inconsistent information encountered in decision-making formulations [6]. Fuzzy-rough sets have been applied in various fields, such as expert systems, knowledge discovery, information system, inductive reasoning, intelligent systems, data mining, pattern recognition, decision-making, and machine learning [7].

Additionally, fuzzy sets have been intensively applied in decision-making problems modelled within the aggregation operator framework. Various aggregation operators under different fuzzy sets are reviewed in [8].

The methodological aspects of the decision-making problems in civil engineering, concerning the combination and integration of fuzzy and probabilistic models to deal with the uncertainties, were discussed in [9].

Research into the development of new MCDM methods has been directed towards hybrid MCDM approaches. The most popular hybrid MCDM methods demonstrate advantages over the traditional ones in solving complicated problems, which involve stakeholder preferences, interconnected or contradictory criteria, and uncertain environments. The evolution of the new hybrid MCDM approaches, such as multiple rule-based decision-making (MRDM), which can be characterized by relevant knowledge for supporting systematic improvements based on influential network relation maps (INRM), has been studied in [10–12].

Extensive reviews dedicated to the application of the MCDM methods in different fields of human activity are presented in [13–17]. Applications in areas such as transportation, supplier evaluation and selection, the tourism and hospitality industries, service quality evaluation, and the circular economy in the context of the supply chains are discussed in these publications. The particular aspects concerning cultural heritage object preservation, including economic, historical, archaeological, religious, technological, and research indicators are considered in [18]. For the solution of this problem, analytic hierarchy process (AHP) and evaluation based on distance from average solution (EDAS) methods are applied. The issue of the conceptual design of a bridge structure by a modified fuzzy Technique for Order of Preference by Similarity to Ideal Solution method, under uncertainty, is solved in [19]. The effective material selection for civil engineering objects is performed by an Analytic Hierarchy Process (AHP) and a fuzzy Multi-Objective Optimisation on the Basis of Ratio Analysis (MOORA) [20]. The distinctive features of the application of hesitant fuzzy and single-valued neutrosophic sets are taken into consideration in [21,22].

2. Contributions

The current Special Issue collects eight articles. They all are original research articles; no review articles or technical reports have been published in the current issue.

The papers contribute to decision-making techniques for civil engineering problems involving symmetric, asymmetric, or non-symmetrical information. The suggested methodologies and tools mainly include novel or extended multiple-criteria decision-making models and methods under uncertain environments. Additionally, three papers published in the current issue do not apply MCDM methods. They contribute to problems related to symmetry by offering other solution methods.

The topics of the Special Issue gained attention mostly in Europe, as well as in Asia. Thirty-four authors from eight countries contributed to the Issue (see Figure 1).

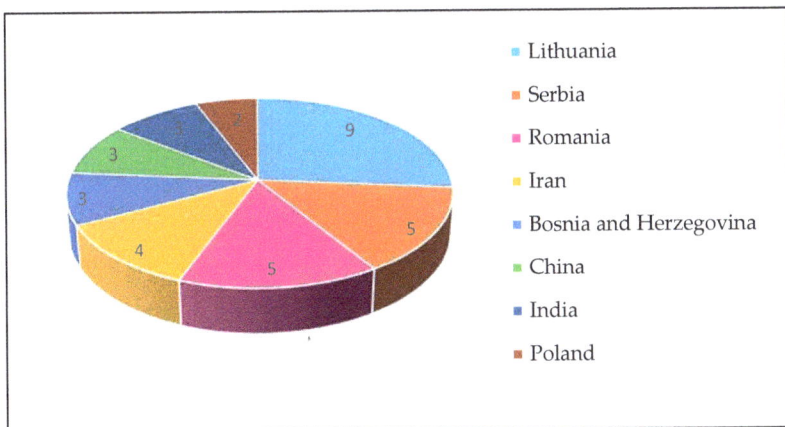

Figure 1. Distribution of authors by countries.

The distribution of papers, according to author affiliations, is presented in Table 1. Co-authors from Lithuania contributed to four papers, together with co-authors from Serbia, Bosnia and Herzegovina, India, and Iran. Authors from Poland, Rumania, and China contributed with a single paper each. Researchers from Serbia, together with co-authors from Bosnia and Herzegovina, contributed one more paper.

Table 1. Publications by countries.

Countries	Number of Papers
Poland	1
Rumania	1
China	1
Bosnia and Herzegovina–Serbia	1
Iran–Lithuania	2
India–Lithuania	1
Bosnia and Herzegovina–Serbia–Lithuania	1

The papers concerning related analysis methods or decision-making approaches are classified into several groups, as presented in Figure 2. Five of the eight papers apply MCDM methods and, mostly, they propose models and techniques under uncertain environments (i.e., fuzzy or rough models). Individual articles that do not deal with multiple-criteria decisions apply other approaches: Experimental testing, image processing, and 3D modelling.

The case studies and application examples of the proposed approaches dealing with symmetrical, asymmetrical, or non-symmetrical problems and presented in the current special issue, can be grouped into three research areas, consisting of 2,3 papers each (see Figure 3).

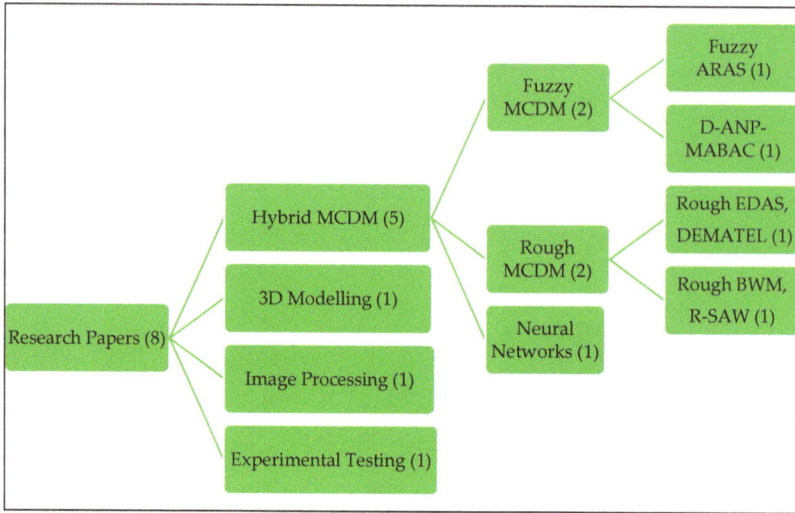

Figure 2. Decision-making approaches.

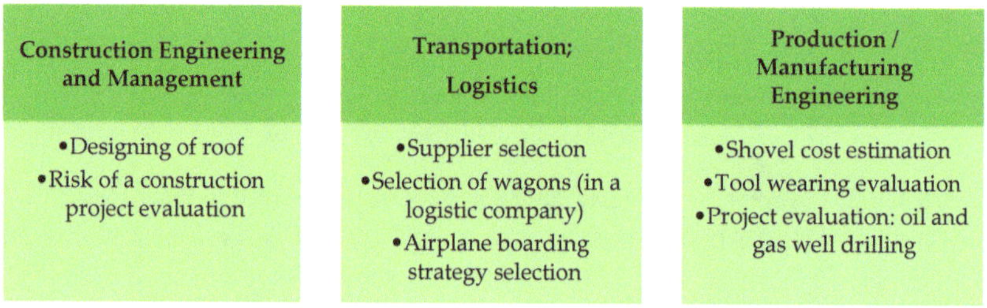

Figure 3. Research areas of case studies.

One group of papers is related to construction engineering and management. In one of the papers, the crucial problem, concerning construction management, is studied by applying a hybrid fuzzy D-ANP-MABAC model for risk evaluation of the construction projects, involving a combination of D-numbers, an analytical network process (ANP), and a multi-attributive border approximation area comparison (MABAC) method [23]. Another construction engineering paper does not apply MCDM methodology. The article aims to develop a systematic and practical approach to the early stages of the parametric design of roof shells; a compound of four concrete elements [24].

One of the most numerous research areas is Production/Manufacturing Engineering. Two papers in the area apply MCDM methodology. An oil and gas well drilling project evaluation is made by a proposed novel approach using an interval-valued fuzzy Additive Ratio Assessment (ARAS) method [25]. The paper makes a significant contribution to the literature, as an alternative method for the evaluation of this type of project. Next, for the mining industry, a hybrid multi-criteria model for shovel capital cost estimation, using multivariate regression and neural networks, is proposed [26]. One more proposal for flexible manufacturing systems is presented in a paper [27], which is aimed at a tool-wear analysis of the tool flank by applying image processing.

Another research area is Transportation and Logistics, involving three research papers. It is nice to mention that two of the papers suggest applying Rough MCDM methods. The supplier selection in a construction company is recommended to be made by utilising a combination of two extended methods: DEMATEL (Decision Making Trial and Evaluation Laboratory) for obtaining the relative weights of important criteria, and an EDAS (Evaluation based on Distance from Average Solution) method for the supplier evaluation and selection [28]. A combination of another two rough methods, rough BWM (Best–Worst Method) and rough SAW (Simple Additive Weighting), is provided for the selection of wagons in a logistics company [29]. However, the last paper from the special issue is not related to MCDM. It analyses the optimal organising of airplane passenger boarding/deboarding strategies as one of the potential possibilities to reduce the airplane turnover time by experimental testing [30].

3. Conclusions

The topics of this Special Issue have raised the interest of researchers both in Europa and in Asia; researchers from eight countries, including international collaborations, authored and co-authored papers published in this Issue.

Although multiple-criteria decision-making was one out of many announced topics, more than half of the papers (five papers out of the eight published papers) applied MCDM methods in their research. Therefore, multiple-criteria decision-making techniques proved to be well-applicable to symmetric/asymmetric information management.

Most approaches suggested decision models under uncertainty, proposing hybrid MCDM methods in combination with fuzzy or rough set theory, as well as D-numbers.

The application areas of the proposed MCDM techniques mainly covered production/manufacturing engineering, logistics and transportation, and construction engineering and management.

Author Contributions: All authors contributed equally to this work.

Acknowledgments: Authors express their gratitude to the journal Symmetry to offer an academic platform for researchers to contribute and exchange their recent findings in civil engineering and symmetry.

Conflicts of Interest: The authors declare no conflict of interest.

References

1. Mardani, A.; Jusoh, A.; Nor, K.M.D.; Khalifah, Z.; Zakwan, N.; Valipour, A. Multiple criteria decision-making techniques and their applications—A review of the literature from 2000 to 2014. *Econ. Res.-Ekon. Istraz.* **2015**, *28*, 516–571. [CrossRef]
2. Mardani, A.; Jusoh, A.; Zavadskas, E.K. Fuzzy multiple criteria decision-making techniques and applications—Two decades review from 1994 to 2014. *Expert Syst. Appl.* **2015**, *42*, 4126–4148. [CrossRef]
3. Kahraman, C.; Onar, S.C.; Oztaysi, B. Fuzzy Multicriteria Decision-Making: A Literature Review. *Int. J. Comput. Intell. Syst.* **2015**, *8*, 637–666. [CrossRef]
4. Yazdanbakhsh, O.; Dick, S. A systematic review of complex fuzzy sets and logic. *Fuzzy Sets Syst.* **2018**, *338*, 1–22. [CrossRef]
5. Rajab, S.; Sharma, V. A review on the applications of neuro-fuzzy systems in business. *Artif. Intell. Rev.* **2018**, *49*, 481–510. [CrossRef]
6. Khan, M.; Son, L.H.; Ali, M.; Chau, H.T.M.; Na, N.T.N.; Smarandache, F. Systematic review of decision making algorithms in extended neutrosophic sets. *Symmetry* **2018**, *10*, 314. [CrossRef]
7. Mardani, A.; Nilashi, M.; Antucheviciene, J.; Tavana, M.; Bausys, R.; Ibrahim, O. Recent fuzzy generalisations of rough sets theory: A systematic review and methodological critique of the literature. *Complexity* **2017**, *2017*, 1608147. [CrossRef]
8. Mardani, A.; Nilashi, M.; Zavadskas, E.K.; Awang, S.R.; Zare, H.; Jamal, N.M. Decision making methods based on fuzzy aggregation operators: Three decades review from 1986 to 2017. *Int. J. Inf. Technol. Decis. Mak.* **2018**, *17*, 391–466. [CrossRef]
9. Antucheviciene, J.; Kala, Z.; Marzouk, M.; Vaidogas, E.R. Solving Civil Engineering Problems by Means of Fuzzy and Stochastic MCDM Methods: Current State and Future Research. *Math. Probl. Eng.* **2015**, *2015*, 362579. [CrossRef]
10. Zavadskas, E.K.; Antucheviciene, J.; Turskis, Z.; Adeli, H. Hybrid multiple-criteria decision-making methods: A review of applications in engineering. *Sci. Iran.* **2016**, *23*, 1–20.

11. Zavadskas, E.K.; Govindan, K.; Antucheviciene, J.; Turskis, Z. Hybrid multiple criteria decision-making methods: A review of applications for sustainability issues. *Econ. Res.-Ekon. Istraz.* **2016**, *29*, 857–887. [CrossRef]

12. Shen, K.Y.; Zavadskas, E.K.; Tzeng, G.H. Updated discussions on "Hybrid multiple criteria decision-making methods: A review of applications for sustainability issues". *Econ. Res.-Ekon. Istraz.* **2018**, *31*, 1437–1452. [CrossRef]

13. Mardani, A.; Zavadskas, E.K.; Khalifah, Z.; Jusoh, A.; Nor, K. Multiple criteria decision-making techniques in transportation systems: A systematic review of the state of the art literature. *Transport* **2016**, *31*, 359–385. [CrossRef]

14. Keshavarz Ghorabaee, M.; Amiri, M.; Zavadskas, E.K.; Antucheviciene, J. Supplier evaluation and selection in fuzzy environments: A review of MADM approaches. *Econ. Res.-Ekon. Istraz.* **2017**, *30*, 1073–1118. [CrossRef]

15. Govindan, K.; Soleimani, H. A review of reverse logistics and closed-loop supply chains: A Journal of Cleaner Production focus. *J. Clean. Prod.* **2017**, *142*, 371–384. [CrossRef]

16. Correia, E.; Carvalho, H.; Azevedo, S.G.; Govindan, K. Maturity models in supply chain sustainability: A systematic literature review. *Sustainability* **2017**, *9*, 64. [CrossRef]

17. Mardani, A.; Jusoh, A.; Zavadskas, E.K.; Khalifah, Z.; Nor, K.M. Application of multiple-criteria decision-making techniques and approaches to evaluating of service quality: A systematic review of the literature. *J. Bus. Econ. Manag.* **2015**, *16*, 1034–1068. [CrossRef]

18. Turskis, Z.; Morkunaite, Z.; Kutut, V. A hybrid multiple criteria evaluation method of ranking of cultural heritage structures for renovation projects. *Int. J. Strateg. Prop. Manag.* **2017**, *21*, 318–329. [CrossRef]

19. Keshavarz Ghorabaee, M.; Amiri, M.; Zavadskas, E.K.; Turskis, Z.; Antucheviciene, J. Ranking of Bridge Design Alternatives: A TOPSIS-FADR Method. *Balt. J. Road Bridge Eng.* **2018**, *13*, 209–237. [CrossRef]

20. Ilce, A.C.; Ozkaya, K. An integrated intelligent system for construction industry: A case study of raised floor material. *Technol. Econ. Dev. Econ.* **2018**, *24*, 1866–1884. [CrossRef]

21. Yu, D. Hesitant fuzzy multi-criteria decision making methods based on Heronian mean. *Technol. Econ. Dev. Econ.* **2017**, *23*, 296–315. [CrossRef]

22. Bausys, R.; Juodagalviene, B. Garage location selection for residential house by WASPAS-SVNS method. *J. Civ. Eng. Manag.* **2017**, *23*, 421–429. [CrossRef]

23. Chatterjee, K.; Zavadskas, E.K.; Tamošaitienė, J.; Adhikary, K.; Kar, S. A Hybrid MCDM Technique for Risk Management in Construction Projects. *Symmetry* **2018**, *10*, 46. [CrossRef]

24. Dzwierzynska, J.; Prokopska, A. Pre-Rationalized Parametric Designing of Roof Shells Formed by Repetitive Modules of Catalan Surfaces. *Symmetry* **2018**, *10*, 105. [CrossRef]

25. Dahooie, J.H.; Zavadskas, E.K.; Abolhasani, M.; Vanaki, A.; Turskis, Z. A Novel Approach for Evaluation of Projects Using an Interval-Valued Fuzzy Additive Ratio Assessment (ARAS) Method: A Case Study of Oil and Gas Well Drilling Projects. *Symmetry* **2018**, *10*, 45. [CrossRef]

26. Yazdani-Chamzini, A.; Zavadskas, E.K.; Antucheviciene, J.; Bausys, R. A Model for Shovel Capital Cost Estimation, Using a Hybrid Model of Multivariate Regression and Neural Networks. *Symmetry* **2017**, *9*, 298. [CrossRef]

27. Moldovan, O.G.; Dzitac, S.; Moga, I.; Vesselenyi, T.; Dzitac, I. Tool-Wear Analysis Using Image Processing of the Tool Flank. *Symmetry* **2017**, *9*, 296. [CrossRef]

28. Stević, Ž.; Pamučar, D.; Vasiljević, M.; Stojić, G.; Korica, S. Novel Integrated Multi-Criteria Model for Supplier Selection: Case Study Construction Company. *Symmetry* **2017**, *9*, 279. [CrossRef]

29. Stević, Ž.; Pamučar, D.; Zavadskas, E.K.; Ćirović, G.; Prentkovskis, O. The Selection of Wagons for the Internal Transport of a Logistics Company: A Novel Approach Based on Rough BWM and Rough SAW Methods. *Symmetry* **2017**, *9*, 264. [CrossRef]

30. Qiang, S.; Jia, B.; Huang, Q. Evaluation of Airplane Boarding/Deboarding Strategies: A Surrogate Experimental Test. *Symmetry* **2017**, *9*, 222. [CrossRef]

symmetry

MDPI

Article

A Hybrid MCDM Technique for Risk Management in Construction Projects

Kajal Chatterjee [1,*] , **Edmundas Kazimieras Zavadskas** [2] , **Jolanta Tamošaitienė** [2] ,
Krishnendu Adhikary [1] and **Samarjit Kar** [1]

[1] Department of Mathematics, National Institute of Technology, Durgapur 713209, India;
   krish.math23@gmail.com (K.A.); dr.samarjitkar@gmail.com (S.K.)
[2] Faculty of Civil Engineering, Vilnius Gediminas Technical University, Sauletekio al. 11, LT-1022 Vilnius,
   Lithuania; edmundas.zavadskas@vgtu.lt (E.K.Z.); jolanta.tamosaitiene@vgtu.lt (J.T.)
*   Correspondence: chatterjeekajal7@gmail.com

Received: 24 November 2017; Accepted: 9 February 2018; Published: 13 February 2018

Abstract: Multi-stakeholder based construction projects are subject to potential risk factors due to dynamic business environment and stakeholders' lack of knowledge. When solving project management tasks, it is necessary to quantify the main risk indicators of the projects. Managing these requires suitable risk mitigation strategies to evaluate and analyse their severity. The existence of information asymmetry also causes difficulties with achieving Pareto efficiency. Hence, to ensure balanced satisfaction of all participants, risk evaluation of these projects can be considered as an important part of the multi-criteria decision-making (MCDM) process. In real-life problems, evaluation of project risks is often uncertain and even incomplete, and the prevailing methodologies fail to handle such situations. To address the problem, this paper extends the analytical network process (ANP) methodology in the D numbers domain to handle three types of ambiguous information's, viz. complete, uncertain, and incomplete, and assesses the weight of risk criteria. The D numbers based approach overcomes the deficiencies of the exclusiveness hypothesis and completeness constraint of Dempster–Shafer (D–S) theory. Here, preference ratings of the decision matrix for each decision-maker are determined using a D numbers extended consistent fuzzy preference relation (D-CFPR). An extended multi-attributive border approximation area comparison (MABAC) method in D numbers is then developed to rank and select the best alternative risk response strategy. Finally, an illustrative example from construction sector is presented to check the feasibility of the proposed approach. For checking the reliability of alternative ranking, a comparative analysis is performed with different MCDM approaches in D numbers domain. Based on different criteria weights, a sensitivity analysis of obtained ranking of the hybrid D-ANP-MABAC model is performed to verify the robustness of the proposed method.

Keywords: D number; analytical network process (ANP); multi-attributive border approximation area comparison (MABAC); multi-criteria decision-making (MCDM); consistent fuzzy preference relation (CFPR); construction project risk; risk management

1. Introduction

   In recent decades, projects in the construction sector have become more complex and risky due to the diverse nature of activities among global companies [1–8]. In comparison to other sectors, construction projects encounter more risks due to uncertainties occurring because of various construction practices, working conditions, mixed cultures and political conditions between host and home countries [9–12]. Thus, in this scenario, risk management can be considered a vital part of the decision-making process in construction projects. These projects may involve many stakeholders, in addition to uncertain socio-economic conditions at the project site, bringing big challenges to practitioners of the industry in recent decades [13,14]. Construction project failure may cause higher costs and time over-runs, requiring a systematic risk assessment and evaluation procedure to classify and respond to changes [15,16].

Thus, prioritisation among construction based risk portfolios, and finding suitable risk mitigation strategies for construction projects, can be introduced as multi-criteria decision making (MCDM) problems. Researchers have recently proposed new methods for prioritising risks in construction based projects [10,17–19]. In addition, the increasing dynamism of construction projects has resulted in extensive impreciseness and subjectivities in this risk investigation procedure. With respect to the identification of risk criteria, a methodology is needed to sort and prioritise criteria weights, based on specific environments and domain experts' judgment. In the real world, since various uncertainties occur in the decision-making process due to subjective and qualitative judgment of decision-makers (DMs), so it is essential to develop a more optimised technique that can handle various types of uncertainties [15,20–24].

In this scenario, a proper decision-making methodology is required to solve multiple conflicting interdependent criteria when evaluating risks in construction projects. In recent years, the number of papers related to analytic hierarchy process (AHP) and analytic network process (ANP) methods, considering supplier selection procedure, has increased substantially [25–29]. However, increasing the number of criteria or comparison levels causes confusion in decision makers' (DMs) judgments, resulting in incomplete decision-making and inconsistency in the evaluator's judgments, thus reducing strategy selection ability. Under these circumstances, Herrera-Viedma et al. [30] proposed consistent fuzzy preference relations (CFPR) to evade inconsistency in the decision-making process, and thus models have been developed with CFPR structuring of the problem of multi-criteria knowledge based strategy selection both with AHP [31] and ANP [32,33]. The CFPR methodology requires less time, has computational simplicity, and also guarantees consistency of decision matrices. However, as CFPR is based on complete and certain information, there always exists a possible inconsistency risk due to the inability of DMs to deal with overcomplicated objects [34].

It is obvious that the approaches mentioned above can play a vital role under some special circumstances, but it also reveals more uncertainties due to the subjective judgment of experts' assessment. For fuzzy set theory it is difficult to determine in advance membership function values before making any decision in an uncertain and vague environment [8]. Interval theory has the same deficiency [23]. In addition, the frame of discernment and basic probability assignment (BPS) present in dempster–shafer (D–S) evidence theory limits its ability to represent incomplete information in uncertain situations [34,35]. In order to overcome the above shortcomings and effectively handle various uncertain and incomplete information, D numbers [36], a special kind of random set, is applied in the construction of CFPR. The D-CFPR [34] method expresses the expert's linguistic preference values using D numbers, and can also be converted to traditional CFPR. In recent years, papers related to D numbers based MCDM methods have begun to appear, viz. D numbers based vlse kriterijumska optimizacija i kompromisno resenje (D-VIKOR) [37], D number based grey relational projection (D-GRP) [38], D numbers based analytic hierarchy process (D-AHP) [39,40], D numbers based technique for order of preference by similarity to ideal solution (D-TOPSIS) [41], and D numbers based decision-making trial and evaluation laboratory (D-DEMATEL) [42].

Besides managing the various risks associated with construction projects, this study also attempts to categorise and assess their risk mitigation strategies, thus setting up a proper framework that is accountable to investors in the construction sector. Thus, a survey on risk mitigation strategies is performed, based on the recent works [9,23,43], to alleviate construction projects risks. Based on the above considerations, this paper also develops an extended version of the MABAC methodology in an uncertain and incomplete decision environment, in order to evaluate risk mitigation strategies in construction based projects. Some papers related to MABAC have been published in recent years, viz. traditional DEMATEL-MABAC [44], pythagorean fuzzy MABAC [45], interval type 2 fuzzy MABAC [46], interval-valued intuitionistic fuzzy MABAC [47], and the fuzzy AHP-MABAC model [48].

Thus, the key motivation of this paper is to develop a D numbers based ANP-MABAC decision-making methodology for prioritising construction project risks, and to find suitable mitigating strategies in uncertain and incomplete decision environments. In this paper, we choose the ANP methodology due to its ability to represent the potential interactions, interdependences, and feedback among the risk based criteria and sub-criteria. Hesamamiri et al. [33] developed a systematic framework combining ANP and CFPR to properly assess and select a knowledge management strategy.

In this paper, CFPR is taken into the decision matrix under D-ANP to find the priority vectors of the criteria for uncertain and incomplete environments. Thus, the main objectives of this research are as follows:

- Categorise and describe proper risk issues concerning various socio-economic, technical, and geo-political based sectors correlated to construction projects.
- Develop a logical structure incorporating a D-CFPR based ANP model for risk prioritisation during construction.
- Identify and prioritise risk response factors in the construction industry based on the D-MABAC methodology.

This paper is organised as follows: Section 2 offers an outline of construction projects along with a review of risk assessment in the sector. Section 3 briefly discusses the preliminaries of D–S evidence theory and D numbers theory, along with their properties. In Section 4, algorithmic methodologies of D-CFPR, D-ANP, and D-MABAC are discussed. Section 5 demonstrates the effectiveness of the above methodology by presenting a numerical example on risk prioritisation and responses in the construction sector. A discussion based on sensitivity analysis is given in Section 6. Finally, the conclusions and future direction of the present research are given in Section 7.

2. Risks in Construction Projects

2.1. An Overview

The construction industry in general, as well as individual construction projects, deals with various threats, which are called risks. Risk assessment in construction projects has been applied differently from project to project (using various models of risk assessment) to evaluate the risk in certain activities of the project [49]. However, the socio-economic complexity involved in construction events makes it more risk prone, so that there may be negative effects on project sustainability [23]. Due to various complex factors, the construction industry is highly diverse and heterogeneous, experiencing a great deal of dynamic change with global sourcing and increasing price competition [50].

To overcome these risks, contractors have generally used high mark-ups, but this approach is no longer effective, as their margins have become smaller [51]. In recent decades, the stages involved in construction projects have become far more complex in nature, due to technological upgrading and stakeholder pressure, and are characterised by a number of uncertainties, which have a negative influence on the projects [52]. As risk free construction projects are impossible in real life, a controlled risk assessment procedure is required to manage various risks in the projects. The aim of this study is to highlight the main risks that construction projects face and the risk mitigation strategies used to manage them.

2.2. Previous Studies on Risk Assessment in Construction Projects

There have been substantial developments over the last four decades in research related to construction project management. Projects have been considered that are either exposed to risks, or have apparently inherited risk due to the participation of several stakeholders (as owners, contractors and designers); see [53] among others. For classifying and managing risks effectively, many methodologies have been suggested in the literature. In this section, a list of the existing approaches to construction project risk management is now briefly summarised.

Wang et al. [4] developed an alien eyes' risk (AER) model, categorising the project risks and their mutual relationships, and proposing a qualitative risk mitigation framework. Schieg [17] adopted a risk management process in construction project management, concentrating more on personal area risks. Zavadskas et al. [23] presented risk assessment based on the MCDM method and applying TOPSIS grey and grey number based complex proportional assessment (COPRAS-G). Wen [54] integrated rough sets and artificial neural networks (ANN) for risk evaluation of construction projects. Fouladgar et al. [55] proposed a risk evaluation outline faced during tunneling operations based on fuzzy TOPSIS methodology. Taroun and Yang [56] hybridised the dempster–shafer (D–S) theory of evidence and evidential reasoning algorithm for structuring personal experience and professional judgment with a spreadsheet-based decision support system. Mohammadi and Tavakolan [25] combined fuzzy logic and AHP in traditional failure mode effect analysis (FMEA) for construction project risk assessment. Serpella et al. [18] applied a knowledge-based approach, based on a three-fold arrangement and risk management function, to address project risks in the construction management sector. Ebrat and Ghodsi [14] identified the risks in construction projects based on the adaptive neuro fuzzy inference system and stepwise regression model for the evaluation of project risks. Taylan et al. [15] proposed hybrid methodologies combining fuzzy AHP and fuzzy TOPSIS and applied the relative importance index method to prioritise the project risks based on the data obtained. Iqbal et al. [9] considered two types of risk management technique during project execution, viz. preventive techniques (to manage risk before the start of the project) and remedial techniques (after occurrence of the risk). Vafadarnikjoo et al. [24] developed an intuitive fuzzy decision-making trial and evaluation laboratory (DEMATEL) to prioritise risks associated with construction projects by using the risk breakdown structure. Ahmadi et al. [27] analysed the criteria for prioritising potential risk events and quantified it using fuzzy AHP. The best response action for a risk event is then identified with respect to the same criteria using a scope expected deviation index. Santos and Jungles [19] evaluate the completion of construction project risk by considering the correlation of delay and the schedule performance index along with any time over-run. Shin et al. [28] made a comparative analysis of AHP and fuzzy AHP to evaluate the potential risk factors at the construction site of a nuclear power plant. Kao et al. [57] proposed an integrated fuzzy ANP based balanced scorecard system for evaluation of relevant bilateral factors in Taiwanese construction projects. Burcar Dunovic et al. [58] assessed large infrastructure projects by integrating the risk impact cumulative distribution curve based MCDM approach. Yousefi et al. [59] proposed a neural network model for predicting emerging time and cost claims applied to Iranian construction projects. Valipour et al. [60] presented a fuzzy cybernetic ANP model for proper identification of public-private partnership project based risks. Ulubeyli and Kazaz [61] developed a fuzzy based sub-contractor selection model (CoSMo) for global based construction projects. Rajakallio et al. [62] analysed the solution delivery from a network perspective in integrated business model renewal. For assessing risk in deep foundation excavation, Valipour et al. [63] developed a step-wise weight assessment ratio analysis (SWARA) based complex proportional assessment (COPRAS) framework to analyse Iranian construction project uncertainty. Khanzadi et al. [64] solved the dispute resolution problem in the construction sector by considering a grey number based discrete zero-sum two-person matrix game model. Keshavarz Ghorabaee et al. [65] presented the hybrid fuzzy step-wise weight assessment ratio analysis (SWARA) based 'evaluation based on distance from average solution' (EDAS) model for assessment of the construction equipment by taking sustainability into account.

3. Preliminaries

3.1. Dempster–Shafer (D–S) Evidence Theory

To deal with real world information, which may be ambiguous, various imprecise decision-making models based on probability theory, fuzzy set theory, and D–S evidence theory have been developed [66]. D–S evidence theory directly expresses uncertain information by allocating the probability to the multiple object based subsets in lieu of individual items [67]. Due to its ability to compare pairs of evidence or belief functions when deriving new ones, and its superiority in an uncertain environment, evidence theory has been widely applied in various domains [22,34,39,68–70].

Certain basic definitions are now presented. Let U denote the frame of discernment representing a collective set of exhaustive and mutually exclusive events, and each element of 2^U (a power set in U) represents a proposition [22]. Based on these, basic probability assignment (BPA) is defined as follows:

Definition 1. *Basic Probability Assignment (BPA)*

A finite non-empty set of mutually exclusive and exhaustive hypotheses for any problem domain is called *its frame of discernment*. In lieu of a D–S evidence framework, the BPA is defined on a frame of discernment to express the imprecise judgements of experts [39]. Thus, if 2^U denotes the power set of a finite non-empty set U, then a BPA is defined as a mapping m satisfying (1) and (2):

$$m : 2^U \rightarrow [0,1], \tag{1}$$

$$m(\phi) = 0 \ and \ \sum_{A \in 2^U} m(A) = 1, \tag{2}$$

where ϕ is an empty set and A is any element of 2^U. If $m(A) > 0$, A is called its focal element, and the union of all focal elements is the core of the mass function.

Example: Suppose there exists a task to assess a project. In the frame of D–S theory, the frame of discernment is the proposed project set $B = \{b_1, b_2, b_3\}$.

As per the expert view, a BPA can be created to express their assessment result: $m(\{b_1\}) = 0.2$; $m(\{b_2\}) = 0.7$; $m(\{b_2, b_3\}) = 0.1$, where $b_1 = [0, 40]$, $b_2 = [41, 70]$, $b_3 = [71, 100]$. In addition, the set $B = \{b_1, b_2, b_3\}$ is a frame of discernment in D–S theory with $m(\{b_1\}) + m(\{b_2\}) + m(\{b_2, b_3\}) = 1$.

3.2. D Numbers Theory

Although D–S evidence theory is operative in the data fusion problem, it is restricted to handling semantic information where linguistic variables are not mutually exclusive [39,70]. Intuitively, if some hypotheses of D–S theory are removed reasonably, the ability to represent and handle uncertain information may greatly improve. Overall, in many practical scenarios, the experts fail to have complete information, and the assessment is done solely on the basis of partial information, resulting in an incomplete BPA. Based on this idea, Deng [36] proposed D numbers theory, a generalisation of evidence theory, in order to characterise ambiguous data, as defined in (3) and (4) below:

Definition 2. *D Numbers [36]*

Let Ω be a finite nonempty set, D numbers is defined by a mapping:

$$D : \Omega \rightarrow [0,1], \tag{3}$$

$$with \ \sum_{B \subseteq \Omega} D(B) \leq 1 \ and \ D(\phi) = 0, \tag{4}$$

where ϕ is an empty set and B is a subset of Ω.

Compared with D–S evidence theory, the D numbers concept has the advantages that [34,39,42,71]:

- *Firstly*, D numbers with *nonexclusive hypothesis* in each element of the frame of discernment is more applicable for linguistic assessment.
- *Secondly*, in an evidence theory, a normal BPA must be *complete*, implying that the sum of all focal length elements in BPA is 1. D numbers allows the experts to input *incomplete and uncertain information* to the framework resulting in an incomplete BPA, thus releasing the completeness

constraint. Thus, if $\sum\limits_{B \subseteq \Omega} D(B) = 1$, the information is said to be complete, and for $\sum\limits_{B \subseteq \Omega} D(B) < 1$, the information is said to be incomplete.

Example 1. *Let an expert give his assessment on the condition of a bridge, using D numbers, in a scale interval of [0,100]. The expert express his assessment in an incomplete BPA framework as follows:*

$$D(\{a_1\}) = 0.5; \ D(\{a_2\}) = 0.3; \ D(\{a_1, \ a_2, \ a_3\}) = 0.1,$$

where $a_1 = [0, \ 60]$, $a_2 = [45, \ 75]$, and $a_3 = [65, \ 100]$.

Hence, the set of $\{a_1, a_2, a_3\}$ is not a frame of discernment, because the elements in the above set are not mutually exclusive. In addition, since $D(\{a_1\}) + D(\{a_2\}) + D(\{a_1, a_2, a_3\}) = 0.9 < 1$, the above information is incomplete. This example shows that the definition of D numbers is similar to the definition of the mass function.

Definition 3. *D numbers for a discrete set [36]*

For a discrete set $\Omega = \{b_1, b_2, \ldots, b_i, \ldots, b_n\}$, where $b_i \in R$ and $b_i \neq b_j (i \neq j)$, a special form of D numbers can be expressed as:

$$\begin{cases} D(\{b_i\}) = v_i \ (i = 1, 2, \ldots, n) \\ \text{or denoted as,} \\ D = \{(b_1, v_1), \ (b_2, v_2), \ldots, (b_i, v_i), \ldots, (b_n, v_n)\}, \\ \text{where } v_i > 0 \text{ and } \sum\limits_{i=1}^{n} v_i \leq 1. \end{cases} \quad (5)$$

Definition 4. *Permutation invariability [22,71].*

Two D numbers $D_1 = \{(b_1, v_1), ,\ldots, (b_i, v_i), \ldots, (b_n, v_n)\}$ & $D_2 = \{(b_n, v_n), ,\ldots, (b_i, v_i), \ldots, (b_1, v_1)\}$ are said to be invariable if:

$$D_1 \Leftrightarrow D_2 . \quad (6)$$

Example 2. *If there are two D numbers:*

$D_1 = \{(0.0, \ 0.7), \ (1.0, \ 0.3)\}$ *and* $D_2 = \{(1.0, \ 0.3), (0.0, \ 0.7)\}$, *then,* $D_1 \Leftrightarrow D_2$.

Definition 5. *D numbers integration [6,36].*

Let $D = \{(b_1, v_1), \ (b_2, v_2), \ldots, (b_i, v_i), \ldots, (b_n, v_n)\}$ be a D number; then, the integration representation $I(D)$ of the D number is defined as:

$$I(D) = \sum\limits_{i=1}^{n} b_i v_i . \quad (7)$$

Example 3. *Let* $D = \{(1, \ 0.3), \ (2, \ 0.4), (3, \ 0.1), (4, 0.2), (5, 0.2)\}$, *then* $I(D) = 1 \times 0.3 + 2 \times 0.4 + 3 \times 0.1 + 4 \times 0.2 + 5 \times 0.2 = 3.2$.

4. Methodology

4.1. D-CFPR: D Numbers Extended CFPR

Based on the additive transitive property, Herrera-Viedma et al. [30] put forward the consistent fuzzy preference relation (CFPR) for structuring an $n \times n$ decision matrix, which requires only $(n-1)$

pair wise comparisons, in lieu of $n(n-1)/2$ comparisons needed for constructing a fuzzy preference relation. Preserving both the reciprocity and transitivity properties, CFPR is constructed only on complete and certain information but fails to deal with cases containing insufficient information [34,36]. Citing the same deficiency as in the case of fuzzy preference relations, Deng et al. [39] proposed the concept of a D numbers preference relation encompassing the fuzzy preference relations of the D numbers domain. In addition, considering the insufficiency in traditional CFPRs, Zuo et al. [42] proposed a D numbers consistent fuzzy preference relation matrix, abbreviated as D-CFPR matrix using D numbers to express linguistic preferences given by experts to construct the CFPR. In the following steps, we will discuss the algorithmic methodology of D-CFPR, where distribution assessments are assessed under incomplete or missing information.

Step 1: Construct a D-CFPR matrix (R_D). The D numbers extended CFPR is advocated to reinforce the capability of CFPR to express uncertain and incomplete information. A D numbers based preference relation on a set of criteria $C = \{C_1, C_2, \ldots, C_n\}$ is characterized by a $n \times n$ matrix $R_D = [D_{ij}]_{n \times n}$ on the product set $C \times C$, whose elements are formulated as per Equation (8), and represented in matrix form (9):

$$R_D : C \times C \to D, \qquad (8)$$

$$R_D = \begin{array}{c} \\ C_1 \\ C_2 \\ \vdots \\ C_n \end{array} \begin{array}{cccc} C_1 & C_2 & \cdots & C_n \\ \left[\begin{array}{cccc} D_{11} & D_{12} & \cdots & D_{1n} \\ D_{21} & D_{22} & \cdots & D_{2n} \\ \vdots & \vdots & \ddots & \vdots \\ D_{n1} & D_{n2} & \cdots & D_{nn} \end{array} \right] \end{array}, \qquad (9)$$

where $D_{ij} = \left\{ \left(b_{ij}^1, v_{ij}^1\right), \left(b_{ij}^2, v_{ij}^2\right), \ldots, \left(b_{ij}^p, v_{ij}^p\right), \ldots \right\}$, $D_{ji} = \neg D_{ij} = \left\{ \left(1 - b_{ij}^1, v_{ij}^1\right), \left(1 - b_{ij}^2, v_{ij}^2\right), \ldots, \left(1 - b_{ij}^p, v_{ij}^p\right), \ldots \right\}$, $\forall i, j \in \{1, 2, \ldots, n\}$, $b_{ij}^p \in [0,1]$, $v_{ij}^p > 0$, $\sum_p v_{ij}^p = 1$. Obviously, $D_{ii} = \{(0.5, 1.0)\}$ $\forall i \in \{1, 2, \ldots, n\}$ in R_D.

In Equation (9), is R_D called a D-CFPR as it is constructed based on $(n-1)$ pair wise comparisons, denoted as $\left\{ D_{12}, D_{23}, \ldots, D_{(n-1)n} \right\}$ representing a set of D numbers. Considering the set of preference values $B = \{D_{ji}, 1 \le i < j \le n\}$, the elements of which are denoted $D_{ji} = \left\{ \left(b_{ji}^1, v_{ji}^1\right), \left(b_{ji}^2, v_{ji}^2\right), \ldots, \left(b_{ji}^k, v_{ji}^k\right), \ldots \right\}$, which can also be represented as follows [42]:

$$D_{ji} = \frac{j - i + 1}{2} - D_{i(i+1)} - D_{(i+1)(i+2)} - \cdots - D_{(j-1)j},$$

where each component $\left(b_{ji}^k, v_{ji}^k\right)$ in D_{ji} is obtained by:

$$\left. \begin{array}{l} b_{ji}^k = \frac{j - i + 1}{2} - b_{i(i+1)}^x - b_{(i+1)(i+2)}^y - \cdots - b_{(j-1)j}^k, \quad \forall (x, y, \ldots, z) \\ v_{ji}^k = \sum\limits_{(x,y,\ldots z) \in \Omega} v_{i(i+1)}^x \times v_{(i+1)(i+2)}^y \times \cdots \times v_{(j-1)j}^z, \end{array} \right\} \qquad (10)$$

where $\Omega = \left\{ (x, y, \ldots, z) \middle| b_{ji}^k = \frac{j - i + 1}{2} - b_{i(i+1)}^x - b_{(i+1)(i+2)}^y - \cdots - b_{(j-1)j}^k \right\}$, $\left(b_{i(i+1)}^x, v_{i(i+1)}^x\right)$ is the xth component of $D_{i(i+1)}$, $\left(b_{(i+1)(i+2)}^y, v_{(i+1)(i+2)}^y\right)$ is the yth component of $D_{(i+1)(i+2)}$, $\left(b_{(j-1)j}^z, v_{(j-1)j}^z\right)$ is the zth component of $D_{(j-1)j}$.

Thus, based on the reciprocal property, the rest of the entries in D-CFPR (9) are calculated as per Equation (11) as follows:

$$D_{ji} = \neg D_{ij}, \quad \forall i, j \in \{1, 2, \ldots, n\}. \qquad (11)$$

As per Equations (9)–(11), a D-CFPR can be constructed based on $(n-1)$ pair wise comparisons $\left\{ D_{12}, D_{23}, \ldots, D_{(n-1)n} \right\}$. In the D-CFPR generated above, some values of the $b_{ij}'s$ in D_{ij}, $i, j \in (1, 2, \ldots, n)$ that fall in the interval $[-a, 1 + a]$, $a > 0$ need to be transformed using Equation (12):

$$f : [-a, 1+a] \rightarrow [0,1], \ f(r) = \frac{r+a}{1+2a}. \tag{12}$$

The transformation function (12) works as a normalisation, which transforms the values of $b'_{ij}s$ from $[-a, 1+a]$ to the interval $[0,1]$.

Step 2: Determine the crisp matrix R_c. Convert the matrix $R_D = [D_{ij}]_{n \times n}$ in (9) to a *crisp matrix* $R_C = [c_{ij}]_{n \times n}$ (13) using the integration representation of D numbers (as per Equation (7)), where each element $c_{ij} = I(D_{ij}) \ i, j = 1, 2, \ldots, n$:

$$R_c = [c_{ij}]_{n \times n} = \begin{bmatrix} c_{11} & c_{12} & \cdots & c_{1n} \\ c_{21} & c_{22} & \cdots & c_{2n} \\ \vdots & \vdots & \ddots & \vdots \\ c_{n1} & c_{n2} & \cdots & c_{nn} \end{bmatrix}. \tag{13}$$

Step 3: Construct a probability matrix R_p. Based on the crisp matrix R_C (13), construct a probability matrix $R_p = [p_{ij}]_{n \times n}$ (14), representing the preference probability between a pair wise set of n criteria $C = \{C_1, C_2, \ldots, C_n\}$:

$$R_p = [p_{ij}]_{n \times n} = \begin{bmatrix} p_{11} & p_{12} & \cdots & p_{1n} \\ p_{21} & p_{22} & \cdots & p_{2n} \\ \vdots & \vdots & \ddots & \vdots \\ p_{n1} & p_{n2} & \cdots & p_{nn} \end{bmatrix}. \tag{14}$$

In R_p, the element $p_{ij} = \Pr(C_i \succ C_j), \forall i, j \in \{1, 2, \ldots, n\}$, where " $\succ$ " represents *"prefer to"*. Based on (13), a set of rules is suggested to generate R_p.

- For the elements satisfying $c_{ij} + c_{ji} = 1.0$,

  ○ If $c_{ij} > 0.5$, then $\Pr(C_i \succ C_j) = 1$ and $\Pr(C_i \prec C_j) = 0$,
  ○ If $c_{ij} \leq 0.5$, then $\Pr(C_j > C_i) = 1$ and $\Pr(C_i > C_j) = 0$.

- When $c_{ij} + c_{ji} > 1.0 : c_{ij} \geq 0.5 \ and \ c_{ji} \geq 0.5$,

  ○ If $c_{ij} \geq 0.5$, then $\Pr(C_i \succ C_j) = 1$ and $\Pr(C_j \succ C_i) = 0$,
  ○ If $c_{ji} \geq 0.5$, then $\Pr(C_j \succ C_i) = 1$ and $\Pr(C_i \succ C_j) = 0$.

- when $c_{ij} + c_{ji} < 1.0 : c_{ij} < 0.5 \ and \ c_{ji} < 0.5$.

In this condition, the unallocated preference *(up)* is $c_{up} = 1 - (c_{ij} + c_{ji})$. The probability of one criteria outperforming another criteria is as follows:

  ○ If $c_{ij} < 0.5$, then $\Pr(C_i \succ C_j) = 1 - \frac{(0.5 - c_{ij})}{c_{up}}$,
  ○ If $c_{ji} < 0.5$, then $\Pr(C_j \succ C_i) = 1 - \frac{(0.5 - c_{ji})}{c_{up}}$.

Step 4: Construct a triangular probability matrix R_p^T. Rank the criteria using the triangularisation procedure, viz. by maximising the sum of the values above, the main diagonal in the $n \times n$ square matrix in the final order is given as:

- First, sum up each row of the $n \times n$ matrix and determine the row number with maximum value.
- Then, assuming the obtained row number is k, delete the k-th row and k-th column in the matrix.
- Replicate the two procedures above until the matrix is empty.

Thus, by applying this defined row deletion order operation on the matrix $R_p = [p_{ij}]_{n \times n}$ (14), construct the triangular probability matrix R_p^T, shown in (15):

$$R_p^T = \left[p_{ij}^T \right]_{n \times n} = \begin{bmatrix} p_{11}^T & p_{12}^T & \cdots & p_{1n}^T \\ p_{21}^T & p_{22}^T & \cdots & p_{2n}^T \\ \vdots & \vdots & \ddots & \vdots \\ p_{n1}^T & p_{n2}^T & \cdots & p_{nn}^T \end{bmatrix}. \tag{15}$$

Step 5: Construct triangulated crisp matrix R_c^T. Firstly, the crisp matrix R_c (13) is triangulated based on the triangular matrix R_p^T (15). From that step, the triangulated crisp matrix R_c^T (16) is derived:

$$R_c^T = \left[c_{ij}^T \right]_{n \times n} = \begin{bmatrix} c_{11}^T & c_{12}^T & \cdots & c_{1n}^T \\ c_{21}^T & c_{22}^T & \cdots & c_{2n}^T \\ \vdots & \vdots & \ddots & \vdots \\ c_{n1}^T & c_{n2}^T & \cdots & c_{nn}^T \end{bmatrix}. \tag{16}$$

Secondly, for elements satisfying $R_c^T(i, j) + R_c^T(j, i) < 1$, a new *"polishing operation"* (17) is implemented to obtain a triangulated crisp matrix R_k^T (18):

$$R_k^T(i, j) = R_c^T(i, j) + \frac{1 - \left[R_c^T(i, j) + R_c^T(j, i) \right]}{2}, \tag{17}$$

$$R_k^T = \left[k_{ij}^T \right]_{n \times n} = \begin{bmatrix} k_{11}^T & k_{12}^T & \cdots & k_{1n}^T \\ k_{21}^T & k_{22}^T & \cdots & k_{2n}^T \\ \vdots & \vdots & \ddots & \vdots \\ k_{n1}^T & k_{n2}^T & \cdots & k_{nn}^T \end{bmatrix}. \tag{18}$$

Step 6: Calculate the relative priority weights of the criteria. Assume that the weight vector is $W = (w_i)^T$, where $w_i (i = 1, 2, \ldots, n)$ are the weights of criteria $C_i (i = 1, 2, \ldots, n)$. The elements above and alongside the main diagonal of the matrix $R_k^T = \left[k_{ij}^T \right]_{n \times n}$ (18) indicate the weight relationship of the criteria. Thus, by adding some necessary constraints, a set of Equation (19) is formed:

$$\begin{cases} \lambda(w_2 - w_1) = k_{22}^T - k_{12}^T \\ \lambda(w_3 - w_2) = k_{33}^T - k_{23}^T \\ \ldots \ldots \ldots \\ \lambda(w_n - w_{n-1}) = k_{nn}^T - k_{(n-1)n}^T \end{cases}, \tag{19}$$

where λ indicates the *granular information* about the pairwise comparison, which reveals the expert's perceptive competency. The value of each weight $w_i (i = 1, 2, \ldots, n)$, subjected to the parameter λ, relates to the expert's cognitive aptitude. The values of λ highly depend on the influence of the experts' judgment and belief function. Thus, a feasible outline (20) is obtained as follows:

$$\lambda = \begin{cases} \lceil \underline{\lambda} \rceil, & \text{The information has high credibility} \\ n, & \text{The information has medium credibility} \\ \frac{n^2}{2}, & \text{The information has low credibility} \end{cases}, \tag{20}$$

where $\underline{\lambda}$ signifies the lower bound of λ, $\lceil \underline{\lambda} \rceil = \min\{k \in \mathbb{Z} | k \geq \underline{\lambda}\}$, and n is the number of alternatives [39]. The concrete priority weight value of each criteria is calculated here when $\lambda = n$.

Step 7: Inconsistency for the D-CFPR. The traditional CFPR is fully consistent when constructed under the transitivity property for $n - 1$ preference values. The D numbers based CFPR is *consistent*, but, when reduced to classical CFPR, it shows inconsistency under *uncertain or incomplete* information. Thus, to express this inconsistency of D-CFPRs, an inconsistency degree (ID) (21) is defined, based on the triangular probability matrix R_p^T (18), as follows:

$$I.D. = \frac{\sum\limits_{i=1, j<i}^{n} R_p^T(i,j)}{n(n-1)/2}, \tag{21}$$

where $R_p^T(i,j)$ is an element of R_p^T, and n is the number of comparison objects. This emphasises that the inconsistency of the D matrix, and its acceptable level is decided by the DMs, whose subjective requirement regulates the tolerance level of the aforementioned inconsistency.

4.2. Evaluating the Risk Criteria Weight Using D-ANP

The evaluation of D-ANP methodology is composed of two phases.

The first phase emphasises the formation of pairwise judgments for every dependent relationship among given criteria $C_i(i = 1, 2, \ldots, n)$, and determination of their priority weight. The priority weights thus obtained are input to the system-with-feedback supermatrix for computing the network influences amongst the different relationships.

The second phase, namely supermatrix evaluation, incorporates five steps: formation of the unweighted supermatrix, formation of the weighted supermatrix, normalisation of the weighted supermatrix (a column stochastic matrix), and finally convergence to a solution using the limited supermatrix.

The converged supermatrix will provide us with the relative priorities for each of the criteria (*or sub-criteria*) considered within clusters of the decision framework. Thus, the following algorithmic steps of D-ANP are as follows:

Step 1: Model construction and problem structuring. First, we properly outline the decision problem with detailed criteria and sub-criteria (*if taken*) and then delimit the cluster's (*dimension's*) network and elements (*criteria set*) within the given clusters. Next, we decide which inter- and inner-dependencies will prevail in the decision problem and clusters of the over-all feedback system.

Step 2: Pairwise comparison matrices and priority vectors. The expert inputs prerequisite for the ANP method are the pairwise judgments of the elements within each cluster, from which inter-and-inner-dependence matrices are formed. Primarily, three categories of pairwise comparisons are formed. First, the inner dependence comparison matrix, based on the criteria of different clusters (dimensions), is used to get the priority weight vector. Then, the pairwise comparison matrices obtained from criteria with respect to other factors (criteria) of the same cluster (dimension). Finally, the pairwise comparison matrix is developed by considering the factor (criteria) of a cluster (dimension) among factors (criteria) of the other cluster (dimension), which affect the criteria.

These pairwise comparison matrices, along with their valuation elicitation, constitute a modified version of D-AHP [22] and D-CFPR [34]. Based on Section 4.1, a brief summary for getting the priority weight of any pairwise comparison matrix is given below, the graphical description of which is shown in Figure 1.

- In the first step, the D-CFPR matrix $R_D = [D_{ij}]_{n \times n}$, is constructed for n criteria, by considering the system as an input using Equations (9)–(12).
- The D-CFPR matrix formed (R_D), is converted to a crisp matrix $R_C = [c_{ij}]_{n \times n}$ using the integration representation of D number, shown in Equation (13).
- The probability matrix $R_p = [p_{ij}]_{n \times n}$ is then constructed based on the derived crisp matrix (R_c) using Equation (14), and it satisfies a set of rules in Step 3 of Section 4.1.
- In the next step, using Equation (15), triangularisation $R_p^T = [p_{ij}^T]_{n \times n}$ is applied to the probability matrix using local information that contains the preference relations of pairwise criteria.
- Lastly, applying Equations (16)–(19), the crisp based triangular matrix $R_c^T = [c_{ij}^T]_{n \times n}$, is obtained, and relative priority weights $w_i(i = 1, 2, \ldots, n)$ of each criteria $C_i(i = 1, 2, \ldots, n)$, based on clusters (*dimensions*) are calculated, thereby checking its inconsistency as per Equation (21).

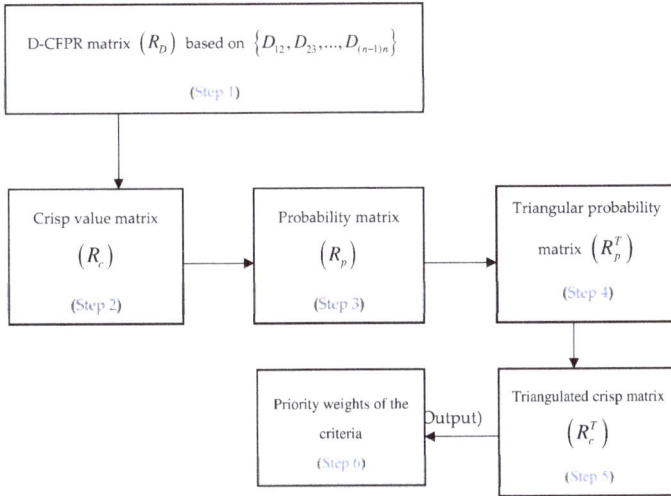

Figure 1. Procedure to obtain priority weights of criteria based on D numbers extended consistent fuzzy preference relation.

Step 3: Formation of the unweighted supermatrix. The local priority weights resulting from the pairwise comparison are used as input in suitable columns of the unweighted constructed supermatrix, to obtain the global priorities in a system. As a result, a supermatrix takes the form of a partitioned matrix, each segment of which represents an association between two clusters in the given system. Thus, the supermatrix W Equation (22) formed is represented by sub-matrices W^{ij} as $W = \left[W^{ij}\right]_{n \times n} (i, j = 1, 2, \dots, n)$. The size of the sub-matrix W^{ij} Equation (23) depends on the compared factors (criteria). Interdependency between clusters is illustrated in Equation (22) by analysing both inter- and intra-relations among clusters.

Arrange all priority vectors, representing the impact of a given set of elements in a cluster on another element in the network, as sub-columns of the corresponding column of an unweighted supermatrix W as in Equations (22) and (23). This is composed of k clusters $\{D_1, D_2, \dots, D_k\}$, and linkages of these clusters $\{e_{k1}, e_{k2}, \dots, e_{kn_k}\}$, are elements of the cluster $D_p (p = 1, 2, \dots, k)$. Each column of the sub-matrix W^{ij} Equation (23) is the priority vector acquired from the identical pairwise judgment, indicating the significance of the elements in the ith cluster with respect to an element in the jth cluster:

$$W = \begin{array}{c} \\ \\ D_1 \\ \\ \\ \\ D_2 \\ \\ \\ \\ D_k \\ \\ \end{array} \begin{array}{c} e_{11} \\ e_{12} \\ \vdots \\ e_{1(n_1)} \\ e_{21} \\ e_{22} \\ \vdots \\ e_{2(n_2)} \\ e_{k1} \\ e_{k2} \\ \vdots \\ e_{k(n_k)} \end{array} \begin{bmatrix} W^{11} & W^{12} & \cdots & W^{1k} \\ & & & \\ W^{21} & W^{22} & \cdots & W^{2k} \\ & & & \\ \cdot & \cdot & \cdot & \cdot \\ \cdot & \cdot & \cdot & \cdot \\ \cdot & \cdot & \cdot & \cdot \\ W^{k1} & W^{k2} & \cdots & W^{kk} \end{bmatrix} \qquad where \ \sum_{j=1}^{k} (n_j) = n, \quad (22)$$

with column headers D_1: $e_{11} e_{12} \cdots e_{1(n_1)}$, D_2: $e_{21} e_{22} \cdots e_{2(n_2)}$, $\cdots$, D_k: $e_{k1} e_{k2} \cdots e_{k(n_k)}$

$$W^{ij} = \begin{bmatrix} W^{ij}_{11} & \cdots & W^{ij}_{1j} & \cdots & W^{ij}_{1n_j} \\ \vdots & & \vdots & \cdots & \vdots \\ W^{ij}_{i1} & \cdots & W^{ij}_{ij} & \cdots & W^{ij}_{in_j} \\ \vdots & & \vdots & & \vdots \\ W^{ij}_{n_i1} & \cdots & W^{ij}_{n_ij} & \cdots & W^{ij}_{n_in_j} \end{bmatrix} \quad i,j = 1,2,\ldots,k,. \tag{23}$$

Step 4: Determine the normalised weighted supermatrix. If there is no linkage between clusters D_k and D_l, then the sub-matrix W^{kl} equals zero. We compute the weighted supermatrix $W^a = \left[\widetilde{W}^{ij}\right]_{n \times n}$, by multiplying the unweighted matrix $W = \left[W^{ij}\right]_{n \times n}$, Equation (23) by priority of dimensions $V_j (j = 1,2,\ldots,n)$, in this stage. As the weighted supermatrix (W^a) needs to be stochastic, we normalise each column of it, and develop meaningful limiting priorities for determining overall cluster influences. This normalisation procedure ensures that the weighted supermatrix is column stochastic; it is finally represented by matrix $\hat{W} = \left[\overline{W}^{ij}\right]_{n \times n}$.

Step 5: Compute the limiting priorities for criteria weights. Finally, the column stochastic weighted supermatrix is raised to an appropriately large power until it converges. Thus, the weighted supermatrix $\hat{W}$ is raised to limiting powers $(\hat{W})^{2m+1}$, m being an arbitrarily large number, to attain a steady-state limiting matrix W^q. Details are shown in Equation (24):

$$W^q = \lim_{k \to \infty} (\hat{W})^{2m+1}. \tag{24}$$

The priority weight of criteria for the corresponding clusters (dimensions) can now be found in the rows of the limiting supermatrix W^q. The limiting supermatrix provides the priority information for the elements of each individual cluster. The strategy outcome with the highest value should be selected from the cluster of criteria. Other priority rankings in different clusters are also provided.

4.3. D-MABAC for Ranking Alternatives

The MABAC methodology was developed by Pamucar and Cirovic [44] to handle problems in MCDM. D numbers is a new representation of uncertain information that can denote the more imprecise based conditions. Thus, the combination of MABAC and D numbers is a new experiment to make decisions in an uncertain environment. The basic setting of the MABAC technique is revealed in the definition of the distance of the criterion function of each of the observed alternatives from the approximate border area [72]. The algorithmic steps displays the execution process for the aforesaid D-MABAC methodology in six steps as follows:

Step 1: Constructing the initial decision matrix M. Initially, the assessment of m alternatives in respect of n criteria is carried out. The alternatives are represented in vector form $A_i = (d_{i1}, d_{i2}, \ldots, d_{in})$, where $d_{ij}(j = 1,2,\ldots,n)$ denotes the value (*in D numbers format*) for the i alternative $(i = 1,2,\ldots,m)$ according to j criteria for matrix $M = \left[d_{ij}\right]_{m \times n}$. Details shown in decision-matrix (25):

$$M = \begin{bmatrix} d_1(\{1\}) = d_{11} & d_1(\{2\}) = d_{12} & \cdots & d_1(\{n\}) = d_{1n} \\ d_2(\{1\}) = d_{11} & d_2(\{2\}) = d_{22} & \cdots & d_2(\{n\}) = d_{2n} \\ \cdots & \cdots & \ddots & \cdots \\ d_m(\{1\}) = d_{m1} & d_m(\{2\}) = d_{m2} & \cdots & d_m(\{n\}) = d_{mn} \end{bmatrix}. \tag{25}$$

Next, applying Equation (8), (i.e., integration representation of D numbers) on the elements d_{ij} $(i = 1,2,\ldots,m; j = 1,2,\ldots,n)$ of decision matrix $M = \left[d_{ij}\right]_{m \times n}$, crisp decision matrix $X = \left[I(d_{ij})\right]_{m \times n}$ is formed as per Equation (26) as follows:

$$X = \begin{bmatrix} I(d_{11}) & I(d_{12}) & \cdots & I(d_{1n}) \\ I(d_{21}) & I(d_{22}) & \cdots & I(d_{2n}) \\ \cdots & \cdots & \ddots & \cdots \\ I(d_{m1}) & I(d_{m2}) & \cdots & I(d_{mn}) \end{bmatrix}_{m \times n} = \begin{bmatrix} x_{11} & x_{12} & \cdots & x_{1n} \\ x_{21} & x_{22} & \cdots & x_{2n} \\ \cdots & \cdots & \ddots & \cdots \\ x_{m1} & x_{m2} & \cdots & x_{mn} \end{bmatrix}_{m \times n} \quad \text{where each } x_{ij} = I(d_{ij}). \quad (26)$$

Step 2. Normalisation of the elements of the initial matrix X. The elements of the normalized matrix $N = [t_{ij}]_{m \times n}$ are obtained using the following expressions:

$$N = \begin{bmatrix} t_{11} & t_{12} & \cdots & t_{1n} \\ t_{21} & t_{22} & \cdots & t_{2n} \\ \cdots & \cdots & \ddots & \cdots \\ t_{m1} & t_{m2} & \cdots & t_{mn} \end{bmatrix}, \quad (27)$$

where

$$t_{ij} = \begin{cases} \frac{x_{ij} - x_i^-}{x_i^+ - x_i^-}, & \text{for "benefit type" criteria elements} \\ \frac{x_{ij} - x_i^+}{x_i^- - x_i^+}, & \text{for "cos t type" criteria elements} \end{cases}, \quad (28)$$

where $x_{ij} = I(d_{ij})$ and the components $x_i^+ = \max(x_1, x_2, \ldots, x_m)$, $x_i^- = \min(x_1, x_2, \ldots, x_m)$ of the decision matrix (X) represent the maximum and minimum values of the criteria $C_j (j = 1, 2, \ldots, n)$, by alternatives $A_i (i = 1, 2, \ldots, m)$.

Step 3. Calculation of the elements of weighted matrix (V). The elements of the weighted matrix $V = [v_{ij}]_{m \times n}$, are calculated using the following equations:

$$V = \begin{bmatrix} v_{11} & v_{12} & \cdots & v_{1n} \\ v_{21} & v_{22} & \cdots & v_{2n} \\ \cdots & \cdots & \ddots & \cdots \\ v_{m1} & v_{m2} & \cdots & v_{mn} \end{bmatrix}, \quad (29)$$

$$v_{ij} = w_j \cdot (t_{ij} + 1), \quad (30)$$

where t_{ij} are the elements of the normalised matrix $N = [t_{ij}]_{m \times n}$ and $w_i (j = 1, 2, \ldots, n)$ the weight coefficients of criteria, respectively.

Step 4. Determine the border approximation area matrix (G). The elements of the border approximation area (BAA) for each criterion $C_j (j = 1, 2, \ldots, n)$ are determined as follows:

$$g_j = \left(\prod_{i=1}^{m} v_{ij} \right)^{1/m}, \quad (31)$$

where v_{ij} are the elements of the weighted matrix $V = [v_{ij}]_{m \times n}$, and m is the total number of alternatives. After calculating the value $g_j (j = 1, 2, \ldots, n)$, for each criterion, the BAA matrix (G) is formed with format $n \times 1$ (n = the total number of criteria, according to which the selection is made from the alternatives).

$$G = [g_1 \ g_2 \ \cdots \ g_n]_{1 \times n}. \quad (32)$$

Step 5. Calculation of the distance of the alternative from the BAA for the matrix elements (Q). The distance (q_{ij}) of the alternatives $A_i (i = 1, 2, \ldots, m)$ from the BAA matrix $G = [g_1 \ g_2 \ \cdots \ g_n]$ is determined as the difference between the elements $v_{ij} (i = 1, 2, \ldots, m; \ j = 1, 2, \ldots, n)$ in the weighted matrix $V = [v_{ij}]_{m \times n}$, and the value $g_j (j = 1, 2, \ldots, n)$ of the BAA (G). The distance $q_{ij} (i = 1, 2, \ldots, m; \ j = 1, 2, \ldots, n)$ are elements of a matrix $Q = [q_{ij}]_{m \times n}$ and are shown in Equation (33):

$$Q = \begin{bmatrix} v_{11} - g_1 & v_{12} - g_2 & \cdots & v_{1n} - g_n \\ v_{21} - g_1 & v_{22} - g_2 & \cdots & v_{2n} - g_n \\ \cdots & \cdots & \ddots & \cdots \\ v_{m1} - g_1 & v_{m2} - g_2 & \cdots & v_{mn} - g_n \end{bmatrix} = \begin{bmatrix} q_{11} & q_{12} & \cdots & q_{1n} \\ q_{21} & q_{22} & \cdots & q_{2n} \\ \cdots & \cdots & \ddots & \cdots \\ q_{m1} & q_{m2} & \cdots & q_{mn} \end{bmatrix}. \tag{33}$$

Alternative $A_i (i = 1, 2, \ldots, m)$ can belong to BAA (G), upper approximation area (G^+), or lower approximation area (G^-), i.e., $A_i \in \{G \vee G^+ \vee G^-\}$ and are determined as per Equation (34):

$$A_i \in \begin{cases} G^+ \text{ if } q_{ij} > 0 \\ G \text{ if } q_{ij} = 0 \quad , \text{ where each } q_{ij} = v_{ij} - q_j. \\ G^- \text{ if } q_{ij} < 0 \end{cases} \tag{34}$$

Step 6. Ranking of the alternatives. A calculation of the values of the criterion functions for the alternatives is obtained as the sum of the distance of the alternatives from the BAA (q_{ij}). Using Equation (35), we calculate the sum of the elements q_{ij} of matrix (Q) by rows, and obtain the final values of the criterion functions $\hat{S}_i$ $(i = 1, 2, \ldots, m)$, for the alternatives $A_i (i = 1, 2, \ldots, m)$:

$$\hat{S}_i = \sum_{j=1}^{n} q_{ij}, \ i = 1, 2, \ldots, m; \ j = 1, 2, \ldots, n. \tag{35}$$

5. Numerical Example: Risk Assessment in a Construction Project

5.1. Identification of Construction Projects Risk Indicators and Their Mitigation Strategies

For categorising and managing construction project risks effectually, several methodologies are recommended in the literature [43,73–75]. Keeping this in mind, we proposed the hybrid D-ANP-MABAC approach for the risk assessment of projects in the construction sector involving uncertain and incomplete information data. The authors employed a combination of questionnaire surveys involving literature reviews and subjective judgments of highly proficient experts to detect different risk response strategies that optimise the performance of construction projects. The proposed methodology incorporates the knowledge and experience of ten experts (five *technical* and five *management* based experts) for risk identification and structuring, along with proper risk mitigation strategies. The demographic profile of the respondents is given in Table 1.

Table 1. Summary of demographic profile of respondents.

	Characteristics	Frequency	Percentage (%)
Age group	21–31	2	20
	31–39	4	40
	39–45	3	30
	45–58	1	10
Gender	Female	4	40
	Male	6	60
Level of Education	Bachelor's degree	4	40
	Master's degree	5	50
	Higher	1	10
Role of respondents	Chief personal officer	1	10
	Manager or general manager	2	20
	Staff or assistant manager	1	10
	Project risks analyst	2	20
	Purchasing manager	1	10
	Construction site engineer	3	30
Years of experience in construction sector	Above 15 years	2	20
	10 years~15 years	4	40
	5 years~10 years	3	30
	Less than 5 years	1	10
	Total available number	10	

Based on the literature review of papers discussed in Table 2, the DMs consider *nine construction project risk criteria* identified under *three dimensions*: political instability (C_1), economic risk (C_2), and social risk (C_3). These are defined as environmental based *external risks* (D_1). Technological risk (C_4), work quality risk (C_5), and time and cost risk (C_6) are defined as construction process based *project risks* (D_2). Resource risk (C_7), documents and information risk (C_8), and stakeholder's risk (C_9) are defined under intrinsic criteria based *internal risks* (D_3). Details are given below in Table 3.

Table 2. Risk factors involved in construction projects.

Risk Indicators in Project Based Construction Management	References
Environmental risk; political, social and economic risk; contractual agreement risk; financial risk; construction risk; project design risk; market risk.	[1]
Safety risk, quality risk, environmental risk, political risk, project site risk, project complexity risk.	[53]
Quality risks, personnel risks, cost risks, deadline risks, strategic decision risks, external risks.	[17]
Operational risk, economic risk, political risk, financial risk, legal risk, currency and inflation risk, corruption risk, tendering procedures.	[3]
Political risks, economic risk, social risk, weather risk, cost, quality risk, technical risk, construction risk, resources risk, project member risk, information risk, construction site risks.	[23]
Resources risk, inexperience of project members, lack of motivational approach, design errors risk, efficiency risk, technical risk, quality risk.	[21]
Inflation risk, Payment security risk, Programme overrun risk, subcontractor pricing risk.	[36]
Political risk, economic risk, natural risk, legal risk, contractor risk, financial risk, management risk, equipment risk, designer risk.	[25]
Management risk, project risk, design risk, financial risk, operational risk, external risk.	[14]
Information risk, cost risks, lack of coordination, project schedule risk, lack of professional planning, legal dispute risk.	[15]
Designing risk, time risk, budget risk, labour risk, political risk.	[16]
Design risk, payment delay risk, funding risk, quality risk, labour dispute risks, natural disaster risk, exchange rate fluctuation risk, political instability, site condition risks, insurance inadequacy risk.	[9]
Technical risks, organisational risks, socio-political risks, environmental risks, financial risks.	[6]
Inflation (economic) risk, environmental and geological risk, design risk, construction delay risk, inadequate managerial skills risk, resource risk.	[29]

Table 3. Dimensions and risk criteria involved in construction projects.

Risk Dimension	Risk Criteria *	Brief Descriptions of Causes of the Mentioned Criteria Risks
External risks (D_1)	Political instability (C_1)	Frequent changes in government due to disputes among political parties, change in law due to local government's unpredictable new regulations, needless influence by local government on court proceedings regarding project disputes.
	Economic risk (C_2)	Fluctuation in currency exchange rate, unpredictable inflation due to immature banking systems, payment delays due to poor funding for project, inadequate forecasting about market demand.
	Social risk (C_3)	Racial tension and differences in work culture and language between foreign and local partners.
Project risk (D_2)	Technological risk (C_4)	Risk of insufficient technology, improper design, unexpected design changes; inadequate site investigation; change in construction procedures and insufficient resource availability.
	Work quality risk (C_5)	Corruption, including bribery, at sites; obsolete technology and practices by the local partner; low local workforce labor productivity due to poor skills or inadequate supervision; improper quality control; local partner tolerance of defects and inferior quality.
	Time and cost risk (C_6)	Delays due to disputes with contractors, natural disasters, and lack of availability of utilities; risk of labor disputes and strikes; insufficient cash flow, improper measurements, ill planned schedules, and delays in payment; lack of proper benchmarking and monitoring of construction activities.
Internal risks (D_3)	Resource risk (C_7)	Difficulty in hiring suitable skilled employees; risk of defective material from suppliers; risk of labor, materials, and equipment availability; poor competence and productivity of labor *.
	Documents and information risk (C_8)	Intellectual property protection risk from former local employees, partners, and third parties; corporate fraud including unexpected increases in turnover, unexpected resignations of financial advisers, intentional or unintentional negligence by auditors, bankers, or creditors.
	Stakeholder's risk (C_9)	Local partner's creditworthiness: Information on local partner's accounts lucidity, financial soundness, foreign exchange liquidity, staff reliability. Termination of joint ventures (JV): unfair dividends, e.g., assets, shares, and benefits, to foreign firms by local partner upon termination of JV contract.

* Sources: [4,6,9,23,27,52,75].

In past decades, construction businesses were constrained to use only a limited number of risk management procedures, even though they were not suitable for all situations. For example, Lyons and Skitmore [50] found that brainstorming is the most common risk identification technique used in the Queensland engineering construction industry. Forbes et al. [74] developed a matrix for selecting appropriate risk management techniques such as artificial intelligence, probabilistic analysis, sensitivity analysis, and decision trees in the built environment for each stage of risk management. In this paper, based on the brainstorming method, the experts sought five alternative construction project risk response mitigation strategies, as detailed in Table 4.

Table 4. Construction risks response strategies.

Alternative (s)	Preventive Management Techniques	References
A_1	Proper scheduling for getting updated project information.	[9]
A_2	Adjust plans for scope of work and estimates to counter risk implications.	[2]
A_3	Get information about local partner's credibility from present and past business partners.	[4]
A_4	Transfer or share risks to/with other parties.	[6]
A_5	Merger and diversification of projects.	[23]

5.2. Calculating Risk Based Criteria Weight Using D-ANP Framework

In this section, the weights of the risk based criteria in construction projects are calculated by D-ANP. The ANP has substantial influence in MCDM problems involving a wide range of factors and sub-factors. In the ANP, a decision problem is transformed into a network structure that allows both inter-intra dependency and feedback among the decision clusters, and even amongst elements within the same clusters. In this phase, the decision group is asked to make pairwise comparison matrices for priority weights of three dimensions and nine criteria (as detailed in Table 3).

Using Steps 1–6 (of Section 4.1) of D-CFPR, the priority weights are calculated in the decision matrices. The algorithmic steps of D-ANP are shown below:

Step 1: Construction of the hierarchy of criteria and alternative risks strategies. The clearly defined risk based construction project model is decomposed into a logical system like a network. Based on the hierarchy of Figure 2, we have five risk mitigation strategies $A_i(i = 1, 2, \ldots, 5)$ in Level 4 hierarchical position, three dimensions $D_j(j = 1, 2, 3)$ (external, project, and internal risk) in Level 2, and corresponding to each dimension, a total of nine criteria $C_j(j = 1, 2, \ldots, 9)$ in Level 3.

Step 2: Determination of the pairwise comparison matrices and priority vectors within clusters. The D numbers based preference matrix is first constructed before calculating the weight of the indicators using the ANP supermatrix. Here, we present the process of determining the priority weight of criteria $(C_1, C_2, \text{ and } C_3)$ for risk dimension *external risk* with respect to the criterion C_6 under dimension *Project risk*.

First, we calculate the relative significance of sub-criteria $(C_1, C_2, \text{ and } C_3)$ relative to sub-criterion C_6, to construct the inner dependence matrix built on the D numbers based preference relation $R_D = [D_{ij}]_{3\times 3}$. In this case, the preference modelled by a set of D numbers (*based on DM's choice*) involving both *uncertain entry* (i.e., $D_{12} = \{(0.55, 0.7), (0.65, 0.3)\}$ and *incomplete entry* $D_{23} = \{(0.75, 0.8)\}$, respectively:

$$R_D = \begin{bmatrix} D_{11} & D_{12} & D_{13} \\ D_{21} & D_{22} & D_{23} \\ D_{31} & D_{32} & D_{33} \end{bmatrix} = \begin{bmatrix} \{(0.5, 1.0)\} & \{(0.55, 0.7), (0.65, 0.3)\} & - \\ - & \{(0.5, 1.0)\} & \{(0.75, 0.8)\} \\ - & - & \{(0.5, 1.0)\} \end{bmatrix}.$$

The standard CFPR cannot handle this case, but D-CFPR (Equations (9)–(12) in Section 3.1) is effective to fill up the rest of the matrix elements in $R_D = [D_{ij}]_{3\times 3}$ as follows:

$$R_D = \begin{bmatrix} \{(0.5, 1.0)\} & \{(0.55, 0.7), (0.65, 0.3)\} & \{(0.80, 0.56), (0.9, 0.24)\} \\ \{(0.45, 0.7), (0.35, 0.3)\} & \{(0.5, 1.0)\} & \{(0.75, 0.8)\} \\ \{(0.2, 0.56), (0.1, 0.24)\} & \{(0.25, 0.8)\} & \{(0.5, 1.0)\} \end{bmatrix}.$$

Then, the D numbers based CFPR matrix $R_D = [D_{ij}]_{3\times3}$ is converted to a crisp mode, $R_C = [c_{ij}]_{3\times3}$, using Equation (13), and the integration representation of D numbers Equation (7), shown below:

$$R_c = I(R_D) = \begin{bmatrix} 0.5 & 0.58 & 0.66 \\ 0.42 & 0.5 & 0.60 \\ 0.13 & 0.20 & 0.5 \end{bmatrix}.$$

Applying the preference rules proposed for D-CFPR (in Step 3 of Section 4.1) and using Equation (14), the probability matrix $R_p = [p_{ij}]_{3\times3}$ is constructed:

$$R_p = \begin{bmatrix} 0 & 1 & 1 \\ 0 & 0 & 1 \\ 0 & 0 & 0 \end{bmatrix}.$$

Following the process (as mentioned in Step 4 of Section 4.1) of the D-CFPR methodology, we obtain the triangular matrix $R_p^T = [p_{ij}^T]_{3\times3}$. Using the triangularisation method, the ranking of the indicators is calculated and shown as: $I_1 \succ I_2 \succ I_3$, where the symbol "$\succ$" indicates preference,

$$R_p^T = \begin{bmatrix} 0 & 1 & 1 \\ 0 & 0 & 1 \\ 0 & 0 & 0 \end{bmatrix}.$$

We next evaluate the relative weights of the risk criteria. *First*, based on the ranking of the risk criteria in the triangulated matrix R_p^T, the crisp matrix R_c, is converted to a triangular crisp matrix R_c^T, as per Equation (16):

$$R_c^T = \begin{bmatrix} 0.5 & 0.58 & 0.66 \\ 0.42 & 0.5 & 0.60 \\ 0.13 & 0.20 & 0.5 \end{bmatrix}.$$

Next, for elements satisfying $R_c^T(i, j) + R_c^T(j, i) < 1$, a new polishing operation Equation (17) is executed and, by also applying Equation (18), a novel triangulated crisp matrix $R_k^T = [k_{ij}^T]_{3\times3}$ is obtained:

$$R_k^T = \begin{bmatrix} 0.5 & 0.58 & 0.76 \\ 0.42 & 0.5 & 0.70 \\ 0.21 & 0.24 & 0.5 \end{bmatrix}.$$

Finally, applying Equation (19), a group of equations is built to calculate the priority weight $w_i (i = 1, 2, \ldots, n)$ of each risk-based criterion. Applying the weight relation of the indicators in matrix mode, and incorporating necessary constraints, the weight equations are constructed and shown below:

$$\begin{cases} \lambda(w_1 - w_2) = 0.58 - 0.5 \\ \lambda(w_2 - w_3) = 0.76 - 0.5 \\ w_1 + w_2 + w_3 = 1 \\ \lambda > 0 \\ w_i \geq 0, \ \forall i \in \{1, 2, 3\} \end{cases},$$

where w_i denotes the weight of the ith indicator and λ indicates the granular information about the pairwise evaluation, which is connected to the cognitive aptitude of the experts. Setting $\lambda = 3$ and using Equation (20), the weight of risk criteria C_1, C_2, and C_3 for dimension D_1 relative to C_6 of dimension D_2 are calculated as $(0.373, 0.347, 0.280)^T$, respectively.

For quantifying the consistency of the D-CFPR based matrix R_p^T, an ID defined for the D *numbers preference relation* (as defined in Equation (21)) is used to express such inconsistency, and, for the case study taken, it is found to be consistent.

Similarly, using the same process, the priority weights of remaining criteria, shown in Figure 3 with respect to the same dimensions and criteria of other clusters (dimensions), are calculated.

Step 3: Formation of unweighted supermatrix. Arrange all priority vectors, indicating the influence of pre-set elements, in different cluster elements in the network, as sub-columns of the resultant column in an unweighted supermatrix W, which is composed of k clusters $D_k (k = 1, 2, 3)$, with corresponding linkages (*criteria*) $\{e_{k1}, e_{k2}, e_{k3}\}$ for $k = 1, 2, 3$. Putting $k = 1$, we get the first cluster (*dimension*) D_1 along with three elements $\{e_{11}, e_{12}, e_{13}\}$ representing three risk criteria $\{C_1, C_2, C_3\}$, respectively. For $k = 2$, we get a second cluster (*dimension*) D_2 along with three elements $\{e_{21}, e_{22}, e_{23}\}$ representing three risk criteria $\{C_4, C_5, C_6\}$. Similarly, the remaining cluster (*dimensions*) D_3, along with its risk criteria, can be found. Thus, we get three clusters (*dimensions*) $D_k = \{D_1, D_2, D_3\}$ and nine corresponding risk criteria $C_j \{j = 1, 2, \ldots, 9\}$. Based on the above, the unweighted supermatrix W is formed by placing priority vector elements in the particular column of W, where each criterion influences the other risk criteria. Details shown in Table 5.

Step 4: Calculating the weighted supermatrix. The weighted supermatrix (W^a) is calculated by multiplying unweighted supermatrix W (Table 5) by the inner dependence matrix of risk dimension $D_j (j = 1, 2, 3)$ (Table 6). Details are shown in Table 7.

Step 5: Selecting the weight of criteria based on the limit matrix. To make the matrix column stochastic in Table 7, we normalise the weighted supermatrix (W^a) column wise, and the result is shown in Table 8. The normalized weighted supermatrix $\hat{W}$ (Table 8) is raised to its limiting power using Equation (22), to get the limiting supermatrix W^q (Table 9). The final ranking of risk criteria weight for the construction project is shown in Table 10.

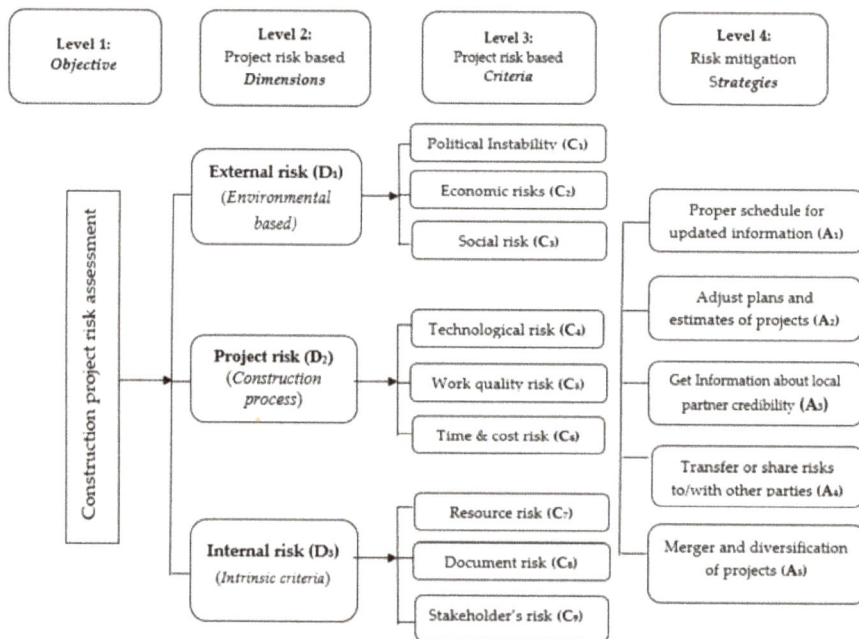

Figure 2. A hierarchical construction project risk breakdown structure.

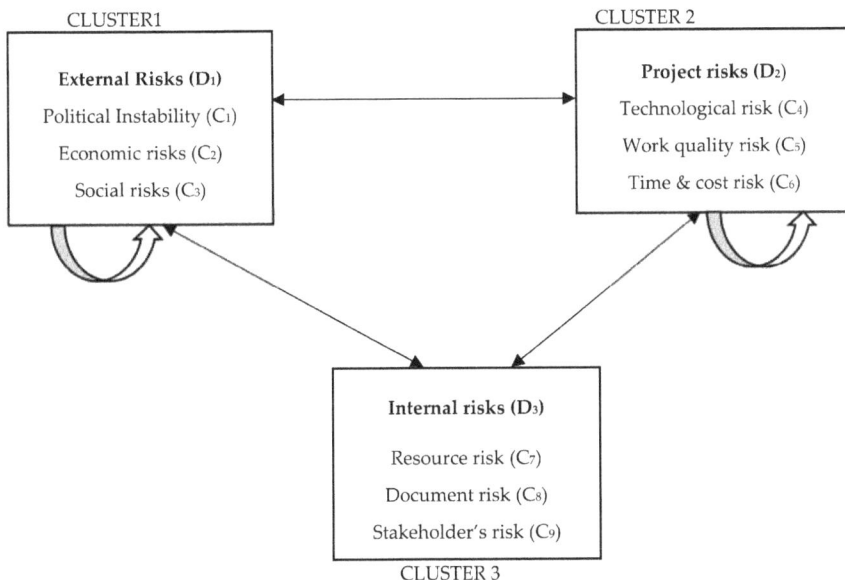

Figure 3. Network structure among construction risk dimensions and criteria.

Table 5. Unweighted supermatrix formed from every risk factor.

		External Risk (D_1)			Project Risk (D_2)			Internal Risk (D_3)		
		C_1	C_2	C_3	C_4	C_5	C_6	C_7	C_8	C_9
External risk (D_1)	C_1	0.000	0.576	0.589	0.315	0.386	0.373	0.332	0.353	0.395
	C_2	0.525	0.000	0.411	0.357	0.351	0.347	0.336	0.321	0.327
	C_3	0.475	0.424	0.000	0.327	0.264	0.280	0.332	0.326	0.278
Project risk (D_2)	C_4	0.288	0.355	0.354	0.000	0.461	0.510	0.348	0.346	0.340
	C_5	0.416	0.320	0.338	0.481	0.000	0.490	0.334	0.361	0.363
	C_6	0.296	0.326	0.308	0.519	0.539	0.000	0.318	0.293	0.298
Internal risk (D_3)	C_7	0.332	0.324	0.357	0.364	0.313	0.370	0.000	1.000	1.000
	C_8	0.351	0.369	0.351	0.343	0.371	0.351	1.000	0.000	1.000
	C_9	0.316	0.308	0.292	0.293	0.315	0.279	1.000	1.000	0.000

Table 6. Inner dependence matrix of construction project factors.

	Dimensions		
	External Risk	**Project Risk**	**Internal Risk**
External risk	1	0.518	0.503
Project risk	0.537	1	0.496
Internal risk	0.462	0.482	1

Table 7. Weighted supermatrix based on supply chain risk factors.

		External Risk			Project Risk			Internal Risk		
		C_1	C_2	C_3	C_4	C_5	C_6	C_7	C_8	C_9
External risk	C_1	0.000	0.576	0.589	0.163	0.200	0.193	0.167	0.178	0.199
	C_2	0.525	0.000	0.411	0.185	0.182	0.180	0.169	0.162	0.165
	C_3	0.475	0.424	0.000	0.170	0.137	0.145	0.167	0.164	0.140
Project risk	C_4	0.155	0.191	0.191	0.000	0.461	0.510	0.173	0.172	0.169
	C_5	0.223	0.172	0.182	0.481	0.000	0.490	0.166	0.179	0.180
	C_6	0.159	0.175	0.165	0.519	0.539	0.000	0.158	0.146	0.148
Internal risk	C_7	0.154	0.150	0.165	0.176	0.151	0.178	0.000	1.000	1.000
	C_8	0.162	0.171	0.162	0.165	0.179	0.169	1.000	0.000	1.000
	C_9	0.146	0.142	0.135	0.141	0.152	0.134	1.000	1.000	0.000

Table 8. Normalised weighted supermatrix based on supply chain risk factors.

		External Risk			Project Risk			Internal Risk		
		C_1	C_2	C_3	C_4	C_5	C_6	C_7	C_8	C_9
External risk	C_1	0.000	0.288	0.294	0.082	0.100	0.097	0.056	0.059	0.066
	C_2	0.263	0.000	0.206	0.093	0.091	0.090	0.056	0.054	0.055
	C_3	0.238	0.212	0.000	0.085	0.068	0.073	0.056	0.055	0.047
Project risk	C_4	0.078	0.095	0.095	0.000	0.231	0.255	0.058	0.057	0.056
	C_5	0.112	0.086	0.091	0.241	0.000	0.245	0.055	0.060	0.060
	C_6	0.079	0.088	0.083	0.260	0.270	0.000	0.053	0.049	0.049
Internal risk	C_7	0.077	0.075	0.083	0.088	0.075	0.089	0.000	0.333	0.333
	C_8	0.081	0.085	0.081	0.083	0.090	0.085	0.333	0.000	0.333
	C_9	0.073	0.071	0.068	0.071	0.076	0.067	0.333	0.333	0.000

Table 9. Limited supermatrix based on supply chain risk factors.

		External Risk			Project Risk			Disruption Risk		
		C_1	C_2	C_3	C_4	C_5	C_6	C_7	C_8	C_9
External risk	C_1	0.1062	0.1062	0.1062	0.1062	0.1062	0.1062	0.1062	0.1062	0.1062
	C_2	0.0958	0.0958	0.0958	0.0958	0.0958	0.0958	0.0958	0.0958	0.0958
	C_3	0.0894	0.0894	0.0894	0.0894	0.0894	0.0894	0.0894	0.0894	0.0894
Project risk	C_4	0.0972	0.0972	0.0972	0.0972	0.0972	0.0972	0.0972	0.0972	0.0972
	C_5	0.0996	0.0996	0.0996	0.0996	0.0996	0.0996	0.0996	0.0996	0.0996
	C_6	0.0971	0.0971	0.0971	0.0971	0.0971	0.0971	0.0971	0.0971	0.0971
Internal risk	C_7	0.1392	0.1392	0.1392	0.1392	0.1392	0.1392	0.1392	0.1392	0.1392
	C_8	0.1406	0.1406	0.1406	0.1406	0.1406	0.1406	0.1406	0.1406	0.1406
	C_9	0.1349	0.1349	0.1349	0.1349	0.1349	0.1349	0.1349	0.1349	0.1349

Table 10. Ranking of construction project risk criteria.

Dimensions	Risk Criteria	Ranking
External risk (D_1)	Political instability (C_1)	4
	Economic risk (C_2)	8
	Social risk (C_3)	9
Project risk (D_2)	Technological risk (C_4)	6
	Work quality risk (C_5)	5
	Time and cost risk (C_6)	7
Internal risk (D_3)	Resource risk (C_7)	2
	Document and information risk (C_8)	1
	Stakeholder's risk (C_9)	3

From Table 10, it is concluded that the third cluster (dimension) *internal risk* (D_3) has a severe risk effect on the construction project sector. *Document and information risk* (C_8) is the most risky, followed by *resource risk* (C_7) and *stakeholder's risk* (C_9). First cluster *External risk* (D_1) has less of a risk effect on construction business. *Economic risk* (C_2) and *social risk* (C_3) are in the 8th and 9th positions, respectively. However, *political instability* attains the 4th position in respect to the risk category. Managers and stakeholders should keep this in view when choosing projects in large construction sectors.

5.3. Determination of Final Alternative Ranking by D-MABAC

In this phase, the evaluation and ranking of risk response alternatives is performed by the application of a D numbers based MABAC (D-MABAC) methodology in construction project risk management. The step-by-step computational procedure is shown below.

Step 1: First, the five *risk response* alternative vectors, with respect to nine risk criteria $C_j (j = 1, 2, \ldots, 9)$, are represented as $A_i = (d_{i1}, d_{i2}, \ldots, d_{i9})$ $(i = 1, 2, \ldots, 5)$ using *incomplete and uncertain* numbers expressed in D numbers. Using Equations (23) and (24), we develop an initial decision matrix $M = [d_{ij}]_{5 \times 9}$ (Table 11) along with its crisp form $X = [I(d_{ij})]_{5 \times 9} = [x_{ij}]_{5 \times 9}$.

Table 11. Comparison of alternatives w.r.t risk criteria using D numbers.

	C_1	C_2	C_3	C_4	C_5	C_6	C_7	C_8	C_9
A_1	(0.62, 0.5)	(0.72, 0.4), (0.48, 0.6)	(0.53, 0.4)	(0.58, 0.5)	(0.72, 0.6)	(0.48, 0.6), (0.64, 0.4)	(0.66, 0.6)	(0.38, 0.9)	(0.84, 0.2)
A_2	(0.68, 0.4)	(0.44, 0.8)	(0.64, 0.6), (0.32, 0.3)	(0.44,.9)	(0.82, 0.9)	(0.88, 0.6)	(0.56, 0.8)	(0.92, 0.8)	(0.69, 0.4)
A_3	(0.54, 0.8), (0.68, 0.2)	(0.68, 0.3)	(0.47, 0.9)	(0.78, 0.8)	(0.38, 0.7), (0.59, 0.3)	(0.68, 0.5)	(0.29, 0.6), (0.39, 0.4)	(0.28, 0.6)	(0.34, 0.6)
A_4	(0.72, 0.9)	(0.49, 0.7)	(0.78, 0.4)	(0.86, 0.4)	(0.88, 0.4)	(0.47, 0.7)	(0.64, 0.2)	(0.62, 1)	(0.56, 0.7)
A_5	(0.48, 1)	(0.56, 0.9)	(0.82, 0.7)	(0.36, 1)	(0.78, 0.7)	(0.59, 0.9)	(0.78, 0.7)	(0.68, 0.3), (0.49, 0.6)	(0.44, 0.8)

Step 2: The elements of the crisp decision matrix $X = [x_{ij}]_{5 \times 9}$ are normalized using Equations (24) and (25) to form a normalized decision matrix $N = [t_{ij}]_{5 \times 9}$, shown in Table 12.

Table 12. Normalised decision matrix of alternatives w.r.t criteria.

	C_1	C_2	C_3	C_4	C_5	C_6	C_7	C_8	C_9
A_1	0.8989	0	1	1	0.7927	0	0.3589	0.5846	1
A_2	1	0.6022	0.2597	0.6826	0	0.0744	0.2344	0	0.5179
A_3	0.2128	1	0.4171	0	0.7642	0.9488	0.5167	0.8427	0.8393
A_4	0	0.6263	0.7238	0.8383	1	1	1	1	0
A_5	0.4468	0.1935	0	0.7904	0.4974	0.0605	0	0.3531	0.1786

Step 3: Using Equations (26) and (27), the elements of the weighted normalized decision matrix $V = [v_{ij}]_{5 \times 9}$ are calculated, and are shown in Table 13.

Table 13. Weighted normalised decision matrix of alternatives w.r.t criteria.

	C_1	C_2	C_3	C_4	C_5	C_6	C_7	C_8	C_9
A_1	0.2017	0.1916	0.1788	0.1944	0.1786	0.0971	0.1892	0.2381	0.2698
A_2	0.2124	0.1304	0.1126	0.1636	0.0996	0.1043	0.1718	0.1406	0.2048
A_3	0.1288	0.1641	0.1267	0.0972	0.1757	0.1892	0.2111	0.2812	0.2481
A_4	0.1062	0.1324	0.1541	0.1787	0.1992	0.1942	0.2784	0.1693	0.1349
A_5	0.1537	0.0958	0.0894	0.1740	0.1491	0.1030	0.1392	0.1995	0.1590

Step 4: Next, using Equations (28) and (29), we determine the BAA, $G = [g_1 \ g_2 \ \cdots \ g_9]$ for each criterion $C_j(j = 1, 2, \ldots, 9)$. This is followed by calculation of the distance (q_{ij}) for matrix elements $Q = [q_{ij}]_{5 \times 9}$ for risk response alternatives $A_i(i = 1, 2, \ldots, 5)$ from BAA $G = [g_1 \ g_2 \ \cdots \ g_9]$ using Equations (30) and (31).

Step 5: Finally, using Equation (32), we calculate the sum function $\hat{S}_i(i = 1, 2, \ldots, 5)$ to obtain the ranking of alternatives $A_i(i = 1, 2, \ldots, 5)$. A graphical representation of the process is shown in Figure 4. Ranking of alternative risk response alternatives (Table 14) is finalised according to values calculated by D-MABAC in descending order. In this paper, the first alternative risk response was selected and implemented.

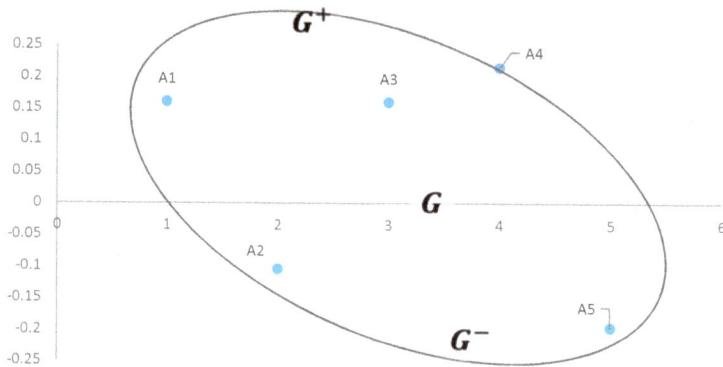

Figure 4. D numbers based Multi-Attributive Border Approximation area Comparison (D-MABAC) in risk mitigation strategy selection.

Table 14. Ranking of alternatives using the D-MABAC method.

	Alternative Risk Responses	Q	Rank
Risk response (A_1)	Proper scheduling for getting updated project information.	0.2830	1
Risk response (A_2)	Adjust plans for scope of work and estimates to counter risk implications.	−0.1161	4
Risk response (A_3)	Get information about local partner's credibility from its present and past business partners.	0.1660	2
Risk response (A_4)	Transfer or share risks to/with other parties.	0.0913	3
Risk response (A_5)	Merger and diversification of projects.	−0.1935	5

6. Results and Discussion

In this section, a detailed comparative analysis of all alternative initiatives (with respect to criteria and dimensions) is conducted.

6.1. Comparison of Alternative Ranking Using Different MCDM Methods

The hybrid MCDM methods in D numbers environment namely, D numbers based multi-attributive border approximation area comparison (D-MABAC), and D numbers based technique for order of preference by similarity to ideal solution (D-TOPSIS), D numbers based complex proportional assessment (D-COPRAS), and D numbers based additive ratio assessment (D-ARAS), were applied to the construction project based case study data to obtain the weighted normalised decision-making matrix $V = [v_{ij}]_{5 \times 9}$ (Table 13). The priority order of the risk response alternatives $A_i(i = 1, 2, \ldots, 5)$ is compared and presented in Table 15.

Table 15. Comparison of MABAC with various existing multi-criteria decision making methods.

Alternative Risk Responses	D-MABAC	D-TOPSIS	D-COPRAS	D-ARAS
A_1	1	1	1	1
A_2	4	4	4	4
A_3	2	2	2	2
A_4	3	3	3	3
A_5	5	5	5	5

Ranking of the risk response alternatives according to the presented MCDM methods concluded that the optimal alternative risk response is A_1 (Proper scheduling for getting updated project information), followed by A_3 (Information about local partner's credibility), A_4 (Transfer risks with other parties), and A_2 (Adjust plans for scope of work). The worst performing risk response alternative is A_5 (Merger and diversification of projects).

Spearman's rank correlation coefficient (r_k) between ranks is applied for determining correlation of ranks obtained by various approaches. Here, this coefficient is applied to demonstrate the statistical importance of difference among the ranking obtained through pairwise correlation analysis of different MCDM methods. Based on the recommendation of Keshavarz Ghorabaee et al. [65], all (r_k) values higher than 0.80 show considerably high correlation. As per Table 16, a strong correlation (1.000) among the MCDM approaches is shown, confirming the credibility of the proposed approach.

Table 16. Rank correlation of various MCDM methods.

Spearman's Coefficient	D-MABAC	D-TOPSIS	D-COPRAS	D-ARAS
r_k	-	1.000	1.000	1.000

6.2. Sensitivity Analysis

Ranking of results in MCDM problems are subject to the distribution of weight coefficients of the criteria. Sometimes, modifying these criteria weight coefficients may change the ranking order of alternatives, generally analysed by sensitivity analysis during the decision-making process. The above weight coefficients are usually based on expert subjective perception, and thus the outcome of probable deviation of these weight values need to be properly assessed.

A sensitivity analysis was executed to measure the level of crosstalk amongst the criteria, revealing the variation in alternative rankings with variation in criteria weight. Outcomes of the sensitivity analysis for prioritising specific project based criteria $C_j(j = 1, 2, \ldots, 9)$ weights are shown in Table 17, and its corresponding effect on ranking of risk response alternatives $A_i(i = 1, 2, \ldots, 5)$ in Table 18.

- The results (Table 17 and Figure 5) shows that assigning various weights to project based criteria $C_j(j = 1, 2, \ldots, 9)$ through different scenarios $\{S_1 - S_8\}$ results in changes to ranking of individual alternatives, thus proving that the model is sensitive to variations in weight coefficients.
- Analysis of the alternative ranking through eight scenarios (Table 18) showed that alternative A_1 retained its rank in five scenarios $\{S_2, S_3, S_4, S_5, S_8\}$ (*best-ranked alternative*), while, in the remaining two scenarios $\{S_1, S_7\}$, it was ranked second, and third in scenario $\{S_6\}$.
- The worst-ranked alternative A_5 retained its rank in six scenarios $\{S_2, S_3, S_4, S_5, S_7, S_8\}$, while in two scenarios $\{S_1, S_6\}$, it was ranked second worst. Therefore, changing the criteria weights through different scenarios resulted in changes to the ranks of the remaining alternatives.
- In addition, from Tables 17 and 18, it is clear that prioritising criteria C_9 has less of an effect on ranking position of alternatives. However, prioritising criteria set $\{C_4, C_6, \text{and } C_7\}$ in scenario S_7, $\{C_2, C_6, \text{and } C_8\}$ in scenario S_1, along with $\{C_4, C_6, \text{and } C_8\}$ in scenario S_6, all altered the positions of risk response alternatives $A_i(i = 1, 2, \ldots, 5)$.

- The prioritising of criteria weight in scenarios $\{S_2, S_3, S_4, S_5, S_8\}$ has no effect on ranking of best or worst risk response alternative A_1 and A_5, respectively, but it does have an effect on the ranking of the second best risk response alternative A_2.

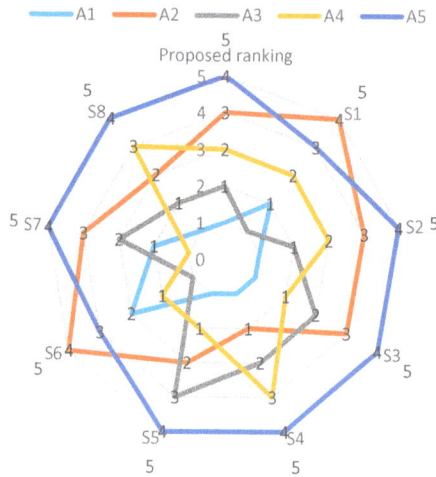

Figure 5. Sensivity analysis of the alternative ranking through different scenarios.

Table 17. Scenarios for different criteria weights.

Criteria	Scenarios *							
	S_1	S_2	S_3	S_4	S_5	S_6	S_7	S_8
C_1	0.0113	**0.1482**	0.0226	**0.2674**	**0.1944**	0.0795	0.0081	**0.1602**
C_2	**0.1578**	0.1121	0.1204	0.1178	0.0914	0.0632	0.0962	0.0537
C_3	0.0576	**0.1789**	**0.2946**	0.0001	**0.1853**	0.0577	0.0308	0.1054
C_4	0.0904	0.1118	**0.2655**	**0.1556**	**0.2219**	0.0519	**0.2032**	0.0958
C_5	0.1172	0.0033	0.0478	0.0598	0.0603	0.0703	0.0875	**0.1805**
C_6	**0.2016**	0.0233	0.0496	0.0631	0.0322	**0.3311**	0.1958	0.0218
C_7	0.0894	**0.1663**	0.0352	0.0937	0.056	0.0055	**0.1485**	0.0223
C_8	**0.2103**	0.0933	0.003	0.0276	0.0871	**0.332**	0.1191	0.0539
C_9	0.0645	**0.1628**	0.1613	0.215	0.0714	0.0089	0.1107	**0.3063**

* Priority criteria are indicated in bold text for different scenarios.

Table 18. Sensitivity in alternative rankings for different scenarios of criteria weighting.

Alternative Risk Responses	Scenarios							
	S_1	S_2	S_3	S_4	S_5	S_6	S_7	S_8
A_1	2	1	1	1	1	3	2	1
A_2	5	4	4	2	3	5	4	3
A_3	1	2	3	3	4	1	3	2
A_4	3	3	2	4	2	2	1	4
A_5	4	5	5	5	5	4	5	5

7. Conclusions

Construction projects present a very complex field involving a large number of stakeholders. From the perspective of information sharing, uncertain information and data cause loss of faith among various stakeholders and increased risk in construction management, including project

Symmetry **2018**, *10*, 46

complexity and decision-making environment conditions. These risks affect project activities, which indirectly impact construction costs, resulting in delays and poor building quality. Thus, in this paper, we rank uncertain risk strategies in construction projects using a D-ANP-MABAC multi-criteria decision-making (MCDM) approach. In the new proposed method, the decision matrix determination from the MCDM problem is transformed to D numbers, which effectively represent the inevitable uncertainties such as incompleteness and imprecision, due to the subjective assessment of decision makers. As the basic element of many decision-making methods especially in analytic network process (ANP) model, the preference relation has attracted interests among researchers and practitioners. Fuzzy preference relation construct pairwise decision matrices based on linguistic values but is inconsistent due to inability of experts in dealing with overcomplicated objects. CFPR methodology [30] removes this inconsistency, but fails to deals with incomplete and uncertain information due to the lack of experts' knowledge and the limitation of cognition. To overcome this weaknesses, D-CFPR methodology is applied to decision matrices allowing all stakeholders (members) of a construction business to address multiple criteria involving various types of uncertainties, such as imprecision and fuzziness in the decision process. The D-CFPR uses D numbers to express the linguistic preference values given by experts, and it can also be reduced to classical CFPR. Based on the D-CFPR based preference relation, the priority vectors of criteria are determined, to be used as inputs for any cluster of matrix formed by the D-ANP method, and obtain the corresponding risk criteria weights. The alternative risk response strategy ranking is achieved using the D-MABAC method.

In view of several categories of uncertainties, including incompleteness and impreciseness, the proposed technique can effectively represent and address uncertain information weighting risk criteria and alternatives in a logical way. An example of selecting risk strategies in construction risk projects is shown here to demonstrate the efficiency of the proposed approach. The assessment of a real world application of D-ANP-MABAC methodology, along with its output result of sensitivity, clearly identifies its potential to provide stable solutions to the problematic choice of laying-up positions. Based on that, the proposed risk strategy alternatives are successfully ranked. Thus, it can be concluded that the above procedure has provided an alternate approach for sustainable risk analysis and decision-making in the construction sector. In our proposed method, we consider both CFPR and ANP methodologies in D numbers domain. As in the present scenario, since the priority vectors for the criteria set are deduced from D-CFPR methodology, which are used as inputs in ANP matrix, it will thus also consume more computational time. Therefore, the computational complexity occurring in present D-CFPR-ANP methodology in future studies needs to be further optimized. In future research, the theoretical framework needs to be modified considering the hidden risks in construction sectors and further be applied to other real-life application areas such as supplier selection problems, project portfolio management, renewable energy selection, etc., to further validate its effectiveness.

Author Contributions: The individual contribution and responsibilities of the authors were as follows: Kajal Chatterjee designed the research, methodology, performed the development of the paper, Krishnendu Adhikary collected and analyzed the data and the obtained results, Edmundas Kazimieras Zavadskas provided extensive advice throughout the study, Jolanta Tamošaitienė assisted with the research design and revised the manuscript. Samarjit Kar assisted with methodology and findings. All of the authors have read and approved the final manuscript.

Conflicts of Interest: The authors declare no conflict of interest.

References

1. Akintoye, A.S.; MacLeod, M.J. Risk analysis and management in construction. *Int. J. Proj. Manag.* **1997**, *15*, 31–38. [CrossRef]
2. Schatteman, D. Methodology for integrated risk management and proactive scheduling of construction projects. *J. Constr. Eng. Manag.* **2008**, *134*, 885–893. [CrossRef]
3. Skorupka, D. Identification and initial risk assessment of construction projects in Poland. *J. Manag. Eng.* **2008**, *24*, 120–127. [CrossRef]

4. Wang, S.; Dulaimi, M.; Aguria, M. Risk management framework for construction projects in developing countries. *Constr. Manag. Econ.* **2004**, *22*, 237–252. [CrossRef]
5. Abdelgawad, M.; Fayek, A. Risk Management in the construction industry using combined fuzzy FMEA and fuzzy AHP. *J. Constr. Eng. Manag.* **2010**, *10*, 1028–1036. [CrossRef]
6. Mhetre, K.; Konnur, B.A.; Landage, A.B. Risk Management in construction industry. *Int. J. Eng. Res.* **2016**, *5*, 153–155.
7. Ribeiro, M.I.F.; Ferreira, F.A.F.; Jalali, M.S.; Meidutė-Kavaliauskienė, I. A fuzzy knowledge-based framework for risk assessment of residential real estate investments. *Technol. Econ. Dev. Econ.* **2017**, *23*, 140–156. [CrossRef]
8. Ribeiro, C.; Ribeiro, A.R.; Maia, A.S.; Tiritan, M.E. Occurrence of Chiral Bioactive Compounds in the Aquatic Environment: A Review. *Symmetry* **2017**, *9*, 215. [CrossRef]
9. Iqbal, S.; Choudhry, R.; Holschemacher, K.; Ali, A.; Tamošaitienė, J. Risk management in construction projects. *Technol. Econ. Dev. Econ.* **2015**, *21*, 65–78. [CrossRef]
10. Hwang, B.-G.; Zhao, X.; Yu, G.S. Risk identification and allocation in underground rail construction joint ventures: Contractors' perspective. *J. Civ. Eng. Manag.* **2016**, *22*, 758–767. [CrossRef]
11. Butaci, C.; Dzitac, S.; Dzitac, I.; Bologa, G. Prudent decisions to estimate the risk of loss in insurance. *Technol. Econ. Dev. Econ.* **2017**, *23*, 428–440. [CrossRef]
12. Pak, D.; Han, C.; Hong, W.-T. Iterative Speedup by Utilizing Symmetric Data in Pricing Options with Two Risky Assets. *Symmetry* **2017**, *9*, 12. [CrossRef]
13. Ravanshadnia, M.; Rajaie, H. Semi-Ideal Bidding via a Fuzzy TOPSIS Project Evaluation Framework in Risky Environments. *J. Civ. Eng. Manag.* **2013**, *19* (Suppl. 1), S106–S115. [CrossRef]
14. Ebrat, M.; Ghodsi, R. Construction project risk assessment by using adaptive-network-based fuzzy inference system: An Empirical Study. *KSCE J. Civ. Eng.* **2014**, *18*, 1213–1227. [CrossRef]
15. Taylan, O.; Bafail, A.; Abdulaal, R.; Kabli, M. Construction projects selection and risk assessment by fuzzy AHP and fuzzy TOPSIS methodologies. *Appl. Soft Comput.* **2014**, *17*, 105–116. [CrossRef]
16. Dziadosz, A.; Rejment, M. Risk analysis in construction project-chosen Methods. *Procedia Eng.* **2015**, *122*, 258–265. [CrossRef]
17. Schieg, M. Risk management in construction project management. *J. Bus. Econ. Manag.* **2006**, *7*, 77–83.
18. Serpella, A.F.; Ferrada, X.; Howard, R.; Rubio, L. Risk management in construction projects: A knowledge-based approach. *Procedia-Soc. Behav. Sci.* **2014**, *119*, 653–662. [CrossRef]
19. Santos, R.; Jungles, A. Risk level assessment in construction projects using the schedule performance index. *J. Constr. Eng.* **2016**, *2016*, 5238416. [CrossRef]
20. Sadeghi, N.; Fayek, A.; Pedrycz, W. Fuzzy Monte Carlo simulation and risk assessment in construction. *Comput.-Aided Civ. Infrastruct. Eng.* **2010**, *25*, 238–252. [CrossRef]
21. Nieto-Morote, A.; Ruz-Vila, F. A fuzzy approach to construction project risk assessment. *Int. J. Proj. Manag.* **2011**, *29*, 220–231. [CrossRef]
22. Deng, X.; Hu, Y.; Deng, Y. Bridge condition assessment using D numbers. *Sci. World J.* **2014**, *2014*, 358057. [CrossRef] [PubMed]
23. Zavadskas, E.K.; Turskis, Z.; Tamošaitienė, J. Risk assessment of construction projects. *J. Civ. Eng. Manag.* **2010**, *16*, 33–46. [CrossRef]
24. Vafadarnikjoo, A.; Mobin, M.; Firouzabadi, S. An intuitionistic fuzzy-based DEMATEL to rank risks of construction projects. In Proceedings of the 2016 International Conference on Industrial Engineering and Operations Management, Detroit, MI, USA, 23–25 September 2016; pp. 1366–1377.
25. Mohammadi, A.; Tavakolan, M. Construction project risk assessment using combined fuzzy and FMEA. In Proceedings of the 2013 Joint IFSA World Congress and NAFIPS Annual Meeting, Edmonton, AB, Canada, 24–28 June 2013; pp. 232–237.
26. Hashemi, S.; Karimi, A.; Tavana, M. An integrated green supplier selection approach with analytic network process and improved grey relational analysis. *Int. J. Prod. Econ.* **2015**, *159*, 178–191. [CrossRef]
27. Ahmadi, M.; Behzadian, K.; Ardeshir, A.; Kapelan, Z. Comprehensive risk management using fuzzy FMEA and MCDA technique in highway construction projects. *J. Civ. Eng. Manag.* **2016**, *23*, 300–310. [CrossRef]
28. Shin, D.; Shin, Y.; Kim, G. Comparison of risk assessment for a nuclear power plant construction project based on analytic hierarchy process and fuzzy analytic hierarchy process. *J. Build. Const. Plan. Res.* **2016**, *4*, 157–171. [CrossRef]

29. Dehdasht, G.; Zin, R.M.; Ferwati, M.S.; Abdullahi, M.M.; Keyvanfar, A.; McCaffer, R. DEMATEL-ANP risk assessment in oil and gas construction projects. *Sustainability* **2017**, *9*, 1420. [CrossRef]
30. Herrera-Viedma, E.; Herrera, F.; Chiclana, F.; Luque, M. Some issues on consistency of fuzzy preference relations. *Eur. J. Oper. Res.* **2004**, *154*, 98–109. [CrossRef]
31. Chen, Y.H.; Chao, R.J. Supplier selection using consistent fuzzy preference relations. *Expert Syst. Appl.* **2012**, *39*, 3233–3240. [CrossRef]
32. Hosseini, L.; Tavakkoli-Moghaddam, R.; Vahdani, B.; Mousavi, S.; Kia, R. Using the analytical network process to select the best strategy for reducing risks in a supply chain. *J. Eng.* **2013**, *2013*, 355628. [CrossRef]
33. Hesamamiri, R.; Mahdavi Mazdeh, M.; Bourouni, A. Knowledge-based strategy selection: A hybrid model and its implementation. *VINE J. Inf. Knowl. Manag. Syst.* **2016**, *46*, 21–44. [CrossRef]
34. Deng, X.; Lu, X.; Chan, F.; Sadiq, R.; Mahadevan, S.; Deng, Y. D-CFPR: D numbers extended consistent fuzzy preference relations. *Knowl.-Based Syst.* **2015**, *73*, 61–68. [CrossRef]
35. Zhang, X.; Deng, Y.; Chan, F.; Adamatzky, A.; Mahadevan, S. Supplier selection based on evidence theory and analytic network process. *Proc. Inst. Mech. Eng. Part B J. Eng. Manuf.* **2014**, *230*, 562–573. [CrossRef]
36. Deng, Y.D. Numbers: Theory and applications. *J. Inf. Comput. Sci.* **2012**, *9*, 2421–2428.
37. Han, X.; Chen, X. D-VIKOR method for medicine provider selection. In Proceedings of the IEEE Seventh International Joint Conference on Computational Sciences and Optimization (CSO), Beijing, China, 4–6 July 2014; pp. 419–423.
38. Liu, H.; You, J.; Fan, X.; Lin, Q. Failure mode and effects analysis using D numbers and grey relational projection method. *Expert Syst. Appl.* **2014**, *41*, 4670–4679. [CrossRef]
39. Deng, X.; Hu, Y.; Deng, Y.; Mahadevan, S. Supplier selection using AHP methodology extended by D numbers. *Expert Syst. Appl.* **2014**, *41*, 156–167. [CrossRef]
40. Fan, G.; Zhong, D.; Yan, F.; Yue, P. A hybrid fuzzy evaluation method for curtain grouting efficiency assessment based on an AHP method extended by D numbers. *Expert Syst. Appl.* **2015**, *44*, 289–303. [CrossRef]
41. Fei, L.; Hu, Y.; Xiao, F.; Chen, L.; Deng, Y. A modified TOPSIS method based on D numbers and its application in human resources selection. *Math. Probl. Eng.* **2016**, *2016*, 6145196. [CrossRef]
42. Zuo, Q.; Qin, X.; Tian, Y.; Wei, D. A multi-attribute decision making for investment decision based on D numbers methods. *Sci. Res.* **2016**, *6*, 765–775. [CrossRef]
43. Renault, B.; Agumba, J. Risk management in the construction industry: A new literature review. *MATEC Web Conf.* **2016**, *66*. [CrossRef]
44. Pamucar, D.; Cirovic, G. The selection of transport and handling resources in logistics centers using Multi-Attribute Border Approximation area Comparison (MABAC). *Expert Syst. Appl.* **2015**, *42*, 3016–3028. [CrossRef]
45. Peng, X.; Yang, Y. Pythagorean fuzzy Choquet integral based MABAC method for multiple attribute group decision making. *Int. J. Intell. Syst.* **2016**, *31*, 989–1020. [CrossRef]
46. Yu, S.; Wang, J.; Wang, J. An interval type-2 fuzzy likelihood-based MABAC approach and its application in selecting hotels on a tourism website. *Int. J. Fuzzy Syst.* **2016**, *9*, 47–61. [CrossRef]
47. Xue, Y.; You, J.; Lai, X.; Liu, H. An interval-valued intuitionistic fuzzy MABAC approach for material selection with incomplete weight information. *Appl. Soft Comput.* **2016**, *38*, 703–713. [CrossRef]
48. Bozanic, D.; Pamucar, D.; Karovic, S. Use of the fuzzy AHP-MABAC hybrid model in ranking potential locations for preparing laying-up positions. *Mil. Tech. Cour.* **2016**, *64*, 705–729. [CrossRef]
49. Salah, A.; Moselhi, O. Risk identification and assessment for engineering procurement construction management projects using fuzzy set theory. *Can. J. Civ. Eng.* **2016**, *43*, 429–442. [CrossRef]
50. Lyons, T.; Skitmore, M. Project risk management in the Queensland engineering construction industry: A survey. *Int. J. Proj. Manag.* **2004**, *22*, 51–61. [CrossRef]
51. Baloi, P.; Price, A. Modelling global risk factors affecting construction cost performance. *Int. J. Proj. Manag.* **2003**, *21*, 261–269. [CrossRef]
52. Jafarnejad, A.; Ebrahimi, M.; Abbaszadeh, M.; Abtahi, S. Risk management in supply chain using consistent fuzzy preference relations. *Int. J. Acad. Res. Bus. Soc. Sci.* **2014**, *4*, 77–89.
53. Tah, J.H.M.; Carr, V. A proposal for construction project risk assessment using fuzzy logic. *Constr. Manag. Econ.* **2000**, *18*, 491–500. [CrossRef]

54.	Wen, G. Construction project risk evaluation based on rough sets and artificial neural networks. In Proceedings of the 2010. IEEE Sixth International Conference on Natural Computation (ICNC), Yantai, China, 10–12 August 2010; pp. 1624–1628.

55.	Fouladgar, M.M.; Yazdani-Chamzini, A.; Zavadskas, E.K. Risk evaluation of tunneling projects. *Arch. Civ. Mech. Eng.* **2012**, *12*, 1–12. [CrossRef]

56.	Taroun, A.; Yang, J. A DST-based approach for construction project risk analysis. *J. Opt. Res. Soc.* **2013**, *64*, 1221–1230. [CrossRef]

57.	Kao, C.H.; Huang, C.H.; Hsu, M.S.C.; Tsai, I.H. Success factors for Taiwanese contractors collaborating with local Chinese contractors in construction projects. *J. Bus. Econ. Manag.* **2016**, *17*, 1007–1021. [CrossRef]

58.	Burcar Dunovic, I.; Radujkovic, M.; Vukomanovic, M. Internal and external risk based assessment and evaluation for the large infrastructure projects. *J. Civ. Eng. Manag.* **2016**, *22*, 673–682. [CrossRef]

59.	Yousefi, V.; Yakhchali, S.H.; Khanzadi, M.; Mehrabanfar, E.; Saparauskas, J. Proposing a neural network model to predict time and cost claims in construction projects. *J. Civ. Eng. Manag.* **2016**, *22*, 967–978. [CrossRef]

60.	Valipour, A.; Yahaya, N.; Noor, N.M.; Mardini, N.; Antucheviciene, J. A new hybrid fuzzy cybernetic analytic network process model to identify shared risks in PPP projects. *Int. J. Strateg. Prop. Manag.* **2016**, *20*, 409–426. [CrossRef]

61.	Ulubeyli, S.; Kazaz, A. Fuzzy multi-criteria decision making model for subcontractor selection in international construction projects. *Technol. Econ. Dev. Econ.* **2016**, *22*, 210–234. [CrossRef]

62.	Rajakallio, K.; Ristimaki, M.; Andelin, M.; Junnila, S. Business model renewal in context of integrated solutions delivery: A network perspective. *Int. J. Strateg. Prop. Manag.* **2017**, *21*, 72–86. [CrossRef]

63.	Valipour, A.; Yahaya, N.; Noor, N.M.; Antucheviciene, J.; Tamošaitienė, J. Hybrid SWARA-COPRAS method for risk assessment in deep foundation excavation project: An Iranian case study. *J. Civ. Eng. Manag.* **2017**, *23*, 524–532. [CrossRef]

64.	Khanzadi, M.; Turskis, Z.; Amiri, G.G.; Chalekaee, A. A model of discrete zero-sum two-person matrix games with grey numbers to solve dispute resolution problems in construction. *J. Civ. Eng. Manag.* **2017**, *23*, 824–835. [CrossRef]

65.	Keshavarz Ghorabaee, M.; Zavadskas, E.K.; Turskis, Z.; Antucheviciene, J. A new combinative distance-based assessment (CODAS) method for multi-criteria decision-making. *Econ. Comput. Econ. Cybern. Stud. Res.* **2016**, *50*, 25–44.

66.	Jiang, W.; Zhuang, M.; Qin, X.; Tang, Y. Conflicting evidence combination based on uncertainty measure and distance of evidence. *Springer Plus* **2016**, *5*, 12–17. [CrossRef] [PubMed]

67.	Li, M.; Hu, Y.; Zhang, Q.; Deng, Y. A novel distance function of D numbers and its application in product engineering. *Eng. Appl. Artif. Intell.* **2015**, *47*, 61–67. [CrossRef]

68.	Jiang, W.; Zhan, J.; Zhou, D.; Li, X. A method to determine generalized basic probability assignment in the open world. *Math. Probl. Eng.* **2016**, *2016*, 3878634. [CrossRef]

69.	Zhou, X.; Shi, Y.; Deng, X.; Deng, Y. D-DEMATEL: A new method to identify critical success factors in emergency management. *Saf. Sci.* **2017**, *91*, 93–104. [CrossRef]

70.	Zhou, D.; Tang, Y.; Jiang, W. An improved belief entropy and its application in decision-making. *Complexity* **2017**, *2017*, 4359195. [CrossRef]

71.	Deng, X.; Hu, Y.; Deng, Y.; Mahadevan, S. Environmental impact based on D numbers. *Expert Syst. Appl.* **2014**, *41*, 635–643. [CrossRef]

72.	Bozanic, D.; Pamucar, D.; Karovic, S. Application the MABAC method in support of decision-making on the use of force in defensive operation. *Tehnika Menadžment* **2016**, *6*, 129–135. [CrossRef]

73.	Lin, J.H.; Yang, C.J. Applying analytic network process to the selection of construction projects. *Open J. Soc. Sci.* **2016**, *4*, 41–47. [CrossRef]

74.	Forbes, D.; Smith, S.; Horner, M. Tools for selecting appropriate risk management techniques in the built environment. *Constr. Manag. Econ.* **2008**, *26*, 1241–1250. [CrossRef]

75.	Stević, Ž.; Pamučar, D.; Vasiljević, M.; Stojić, G.; Korica, S. Novel Integrated Multi-Criteria Model for Supplier Selection: Case Study Construction Company. *Symmetry* **2017**, *9*, 279. [CrossRef]

symmetry

MDPI

Article

Pre-Rationalized Parametric Designing of Roof Shells Formed by Repetitive Modules of Catalan Surfaces

Jolanta Dzwierzynska * and Aleksandra Prokopska

Department of Architectural Design and Engineering Graphics, Rzeszow University of Technology, Poznanska 2, 35-084 Rzeszow, Poland; aprok@prz.edu.pl
* Correspondence: joladz@prz.edu.pl; Tel.: +48-17-865-1507

Received: 13 February 2018; Accepted: 8 April 2018; Published: 11 April 2018

Abstract: The aim of the study is to develop an original, methodical, and practical approach to the early stages of parametric design of roof shells formed by repetitive modules of Catalan surfaces. It is presented on the example of designing the roof shells compound of four concrete elements. The designing process proposed by us consists in linking geometric shaping of roofs' models with their structural analysis and optimization. Contrary to other methods, which use optimization process in order to find free roof forms, we apply it in order to explore and improve design alternatives. It is realized with the application of designing tools working in Rhinoceros 3D software. The flexible scripts elaborated by us, in order to achieve roofs' models of regular and symmetrical shapes, are converted into simulation models to perform structural analysis. It is mainly focused on how the roof shells perform dependently on their geometric characteristics. The simulation enables one to evaluate various roof shells' shapes, as well as to select an optimal design solution. The proposed approach to the conceptual design process may drive the designing to achieve geometric and structural forms which not only follow the design intentions but also target better results.

Keywords: civil engineering; architecture; conceptual design; parametric design; structural analysis; Grasshopper; optimisation; finite element method (FEM); ruled surface; roof shell; multi criteria decision making

---

1. Introduction

The underlying compositional rules for architectural design were established by Vitruvius in The Ten Books on Architecture, in the early ages [1]. Over the years, different design methodologies have emerged according to development of design technology. The classical architecture was determined by order and composition of forms being mostly shaped on the basis of platonic solids arranged on the Cartesian grid [2]. However, no linear architecture as well as organic one historically appeared in the Baroque period. This evolution of design activity understood as the process of form generating was caused by wider scientific search for theory of morphogenesis in the natural word. It was related to the profound study of biological organisms, the structure of matter, and application of this knowledge to the design and construction of built environment. Every element of any form was related to each other, according to the symbiotic ordering systems of nature. On the other hand, Modernism's dominant concepts of architectural design were based on industrial technologies of functionalism and universal models [2]. During the last twenty years, the advancement of digital technologies influenced the whole field of the architectural/civil engineering design [3]. Although at first digital media were applied rather as a representation tool, soon they became the means of conceptual design. Due to this fact, the first stage of the architectural design process moved away from a traditional paper-based process consisting in 3D model creation based on the 2D drawing [4,5]. What is more, due to widespread application of information technology tools in designing, the boundary between physical and virtual models began to blur. It was mostly caused by the hybridization of several methods and techniques for acquiring the model's geometry. Such methods were applied among others in reverse engineering [6]. Moreover, thanks to digital tools, the digital architectural/engineering models could be generated. The fast development of computing technologies brought the need of collaboration in various areas of design [7,8]. Building Information Modeling (BIM) as a 3D model-based approach gave architecture and civil engineering possibilities to streamline the design process. BIM, in a way, bridged the gaps in communication between participants of the design process: owners, architects, engineers, and contractors. The important insight into BIM-based design collaboration in the construction industry is given in [9]. However, the development of methodology to analyze the benefits of BIM is presented in [10].

Architectural/civil engineering designing during the last decade was inspired not only by different possibilities of digital technology, but it was also influenced by other disciplines such as mathematics and physics. It helped to introduce the concepts and software enabling creation of dynamic, parametric, and non-linear forms. Although civil engineering industry was among the last to use the new technologies consisting in smooth modeling, that is, digital modeling based on Non-Uniform Rational B-Splines (NURBS), their application resulted in a significant change of the design approach. That was due to the fact that NURBS curves could be controlled during designing, and their flexible shape became a base for creation of various changeable forms. There was also a significant change in the conception of space which started to be treated as a four-dimensional formation—an intersection of space and time. Architectural design established new computational concepts of architecture; topological and parametric architecture, among others [11]. Topological forms described by parametric functions gave a variety of possibilities for structure creation. In parametric design, a geometrical form is shaped not by declaration of its shape and structure, but by parameters and equations describing them. One can distinguish conceptual parametric design and constructive parametric design [12]. Constructive parametric design refers to additional data embedded for 3D determination of the given object. In this context, the architectural/engineering object being designed is influenced by many variables, and designing can be seen as an multi criteria decision making process (MCDM) [13]. Digital environment, however, offers a new approach to designing. On the one hand, it gives much freedom in shaping free forms; on the other hand, it allows designers to use computer software for optimization and simulation of projects. Due to this fact, it helps to make a right decision at the early stage of designing, whereas a digital model becomes a single source of information which can be generated, controlled, and managed by a designer. Digital architecture has profoundly changed the processes of design and construction. After the first generation of digital design processes and application of new digital design tools, new forms and relationships have been developed, and new processes have been emerged. Design tools have been developed that calibrated the digital form with reality on the construction site. Parallel to this, digital fabrication tools have been added to the design process and designing has got a new quality, which has resulted from so called "parametric design thinking" [14–18]. However, "parametric design thinking" entails ability to understand, read, and construct complex and parametrized operations which make a projected object respond and evolve. Alternate terms, such as "digitally intelligent design", "algorithmic design", "object oriented design", have arisen to describe this trend [19,20].

Along this line of thought, the paper discusses a novel parametric approach to conceptual design of roof shells compound of repetitive units of Catalan surfaces. The Catalan surfaces constitute the subset of the ruled surface class, the class of surfaces generated by straight lines. They are worth considering as, due to their striking shape and relative simplicity of construction, they stand out in the architecture of curvilinearity. Modeling of roofs formed by means of Catalan surfaces gives a variety of forms of aesthetic features [21,22]. The method of shaping free form buildings roofed with profiled steel sheets effectively transformed into strips of screw ruled surfaces is presented in [23–25]. Although many works deal with parametric design of beautiful free curvilinear building forms, we have not found examples of parametric design of compound forms of building covers/shells shaped on the base of Catalan surfaces. Therefore, the focus of this research lies on the parametric geometric description of the compound roof shells of symmetrical and regular shape, as well as how these descriptions can be used at early stages of design. The main goal is to elaborate universal scripts for modeling roof shells, which can be next used in various simulations. Furthermore, our approach is to link geometric designing of roofs' models with their structural analysis and optimization. The flexible scripts describing the geometric form are converted to simulation models to perform structural analysis, which is mainly concentrated on how the shells perform dependently on their geometric characteristics. The simulation enables to evaluate various roof shell shapes, as well as the selection of the optimal design solution meeting established criteria. The paper shows that optimization can support conceptual design being an efficient tool for "form-exploration". It also presents rationalized design strategy, which means both shaping roofs of rationalized surface classes, which have positive characteristics for engineering construction, and incorporating knowledge from engineering to architecture. It presents the advantages and shortcomings in parametric geometric modeling of these roofs, discusses their architectural qualities, and their relationship with construction principles. Our proposal targeting the early stages of the design process of roofs based on Catalan surfaces is an approach to make some contribution to the research results, which have been developed so far.

2. Early Stage of Rationalized Design

A lot of research has tried to outline frameworks for the organization of the decision-making design activities in architecture and civil engineering. The waterfall model proposed in 1970 was a relatively linear sequential design approach [26]. However, this model postulated that the early design process should be organized in a cascading sequence of the designing activities. Another model, closer to design reality, is widely described in [27]. It is a model of different design paths. According to it, the early-stage formation of a building object takes place in the process of successive designing steps with real possibilities of selecting different design ways and paths. These paths form a tree of multivariate opportunity. Comparing a pathfinder model with a waterfall model of design, we can state that a pathfinder model of the conceptual design process has a loose structure. It is characterized by iterations and feedbacks. Iterations and feedbacks assure a practical possibility of multiple modification and verification of the form according to the current need. These are actions in conformity with the design practice. In the architectural design process, it is often assured the possibility of a creative application of a broadly apprehended library of forms, or menu of forms. As a rule, the library of forms is formalized and contains morphems morphemes. They are forms selected and assorted by the designer, which can be ready to use, or can be further modified according to artistic inspiration or need. This permits the designer to think alternatively, by analysis and synthesis, what favors appropriate design decisions. During the design process, the designer examines many possible design decisions and subordinates their selection to his/her proper creative personality, knowledge, and design intuition. The structure of the multivariant decision process permits to perform successive choices according to the vision of the designer. It is in accordance with a statement that "designers work by exploring alternatives", presented in [28,29]. Typically, designers consider several alternatives, and choose the best one comparing the relative benefits across possible alternatives. It permits creative process customization, and does not obstruct thinking and exploration cycles. This is currently an acceptable approach to the problem with conformity with the real aspects of architectural design. Thanks to the development of computer-aided design, the approach based on parametric modeling has found favor in recent years. However, parametric models admit multiple alternatives, each achieved by editing parameters in the digital model, which results in a wider variety of solutions. In this context, parametric design "supports complexity" [15].

The need to develop design frameworks for a complex design process has been noticed by a number of researchers [20]. However, their contribution, although an important one, concerns rather more advanced stages of designing. Nevertheless, the development of the framework for the early stage, responsive and kinetic design of a building skin, is presented in [26]. The authors explore six aspects of designing using diverse means: parametric models, digital simulations, computational analyses, physical models and interactive prototypes. On the other hand, meta-parametric design approach for the concept design stage is proposed in [20]. The purpose of this approach is increasing the number of parametric definitions, which can be implemented at the early stages of design.

The early stage of designing is a conceptual stage when the most important decisions are made for shaping the future of the project. However, at this stage, little is known about objectives that can co-evolve during design development. Therefore, rationalization of this stage of designing is very important. Basically, rationalization in the building industry appears in various forms. It is mostly implemented in order to limit complexity or manufacturing cost. Furthermore, rationalization mostly deals with solving of design problems in relation to construction in the later stages of the design process. This research is focused on rational designing in the sense of incorporating knowledge from engineering to architectural design at the early stage. It is so-called pre-rationalized design insisting on introduction of the active attitude towards geometric design of roof shells.

3. Geometric Properties of Catalan Surfaces

Defining architectural geometric characteristics in an analytical way is becoming, more and more, an area of increasing interest and importance. In general, geometric modeling of surfaces deals with two major aspects: the visual representation focused on aesthetic appearance of surface forms, and analytical representation referring to mathematical descriptions and analysis of their geometric properties. It is also closely linked to the assembly of simple forms into a complex object. In our case, it will be the creation of a multi-shell roof from single shells in order to cover a proper rectangular plan.

The method of geometric modeling of multi-shell roofs depends mostly on the surfaces' properties forming the shell; their curvature, as well as continuity between them. The mathematical development of differential geometry and related general theories provide the needed theoretical basis for understanding surfaces' properties. Catalan surfaces play a specific role, due to their characteristics.

Catalan surfaces are ruled surfaces (scroll surfaces formed by the continuous movement of a straight line) [30]. They are oblique ruled surfaces, which can be divided into two groups:

- oblique ruled surfaces of second order—hyperbolic paraboloid
- oblique ruled surfaces of more than second order—conoids, cylindroids [31,32].

The difference between hyperbolic paraboloid, conoid, and cylindroid results from different path of movement of a surface's ruling during surface's formation. In all cases of Catalan surfaces' creation, the each ruling is parallel to the fixed plane (not containing surface's directrices).

Thin-shell structures are lightweight constructions using shell elements. These elements, typically curved ones, are assembled to make large structures—covers of large buildings free of intermediate supports. The shell material is thin in sections relative to the other dimensions of the roof, and undergoes relatively little deformation under load. All thin shell roofs derive their strength through shape, rather than mass. Considering three various groups of Catalan surfaces, hyperbolic paraboloids are exceptionally stiff, due to their double curvature [33]. What is more, hyperbolic paraboloids exhibit membrane action, wherein internal forces are efficiently transmitted through the surface in an in-plane manner [34]. Due to this fact, they can be analyzed by simple statics.

The design approach to parametric modeling of shell roofs proposed in this research is realized by application of Grasshopper, a parametric plug-in of the 3D modeling software Rhinoceros. The structural analysis is carried out using Karamba 3D, whereas the optimization issues are addressed by means of Galapagos, which work in the same Rhinoceros 3D environment. When it comes to architectural design, Grasshopper is one of the most commonly used generative design editors. It represents geometric shapes, but leaves the mathematical description hidden. However, it allows designers to build form generators from the simple to complicated complex forms.

The goal of our research is to elaborate universal scripts, in order to create digital shell roof models of various forms. Next, the scripts are to be used for modeling four-shell roof composed of repetitive modules of Catalan surfaces. Parametric roofs' models are to be converted to structural models, in order to perform structural analysis. Our comparative criterion to create roofs is the minimum mass and deflection.

4. Parametric Pre-Designing of Roof Shells—Results

4.1. Geometric Modeling by Means of Grasshopper Scripts

Digital modeling always involves the definition of the simple spatial forms, next their transformations and modification. Thanks to Grasshopper plug-in, the form generative algorithm modeling can be applied. Then, the form is generated by means of mathematical operations, functions, and other dependencies, which is shown in a graphical way. Several possibilities exist for defining the motion of a line generating a ruled surface. We can distinguish two methods for ruled surface generation. The first one consists in moving the straight line along generating curves, and the second one consists in connecting the proper corresponding points of two generating curves, which are called surface's directrices [31]. From a geometric point of view, ruled surfaces are infinite surfaces, however, for practical reasons we will consider finite parts of Catalan surfaces composed of rulings' segments.

Each Catalan surface is determined by two directrix lines and the director plane, to which all surface's rulings are parallel. In the case when both directrix lines are curved lines, we obtain a cylindroid; when one directrix is a straight line, we obtain conoid, however, when both directrices are skew straight lines, the Catalan surface is a hyperbolic paraboloid, shown in Figure 1.

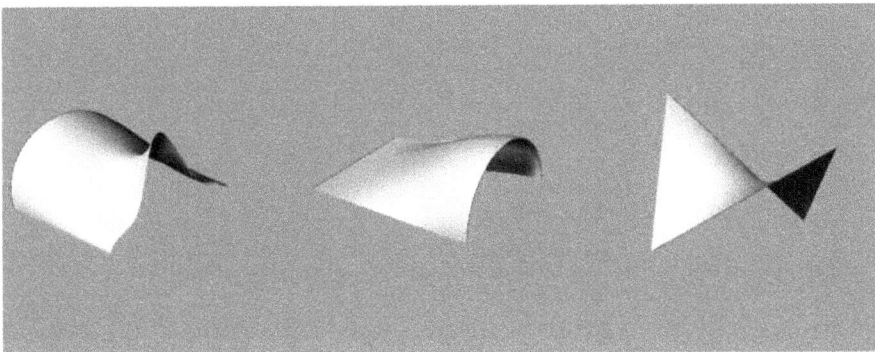

Figure 1. Typical examples of Catalan surfaces; respectively from the left: a cylindroid, a conoid, a hyperbolic paraboloid.

The surfaces as three dimensional objects in three-dimensional space can be described mathematically by a single equation with three space variables (x,y,z). However, to suit the requirements of the Grasshopper's algorithm, the surfaces should be described by two parameters (u,v). The developed Grasshopper's toolbox provides components which perform basic operations. This toolbox allowed us to use a set of functionalities to analyze and generate surfaces within the proposed design approach.

4.1.1. Hyperbolic Paraboloid

We start parametric modeling of a hyperbolic paraboloid from establishing series of points on two arbitrary skew straight lines, however, both contained in vertical parallel planes. The rulings join lines' points, which correspond to the same parameter value along u direction. This guarantees parallelism of all surface's rulings to the same vertical director plane. Application of a graph mapper for one line or for both lines enables steering lines' position in the planes (Figure 2).

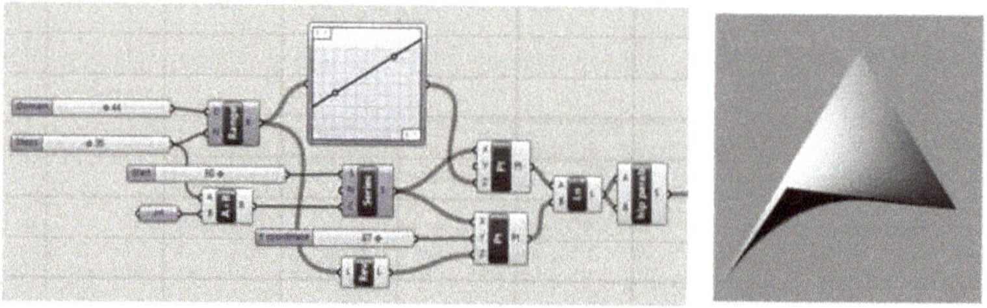

Figure 2. Single hyperboloid paraboloid form creation: (**a**) Grasshopper script for a single form; (**b**) result.

If all proper pairs of points included in the input curves are joined by straight line segments, a strip of a ruled surface connecting the curves is obtained. It is created as a single unit within Grasshopper domain [0, 1] for *u* and *v* variables. Such a unit surface can be a module for complex roof creation. The change of input parameters enables achieving roof forms of different shape, Figure 3.

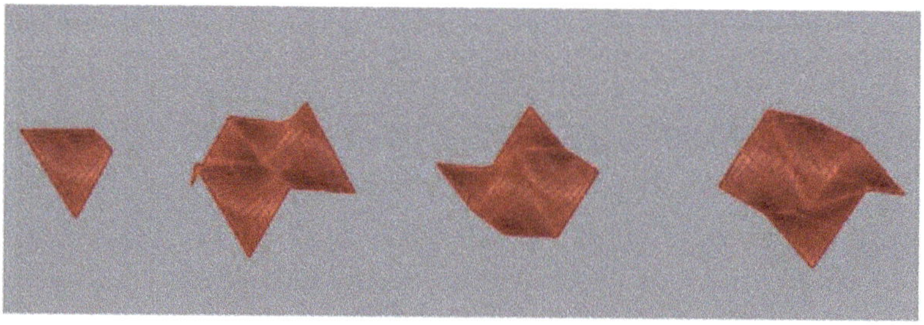

Figure 3. Several compound forms created from four units of the hyperbolic paraboloid surface.

The parametric model of any form can be obtained using different algorithm definitions. That means "that the relationship between a parametric model and its output is many-to-one" [20]. In order to create complex forms from the single one, we have applied various approaches. One of them was morphing box creation on a square surface, which is presented in Figure 3. What is more, thanks to different input parameters which determine both geometric and metric characteristics, it is possible to achieve great amount of complex roofs. They can cover both open and close spaces, Figure 4.

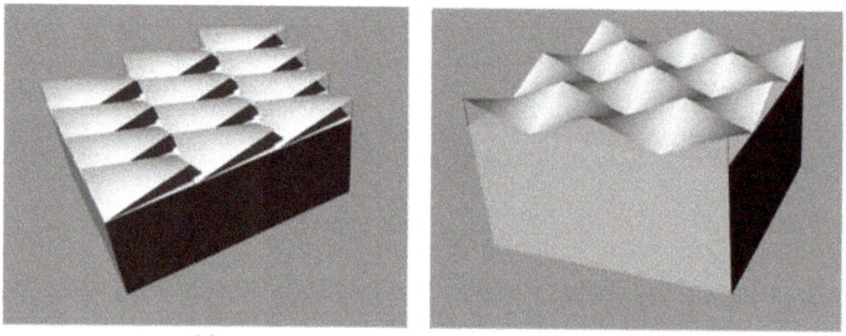

Figure 4. Building covers composed of hyperbolic paraboloid units: (**a**) with the same direction; (**b**) with different directions.

The geometric characteristic describes the shape of roof's repetitive units, their number, and arrangement (Figure 5). However, the metric characteristic determines the roof's span and height.

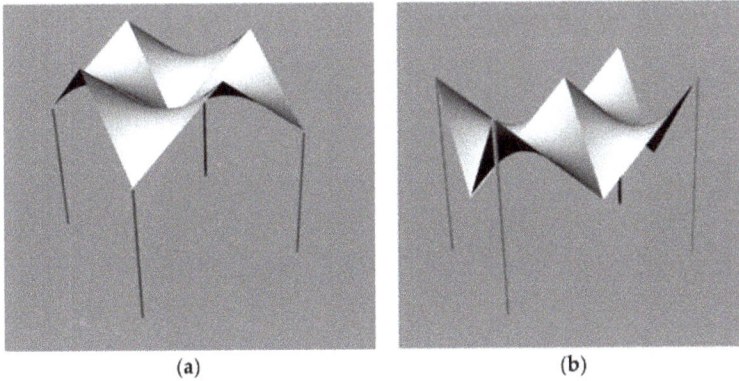

Figure 5. The compound roof shell: (**a**) concave upwards; (**b**) concave downwards.

4.1.2. Cylindroid and Conoid

According to the geometrical definition presented above, the cylindroid's rulings join points located on two curved lines. Therefore, in the script for hyperbolic paraboloid creation, we used curved lines as directrix lines, instead of straight lines. These curves are plane lines included in the vertical parallel planes. The shape of the directrix line can be differentiated dependently on the modification by graph mapper (Figure 6).

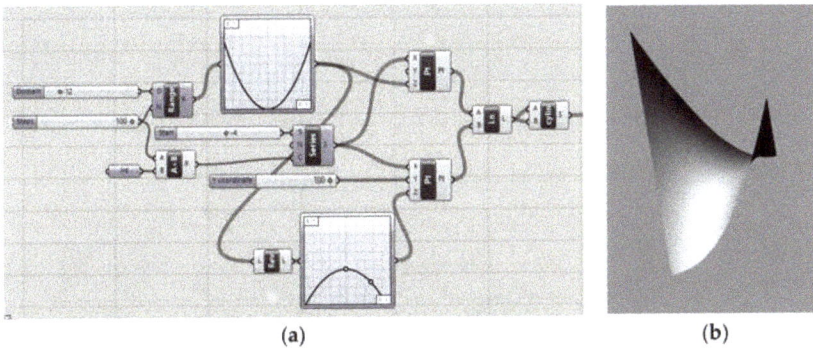

Figure 6. Single cylindroid form creation: (**a**) Grasshopper script for a single form; (**b**) result.

Thanks to it, it is possible to achieve various form creations dependently on the parameters' input, as well as directrices' curvature. However, due to their shape, not all generated forms can be suitable to form a shell roof. Some example forms are presented in Figure 7.

In order to create a parametric conoid surface, only slight modification of our script is necessary to make one directrix line be straight, which gives variety of interesting solutions (Figure 8).

The compound roofs shaped by means of repetitive units of conoid elements are very popular in building industry, and can be also generated by our script (Figure 9).

The principal advantage of parametric designing is the flexibility to perform transformations, which results in various configurations of the same geometrical components, different alternatives.

In order to evaluate roofs' alternatives, an optimization procedure will be applied by means of Galapagos, as well as structural analysis by means of Karamba 3D.

In our further considerations, we take into account the roofs composed of four identical modules of Catalan surfaces covering a rectangular plan.

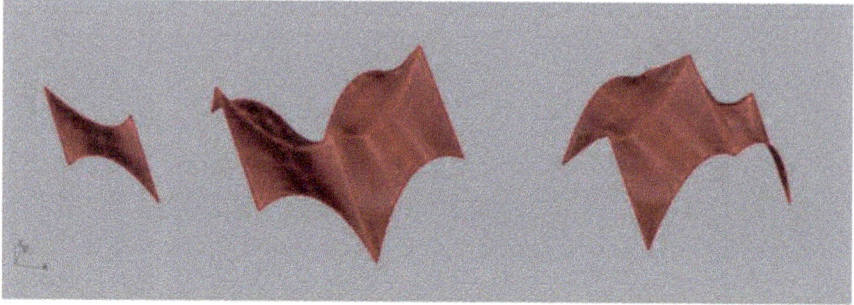

Figure 7. Several compound forms created from the units of a cylindroid surface.

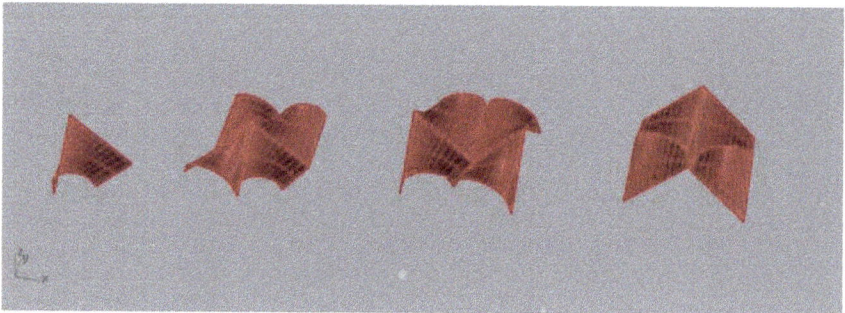

Figure 8. Several compound forms created from the units of a conoid surface.

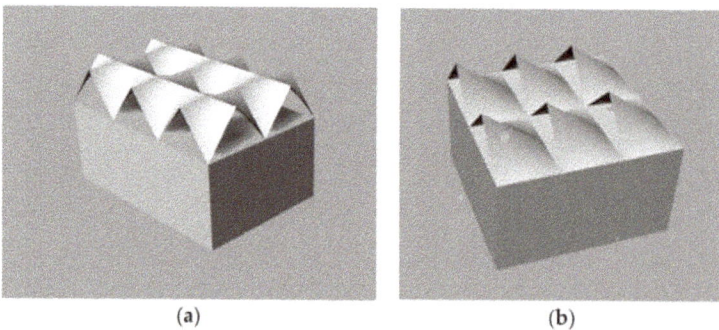

(a)   (b)

Figure 9. Building covers composed of conoid units: (**a**) with the same direction; (**b**) with different directions.

4.2. Form Optimisation by Means of Galapagos

In general, the optimal realization of the designer's idea from both an aesthetic and functional point of view is a difficult and a complex problem. Aesthetics and functionality are mutually dependent issues, as functional optimization cannot be separated from shape optimization, and vice versa. "Any optimum is only an optimum within the conceptualization of the problem space and the boundary conditions applied" [31]. The algorithmic methods and tools proposed by Grasshopper allow for a surfaces' analysis and modeling the rationalized forms, which match the initial ideas of design. Grasshopper enables generation of forms of a predefined logic. However, Galapagos, a module included in Grasshopper, is a numeric approach to drive controlled, suitable, and optimal results within the iterative design process. It enables the user to achieve results that best meet design criteria.

In our case study, we have taken into account the roof shells composed of four identical modules of Catalan surfaces that are four units of a hyperbolic paraboloid surface, a conoid surface, and a cylindroid surface, covering the same rectangular plan. Our optimization criterion was a minimum area of the roof shell, which covers a rectangle of a unit area. The height of the roof was assumed as the integer value within a fixed range of 1 m to 10 m.

Each optimization problem requires defining a fitness function, a sort of problem which is suited for Galapagos Evolutionary Solver. In our function, "each instance"—an alternative roof area (A)—was dependent on the rectangular area (B) covered by this roof, as well as the roof's height. The roof's area (A) was compared further with the area (B) of the covered rectangular plan. Therefore, in our fitness function, the difference, A−B, between the above areas was optimized (Figure 10).

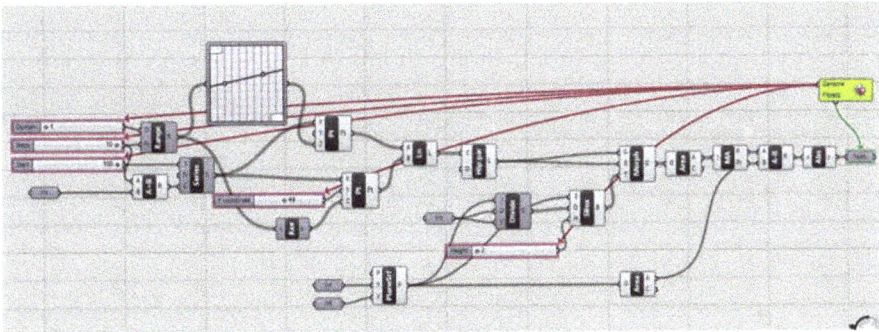

Figure 10. Optimization of a roof shell by Galapagos.

The minimal value of A−B meant the best solution. Thanks to Galapagos Evolutionary Solver, we have found the optimal height for each roof shell, as well as the minimal area of the Catalan surface which covers the unit area. The worst optimization result that is the biggest area of the roof shell was achieved for the roof shell composed from conoid elements. However, the best result that is the smallest roof shell area was obtained in the case of the roof shell composed from cylindroid elements determined by two parabolas as surfaces' directrices. These results were obtained for the height of each roof equal to 1 m.

4.3. Optimization Based on Structural Analysis

The above optimization results mean that the weight of the considered cylindroid shell roof should be the smallest compared to the weight of the other roof types, assuming the same specific height of all roofs equal to 1 m. This issue can be further checked by means of structural analysis. Structural analysis consists in determination of the effects of loads on a designed roof shell. In order to perform an accurate structural analysis, it is necessary to determine geometry of a roof's surface, structural loads, support conditions, as well as material properties. The results of such an analysis typically include support stresses, displacements, deformations, as well as support reactions. This information can be compared to the criteria that indicate the conditions of failure. Structural analysis is thus a key part of the engineering design. As far as the roof shell is concerned, we can state that it is a form-based structure, which means that its shape influences its load carrying capacity. Therefore, shell's geometry along the boundary conditions, as well as the type of loading applied, dictate the way the shell transfers load, or the way it fails. Shell can fail due to increasing deformations, failure of material, or a combination of both. Due to this fact we have taken into consideration shell's deflection as the main criterion for assessing structural stability.

In our research, several representative roof shells have been subjected to structural analysis. Each of them was compound of four identical concrete unit surfaces of the same class Table 1. It was assumed that each roof shell covered a horizontal rectangle with the area of 16 m.

Table 1. The types of the analyzed compound roof shells.

Kind of Thse Urface	Type of the Compoud Roof Shell	
Hyperbolic paraboloid	1	2
Conoid	3	4
Cylindroid	5	6
	7	

Each compound roof shell made up of four concrete surface units has been treated as one coherent shell element for calculation. Therefore, for each roof type, the calculations have been carried out as for a single concrete shell structure supported on four corners with circulation space in the middle. It is a common practice to apply approximate solutions of differential equations as the basis for structural analysis, which can be prepared using numerical approximation techniques. In our structural analysis, we used the finite element method (FEM) as the most commonly used numerical approximation in structural analysis. However, we carried out the analysis by application of Karamba 3D, an interactive parametric finite element program, which works in environment of Rhinoceros 3D. FEM simulation represents physical objects as a collection of discrete components or "elements". Due to this fact, the geometry of shells was presented by meshes, and each mesh face corresponded to the constant strain finite element, Table 2. In order to achieve comparable results each shell's mesh was divided into the same number of 400 quads, which were automatically decomposed to triangles.

Table 2. The scheme of the mesh, supports' location, as well as the obtained simulation results.

Kind of the Surface	The Scheme of the Mesh and Supports' Location	Dimmentions (m)	Mass (kg)	Displacement (m)
Hyperbolic paraboloid 1		$a = b = 4.00$ $h = 1$	3021.93	0.0004
Hyperbolic paraboloid 2		$a = b = 4.00$ $h = 1$	3021.93	0.0000
Conoid 3		$a = 4.75$ $b = 3.35$ $h = 1$	2993.24	0.0042
Conoid 4		$a = 4.75$ $b = 3.35$ $h = 1$	3993.52	0.0035
Cylindroid 5		$a = 3.47$ $b = 4.61$ $h = 1$	2901.03	0.0033
Cylindroid 6		$a = 4.69$ $b = 3.41$ $h = 1$	3113.00	0.0033
Cylindroid 7		$a = 4.67$ $b = 3.47$ $h = 1$	3107.00	0.0008

In order to perform a structural analysis by Karamba 3D, we converted our Grasshopper models given by the scripts to a simulation models (Figure 11).

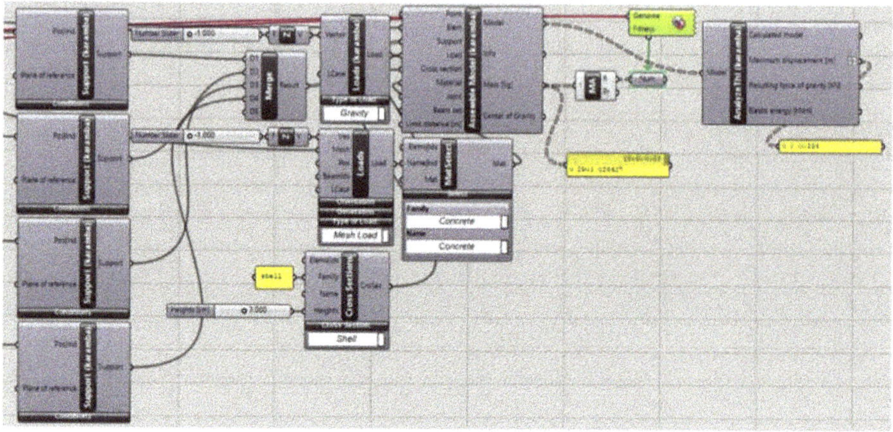

Figure 11. The assembly of load and supports in a simulation model in order to check roof shell behavior under self-load.

The first simulation was performed for the shell submitted to the dead load, which was the self-weight of concrete used in the shell construction. The dead load acted on the surface of the shell in negative z direction. The thickness of each shell was assumed to be of 7 cm. We have examined the behavior of each roof under self-weight, and established both the minimal mass for each roof and deflection assuming that the roof height $h \geq 1$ m. The simulation was performed by means of Galapagos Evolutionary Solver, and a minimum mass was an optimization criterion. The results are presented in Table 2. For each case of roof, there were also established dimensions of the rectangular plan covered by the roof: its length a according to direction x, and its width b according direction y, Table 2. Analyzing the achieved results, we can state that the construction of each roof is stable. However, the minimal mass of roof's construction was obtained in the case of the roof shell number 5, which was composed of the cylindroid units. This result confirmed the results of the previous analysis performed in point 3.2. In order to perform further structural analysis, we have chosen two of the roof shells' alternatives presented in Table 2. The first one was the cylindroid roof number 5 as a roof of a minimal mass. The second one was the roof shell number 2 compound of hyperbolic paraboloid units. It is a regular shaped roof, which covers a square plan. It also has a more favorable shape than other roof of hyperbolic paraboloid shape (the roof number 1), due to less possibility of the accumulation of precipitation (Table 2).

In the second structural analysis performed by us, we considered not only dead loads, but also live loads acting on the construction. Both snow and wind loads can have a considerable effect on shell structures. These loads are calculated in the form of pressure coefficients acting over the surface of the shell. We assumed snow pressure of 1.3 kN/m^2 and wind pressure of 1 kN/m^2, and applied different load combinations, one of them is shown in Figure 12.

We could do this using the load component of the Karamba 3D toolbar. This is a multi-use component which allowed us to specify a number of various types and different load combinations for the model. We have oriented self-load and snow load globally to a system of axes x, y, z, whereas a wind load locally to the mesh. The assemble component gathered all necessary information and created a static model. Based on simulation performed by Karamba 3D, it was possible to predict the behavior of the construction under loads, and the stress that each construction element experienced. Both roof shells were stable under the dead and live loads, however, the best optimization results, that is, the smallest deflection (equal to 0 mm) has been achieved in the case of the roof shell number 2 (Figure 13).

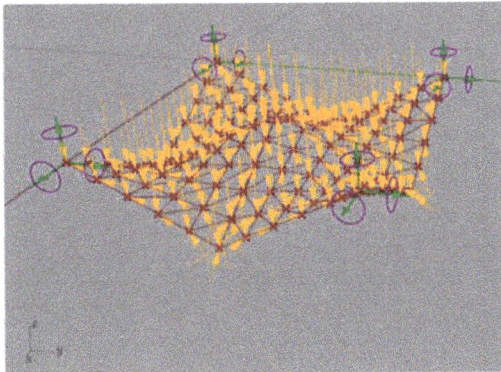

Figure 12. The schema of the application of live loads acting on the structure.

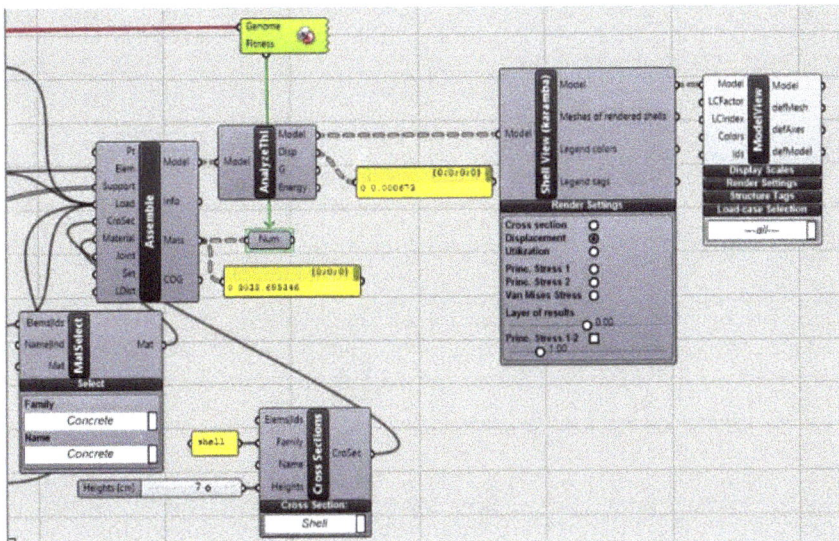

Figure 13. The best simulation result—behavior under dead and live loads for the shell roof number 2.

Due to this fact, the roof shell number 2, which is composed of four parts of a hyperbolic paraboloid surface, has been chosen as the best one from seven alternatives, and can be taken for further consideration in the more advanced process of design.

4.4. Discussion

The elaborated method of conceptual designing of roof shells composed of several parts of Catalan surfaces works well. The scripts for parametric designing developed by us seem to be universal, as they allow for the generation of various roof forms. These flexible scripts are the key for further simulation and optimization processes. Thanks to their application, we could find the roof shell of the best characteristic, that is, the roof compound of four parts of a hyperbolic paraboloid surface, which can be taken into consideration in a further design process. It is evident that in parametric design, shape and metric properties are the most important aspects of the design framework. They should be controlled at all times during the design process. This is due to the fact that they allow automatic evaluation of the performance of various design options in order to target better results. Karamba 3D includes a number of analysis components for performing different types of structural calculations: deflections and mass, among others. We could minimize the mass, deflection, and find out proper parameters to optimize the construction's right shape. However, due to the fact that we propose a structural analysis at the early stages of designing, this analysis is treated rather as an estimated analysis, not an accurate one.

The research shows that uncomplicated scripts give more flexibility in further optimization process. Due to this fact, they are of considerable importance, in particular, for the design of rationalized surfaces. It is worth mentioning, that in all presented plug-in tools for parametric design such as Grasshopper and Karamba 3D, the interactive display of the designed form is generated parallel with an interactive window for visual scripting modification. It helps in generation of complicated geometry and facilitates the re-editing process. What is important is that Karamba 3D and Galapagos work in the same Rhinoceros 3D environment. This minimalizes problems that could occur during application of various toolsets working in different software environments.

5. Conclusions

The research shows that it is possible to deploy structural analysis of the designed form at the early stages of design. What is more, it is very useful as it enables to estimate object's performance in different conditions, as well as generation and testing of the design variants based on various criteria and a specialist input.

However, our aim was not only designing of the optimal roof's form, but also methodology formulation, as well as indication of the approach and tools for the initial designing of roof shells. Due to this fact, we tried to elaborate scripts for further implementation.

The study shows that optimization can be not only a tool to search for optimal form, but can also be applied to check the performance of existing forms. Therefore, it can be applied as an efficient tool for form exploration and improvement in order to support conceptual design.

The rationalized design strategy presented in the paper consists in both application of rational surface classes, such as Catalan surfaces, simple and fundamental to roof design, and the active and complex attitude to the design concept. Such an attitude strengthens and facilitates sharing information, and co-operation between an architect and a civil engineering designer.

Author Contributions: The individual contribution and responsibilities of the authors were as follows: Jolanta Dzwierzynska conceived and designed the research, performed the experiments and analyzed the data; Aleksandra Prokopska contributed in morphological analysis, Jolanta Dzwierzynska wrote the paper. All the authors have read and approved the final manuscript.

Conflicts of Interest: The authors declare no conflict of interest.

References

1. Vitruvius, P. *Vitruvius: The Ten Books on Architecture*, 1st ed.; Harvard University Press: Cambridge, UK, 1914; pp. 13–31.
2. Biermann, V.; Borngasser, B. *Architectural Theory from the Renaissance to the Present*, 1st ed.; Taschen: Koln, Germany, 2003; pp. 6–20.
3. Oxman, R. Theory and design in the first digital age. *Des. Stud.* **2006**, *27*, 229–265. [CrossRef]
4. Unwin, S. Nalyzing architecture through drawing. *Build. Res. Inf.* **2007**, *35*, 101–110. [CrossRef]
5. Hewitt, M. Representational Forms and Modes of Conception; an Approach to the History of Architectural Drawing. *J. Archit. Educ.* **2014**, *39*, 2–9.

6. Biagini, C.; Donato, V. Behind the complexity of a folded paper. In Proceedings of the Mo. Di. Phy. Modeling from Digital to Physical. Innovation in design languages and project procedures, Milano, Italy, 11–12 November 2013; pp. 160–169.

7. Kim, D.-Y.; Lee, S.; Kim, S.-A. Interactive decision Making Environment for the Design optimization of climate Adaptive building Shells. In Proceedings of the International Conference on Cooperative Design, Visualization and Engineering (CDVE 2013), Mallorca, Spain, 22–25 September 2013; Springer: Berlin/Heindelberg, Germany, 2013.

8. Luo, Y.; Dias, J.M. Development of a Cooperative Integration System for AEC Design. In Proceedings of the International Conference on Cooperative Design, Visualization and Engineering (CDVE 2004), Palma de Mallorca, Spain, 19–22 September 2004; Springer: Berlin/Heidelberg, Germany, 2004; Volume 3190.

9. Wang, J.; Chong, H.-Y.; Shou, W.; Wang, X.; Guo, J. BIM- Enabled design Collaboration for Complex Building. In Proceedings of the International Conference on Cooperative Design, Visualization and Engineering (CDVE 2013), Mallorca, Spain, 22–25 September 2013; Springer: Berlin/Heindelberg, Germany, 2013.

10. Barlish, K.; Sullivan, K. How to measure the benefits of BIM- A case study approach. *Autom. Constr.* **2012**, *24*, 149–159. [CrossRef]

11. Kolarevic, B. *Architecture in the Digital Age: Design and Manufacturing*, 1st ed.; Spon Press: London, UK, 2003; pp. 20–98.

12. Stravic, M.; Marina, O. Parametric Modeling for Advanced Architecture. *Int. J. Appl. Math. Inform.* **2011**, *5*, 9–16.

13. Elango, M.; Devadas, M.D. Multi-Criteria Analysis of the Design Decisions In architectural Design Process during the Pre-Design Stage. *Int. J. Eng. Technol.* **2014**, *6*, 1033–1046.

14. Bhooshan, S. Parametric design thinking: A case study of practice-embedded architectural research. *Des. Stud.* **2017**, *52*, 115–143. [CrossRef]

15. Wortmann, T.; Tuncer, B. Differentiating parametric design: Digital Workflows in Contemporary Architecture and Construction. *Des. Stud.* **2017**, *52*, 173–197. [CrossRef]

16. Oxman, R. Thinking difference: Theories and models of parametric design thinking. *Des. Stud.* **2017**, *52*, 4–39. [CrossRef]

17. Woodburg, R.; Arish, R.; Kilian, A. Some Patterns for Parametric Modeling. In Proceedings of the 27th Annual Conference of the Association for Computer Aided Design in Architecture, Halifax, NS, Canada, 1–7 October 2007; pp. 222–229.

18. Turrin, M.; von Buelow, P.; Stouffs, R. Design explorations of performance driven geometry in architectural design using parametric modeling and genetic algorithms. *Adv. Eng. Inform.* **2011**, *25*, 656–675. [CrossRef]

19. Moussavi, F. Parametric software is not substitute for parametric thinking. *Archit. Rev.* **2011**, *230*, 39–40.

20. Hardling, J.E. Meta-parametric Design. *Des. Stud.* **2017**, *52*, 73–95. [CrossRef]

21. Tofli, J. Application of Catalan surface in designing roof structures—An important issue in the education of the future architect engineer. In Proceedings of the International Conference on Engineering Education (ICEE 2007), Coimbra, Portugal, 3–7 September 2007.

22. Przewlocki, S. *Descriptive Geometry in Construction*, 1st ed.; Arkady: Warsaw, Poland, 1982; pp. 57–209.

23. Abramczyk, J. *Shaping Free Forms of Buildings Roofted with Transformed Corrugated Sheeting*; Publishing House of Rzeszow University of Technology: Rzeszow, Poland, 2017.

24. Abramczyk, J. Shaping Innovative Forms of Buildings Roofted with Corrugated Hyperbolic Paraboloid Sheeting. *Procedia Eng.* **2016**, *161*, 60–66. [CrossRef]

25. Abramczyk, J. Building Structures Roofted with Multi-Segment Corrugated Hyperbolic Paraboloid Steel Shells. *Procedia Eng.* **2016**, *161*, 1545–1550. [CrossRef]

26. Zboinska, M.A.; Cudzik, J.; Juchnevic, R.; Radziszewski, K. A Design Framework and Digital Toolset Supporting the Early-Stage Explorations of Responsive Kinetic Building Skin Concepts. In Proceedings of the 33rd Annual eCAADe Conference, Real Time: Extending the Reach of Computation, Vienna, Austria, 16–18 September 2015; Volume 2, pp. 715–724.

27. Prokopska, A. *Methodology of Architectural Design, Preliminary Phases of the Architectural Process*; Publishing House of Rzeszow University of Technology: Rzeszow, Poland, 2015; pp. 39–121. (In Polish)

28. Prokopska, A. Morphology of the Architectural Achievement. A Methodological Analysis of Selected Morphological Systems of the Natural and Architectural Environments. *Syst. J. Transdiscipl. Syst. Sci.* **2002**, *7*, 1–116.

29. Woodburg, R.; Mohiuddin, A. Interactive design galleries: A general approach to interacting with design alternatives. *Des. Stud.* **2017**, *52*, 40–72. [CrossRef]

30. Dzwierzynska, J.; Abramczyk, J. *Engineering Descriptive Geometry*; Publishing House of Rzeszow University of Technology: Rzeszow, Poland, 2015.

31. Pottman, H.; Asperl, A.; Hofer, M.; Kilian, A. *Architectural Geometry*, 1st ed.; Bentley Institute Press: Exton, PA, USA, 2007; pp. 35–194.

32. Krivoshapko, S.; Ivanov, V.N. *Encyclopedia of Analytical Surfaces*; Springer International Publishing: Cham, Switzerland, 2015; pp. 60–77.

33. Chudley, R.; Greeno, R. *Advanced Construction Technology*, 4th ed.; Dorset Press: Dorchester, UK, 2006; pp. 528–529.

34. Min, C.S.; Gupta, A.K. Inelastic behavior of reinforced concrete hyperbolic paraboloid saddle shell. *Eng. Struct.* **1994**, *16*, 227–237. [CrossRef]

symmetry

MDPI

Article

A Novel Approach for Evaluation of Projects Using an Interval–Valued Fuzzy Additive Ratio Assessment (ARAS) Method: A Case Study of Oil and Gas Well Drilling Projects

Jalil Heidary Dahooie [1,*], Edmundas Kazimieras Zavadskas [2], Mahdi Abolhasani [1], Amirsalar Vanaki [1] and Zenonas Turskis [2]

[1] Faculty of Management, University of Tehran, Jalal-e-Al-e-Ahmad Highway, 14155-6311 Tehran, Iran; abolhasani.mahdi@gmail.com (M.A.); amirsalarvanaki@ut.ac.ir (A.V.)
[2] Faculty of Civil Engineering, Vilnius Gediminas Technical University, Sauletekio al. 11, LT-10223 Vilnius, Lithuania; edmundas.zavadskas@vgtu.l (E.K.Z.); zenonas.turskis@vgtu.lt (Z.T.)
* Correspondence: heidaryd@ut.ac.ir; Tel.: +98-021-61117686

Received: 4 November 2017; Accepted: 8 January 2018; Published: 12 February 2018

Abstract: The beginning of the 21st-century resulted in a more developed multi-attribute decision-making (MADM) tool and inspired new application areas that have resulted in discoveries in sustainable construction and building life cycle analysis. Construction and civil engineering stand for the central axis of a body consisting of a multidisciplinary (multi-dimensional) world with ties to disciplines constituting the surface, and with the disciplines, as a consequence, tied to each other. When dealing with multi-attribute decision-making problems generally multiple solutions exist, especially when there is a large number of attributes, and the concept of Pareto-optimality is inefficient. The symmetry and structural regularity are essential concepts in many natural and man-made objects and play a crucial role in the design, engineering, and development of the world. The complexity and risks inherent in projects along with different effective indicators for success and failure may contribute to the difficulties in performance evaluation. In such situations, increasing the importance of uncertainty is observed. This paper proposes a novel integrated tool to find a balance between sustainable development, environmental impact and human well-being, i.e., to find symmetry axe with respect to goals, risks, and constraints (attributes) to cope with the complicated problems. The concept of "optimal solution" as the maximum degree of implemented goals (attributes) is very important. The model is built using the most relevant variables cited in the reviewed project literature and integrates two methods: the Step-Wise Weight Assessment Ratio Analysis (SWARA) method and a novel interval-valued fuzzy extension of the Additive Ratio Assessment (ARAS) method. This model was used to solve real case study of oil and gas well drilling projects evaluation. Despite the importance of oil and gas well drilling projects, there is lack of literature that describes and evaluates performance in this field projects. On the other hand, no structured assessment methodology has been presented for these types of projects. Given the limited research on performance evaluation in oil & gas well-drilling projects, the research identifies a set of performance criteria and proposes an evaluation model using fuzzy Delphi method. An illustrative example shows that the proposed method is a useful and alternative decision-making method.

Keywords: performance evaluation; oil and gas well drilling projects; Step-Wise Weight Assessment Ratio Analysis (SWARA); interval-valued fuzzy Additive Ratio Assessment; Additive Ratio Assessment (ARAS)

1. Introduction

Today, projects have a significant role to play in the success of any company and the integration of activities leading to new products or services can improve its performance [1]. That is why many companies consider the use of project management as a key strategy for their survival in a competitive environment as well as for increasing the possibility of value creation in their businesses [2]. A project is a temporary attempt to produce a unique product or service and project management refers to the application of knowledge, skills, tools and techniques to carry out all activities of the project [3] as well as, project is a complex effort involving interconnected activities, with the purpose of achieving an objective [1]. A project performance evaluation dealing with evaluating and rating all tasks [4] help Decision Makers ('DM') through determining the status of the project and its weaknesses and strengths [5] and will help establish benchmarks of high performance projects for cross-learning and identify inefficiencies of low performance projects for potential improvement [6]. Therefore, the importance of project performance evaluation is inevitable and has long been confirmed by practitioners and academics from a variety of functional disciplines [7]. As a result, several project performance evaluation approach has been advanced that MCDM models is one of them. In a business environment, evaluation of the "best" project can be done by a decision committee, instead of a single DMs. Different DMs can bring their own points of view and knowledge, which must be resolved within a framework of understanding and mutual concessions [8,9]. In such situations, MCDM can help in finding a sufficiently good solution from a collection of alternatives and can address complex problems that involve high uncertainty, conflicting objectives, different forms of information, multiple interests and different perspectives [8]. MCDM methods solve a complex problem by converting it into several small problems and as smaller processes are weighted and re-aggregated, a general picture of decision makers is provided [10]. Therefore, given the inherent complexity, risks and uncertainties of projects and the diversity of success/failure criteria, this paper aims to introduce a novel assessment framework for projects using Interval-Valued Fuzzy Additive Ratio Assessment as a recent MADM approach dealing with uncertainties in decision making. In many practical situations, there exists information which is incomplete and uncertain so that decision makers cannot easily express their judgments on the candidates with exact and crisp values. As well as there are many real-life complex problems that need to involve a wide domain of knowledge. Therefore, fuzzy sets provide generally more adequate description to model real-life decision problems than real numbers [11]. However, it became apparent that the fuzzy sets are not sufficient for uncertain MCDM. Therefore, Zadeh (1975) [12] and Gorzałczany (1987) [13] developed the concept of interval-valued fuzzy sets, whereas Atanassov (1986) [14] proposed intuitionistic fuzzy sets [15]. Consequently, interval-valued fuzzy sets allow us to achieve a better imagination from environmental ambiguity and uncertainty [16].

On the other hand, Over the past two decades, infrastructure projects as an important category of projects have accounted for 3.8% of global GDP and this number is expected to rise to 1.4% by 2030 [5]. These type of projects include general fields of energy, transport, water, communications and social infrastructure such as hospitals [17]. Infrastructure projects are defined as long-term, large-scale and difficult-to-implement projects that can hardly be valued; therefore, evaluation of these projects is a complex and specialized activity [17].

Further, oil and gas well drilling projects are of great importance among major infrastructure projects because of a large volume of investment and the economic benefits of their proper implementation [4,18]. Difficulty in predicting operational costs due to inherent uncertainties in the economic evaluation of oil and gas exploration and development projects [19], costs of renting drilling rigs as well as drilling services required by these projects [20] are major parameters resulting in the need for high-volume investments. Like any costly projects for the highest quality and lowest cost and time, oil and gas drilling projects require appropriate decision-making procedures to face upcoming challenges and achieve the desired level of productivity [21]. A fast, continuous, timely and data-based decision-making process can lead to improved productivity; particularly as today's global demand for energy and environmental constraints have forced oil and gas projects to enhance their activities in terms of efficiency and effectiveness [18]. In addition, similar to other projects, oil and gas well-drilling projects deal with inherent uncertainties [22] that are generally arisen from environmental factors, organizational complexities, as well as changes, deviations and events occurring in a project [23]. Accordingly, oil and gas projects face various risks and complexities that make the decision-making process so difficult [18].

However, a brief review of the literature shows a limited portion of structured performance evaluation models in these field and though a variety of project evaluation approaches like multi-criteria decision-making (MCDM) methods have been applied in other project fields [6] but there is a necessity for structured assessment methodology in the field of oil and gas well drilling projects. Therefore, we choised oil and gas well drilling projects as a case study and tried to identify an initial list of performance criteria based on review of literature and propose an evaluation model using Delphi method.

According to the above discussion, the purpose of this study is to provide an interval-valued MADM-based framework as a novel approach for evaluating the performance of projects that oil and gas well drilling projects was considered in order to remove the limitations noted in this type of projects. To this end, a review on the literature is provided and an initial list of evaluation criteria is extracted. Due to the research limitations in this context, the Fuzzy Delphi technique and expert panels are used to develop a more complete list of effective criteria. Next, the identified criteria are weighted by the SWARA method and finally the Interval-Valued Fuzzy Additive Ratio Assessment is employed to assess and rank active projects of a certain company in the field of oil and gas drilling.

The structure of this paper will be as follows. Section 2 presents the literature review and the initial list of criteria. Section 3 describes the research methodology, while Section 4 provides an experimental example using data from a set of seven oil and gas well drilling projects from a subsidiary of the National Iranian Oil Company (NIOC) in Iran. Finally, Section 5 concludes the study.

2. Literature Review

Decision making in a project context is a complex undertaking. The term complexity is an increasingly important point of reference when we are trying to understand the managerial demands of modern projects in general, and of the various situations encountered in projects [24,25]. On the one hand, a project is a temporary and transient organization surrounded by inherent uncertainty [24,26]. When complexity becomes too great, the possibilities and interrelations become so fuzzy that the system has to be assisted by appropriate tools and skills. Consequently, managers facing complex project need access to a decision-making aid model based on relevant performance evaluation [24]. Therefore the project performance evaluation is inevitable and necessary issue. So far, several decsion approach has been used to help project evaluation such as economic models, mathematical programming, artificial intelligence optimization methods, integrated models, data envelopment analysis (DEA) method, integrated the balanced scorecard (BSC) approach and MCDM models [27]. among these, MCDM models can help in finding a sufficiently good solution from a collection of alternatives and can address complex problems that involve high uncertainty, conflicting objectives, different forms of information, multiple interests and different perspectives [8]. In measuring the overall performance of projects, MCDM models have been used to aggregate multiple performance measures under various application contexts [6] and incidentally are widely used for energy projects [8]. Analytical Hierarchy Process (AHP) [28,29], Analytic Network Process (ANP) [29,30], TOPSIS [31], DEMATEL [32], ELECTRE [33] and some hybrid methods [4,8,34] are number of MCDM methods applied to projects evaluation.

On the other hand in many projects in practical situations, there exists information which is incomplete and uncertain so that decision makers cannot easily express their judgments on the candidates with exact and crisp values. As well as there are many real-life complex problems that need to involve a wide domain of knowledge. These conditions in which decisions are based on obscure and unreliable information or lake of knowledge and personal preferences of the experts can create difficulties in the decision-making process. These difficulties can lead to deceptive and uncertain decisions. Therefore, solving decision-making problems existing in real life and modeling them in the form of multi-criteria decision-making problems is still considered as a challenging topic [35–37] and fuzzy sets provide generally more adequate description to model real-life decision problems than real numbers [11] and presented to fix these challenges and provided a basis for development of a variety of fuzzy decision-making models. The development of the fuzzy concept has led to the provision of models which have the flexibility to control and display uncertainty and low accuracy due to lack of knowledge of experts and inadequate data [38,39]. Many developments have been made to better address inadequate and inaccurate data [14,40]. Atanassov developed intuitive fuzzy sets [14]. These sets included the membership function, the non-membership function and the hesitancy function [41]. Zadeh [12] introduced a type-2 fuzzy set which allowed expressing the membership of the components in the form of a fuzzy set. Further, the type-n fuzzy numbers were defined [42] which were the generalized form of type-2 fuzzy numbers and allowed the membership of elements to be in the form of a type-$(n-1)$ fuzzy set. Since the concept of fuzzy numbers with interval values has been presented, there has been an increasing interest of researchers in this field [43] so that it was successfully used in numerous decision-making issues in conditions of uncertainty. Briefly, there are two approaches to classify studies in this area: (1) Content Approach; (2) Applied Approach [44]. In the content approach, two main steps are considered for decision-making issues: aggregation of opinions [45–47] and method exploitation. In the applied approach, researches on fuzzy numbers with interval values can be classified into five main domains [48]: (1) basic operators in fuzzy space with interval values [49]; (2) group decision [50,51]; (3) combining decision making with linguistic variables [52,53]; (4) matrix of judgment based on priority relations [54] and (5) development of the model of dual interval-valued fuzzy sets [55]. In recent years, researches have been a growing trend in two domains; decision making with linguistic variables and development of basic operators in fuzzy space with interval values.

Given that this paper aims to prioritize the projects, it is related to the method exploitation step. In this category of methods, it is tried to find priority relations in non-preferred alternatives so that a set of options is ranked based on them. So far, various decision-making methods have been combined with fuzzy numbers with interval values [56] including the VIKOR method [57,58], the TOPSIS method [59], the MULTIMOORA method [15] and the TODIM method [60]. In this paper, we tried to use Additive Ratio Assessment (ARAS) method combined with fuzzy numbers with interval values which is addressed when we introduce fuzzy numbers with interval values and combined method steps.

In addition to the performance evaluation methodology, it is also necessary to define the performance evaluation criteria for each project. Evaluation criteria are quantitative or qualitative variables that measure the performances and the impacts of the analyzed alternatives [61]. Through a literature review, this section also seeks to develop an initial list of criteria as inputs for the fuzzy Delphi process. As discussed before, despite the importance of oil and gas well drilling projects, there is a limited research available on performance evaluation of these projects, by using some measures of these criteria. Below provides further details of some relevant research studies:

Dachyar and Pratama (2014) evaluated the efficiency of oil and gas well drilling projects using the MACBETH method [22]. The author introduced criteria such as implementation methods, time, cost, quality, risk and safety. Ahari and Niaki (2014) used a neuro-fuzzy network to assess the quality of oil and gas well drilling projects for a contractor selection problem [4,21]. The authors defined three criteria of time, cost and quality as the basic assessment objectives. Also, a set of five parameters were introduced as inputs; (1) the cost compliance percentage with plans; (2) the time compliance percentage with plans; (3) the percentage of quality failure in all operational failures; (4) the number of HSE incidences; and (5) the number of quality failures without non-productive time. The quality of work measure was defined as the model output. The authors followed the size of drilled holes as an important factor in the work package plan. Exploring the design of drilling contracts, Osmundsen et al. (2006) suggested safety and associated risks from the important indicators of a drilling project [20]. In other study (2010), the authors examined new incentive schemes in offshore drilling contracts to improve project performance. They highlighted higher cost and its rapid growth in drilling activity, especially for renting and drilling services, as well as the need to carefully examine effective criteria in drilling projects [62]. Liu et al. (2013) considered risks in drilling projects [63]. They identified a set of 25 risk factors in six categories of natural risks, R & D risks, management risks, drilling risks, equipment risks and security risk and environmental factors.

In addition to the related studies referred to above, in general, three main traditional criteria have always been considered; including time, cost, and quality. A project would be succeeded when it is implemented at a reasonable cost and quality during a planned period time, and meet stakeholders' satisfaction [1]. This traditional approach—referred as the "Iron Triangle" [4]—merely deals with the economic dimension of projects, while ignoring other major aspects [64]. Further, these criteria are less flexible in evaluating project performance [65]. In order to overcome limitations of the traditional criteria, academic researchers suggest a variety of criteria for project performance evaluations. Some examples are safety of the project site [66–68], geographic location of projects [69], environmental impacts [70] and satisfaction of community, client or customer [71,72]. Moreover, the quality and variety of materials and goods used in a project [73,74], number of active labor force and salaries [75–77] and experience and scientific levels of employees in the project [78,79] are amongst other effective measures of project performance. Risks and operational risks, in particular, are also important indicators that many researchers have taken into account [23,63,80]. In addition, research and development (R & D) and related costs are identified as influential parameters in improving project performance [81,82].

Considering the generality of criteria mentioned above, these can be used in oil and gas wells drilling projects. As noted, there is a limited research available on performance evaluation of oil and gas well-drilling projects, however, a list of initial criteria for evaluating oil and gas well drilling projects can be found by reviewing the literature (Table 1). Obviously, these criteria are precisely related to the drilling of oil and gas wells and do not include all the relevant criteria in the entire oil and gas industry. This list will be completed in Section 4 using the Fuzzy Delphi method.

Table 1. Primary Criteria from Literature Review.

No.	Description	Source
1	Cost spent for drilling	[1,4,21,22]
2	Types of drilled wells in terms of number of holes	[4,32]
3	Time of drilling operations	[1,4,21,22]
4	Number of accidents caused by non-compliance with safety regulation or environmental factors	[4,20–22,63,66–68,70]
5	Actual cost compliance percentage with plans	[4,21]
6	Number of operational experts working on the project	[74–76]
7	Scientific levels of drilling specialists working in the project	[79]
8	Experience of drilling specialists working in Project	[78,79]
9	Average salaries of employees in the project	[74–77]
10	Types of wells drilled in operational risk	[20,22,23,63,69,81]
11	Number of operational failures	[4,21,63]
12	Employer/Senior Manager Satisfaction	[4,21,71,72]
13	Quality of materials and goods	[1,63,73,74]
14	R & D expenditure	[63,81,82]

3. Research Methodology

First, a list of assessment criteria is derived from the literature available on the research subject. Then by using the Fuzzy Delphi method and expert opinions, the criteria are extended and the final list is obtained for oil and gas well-drilling proje cts (Table 4). The Step-Wise Weight Assessment Ratio Analysis (SWARA) method is employed to determine final criteria weights. According to the interval-values fuzzy Additive Ratio Assessment and the decision Table developed for a set of oil and gas projects, project performance is evaluated. Figure 1 presents the flowchart of the research.

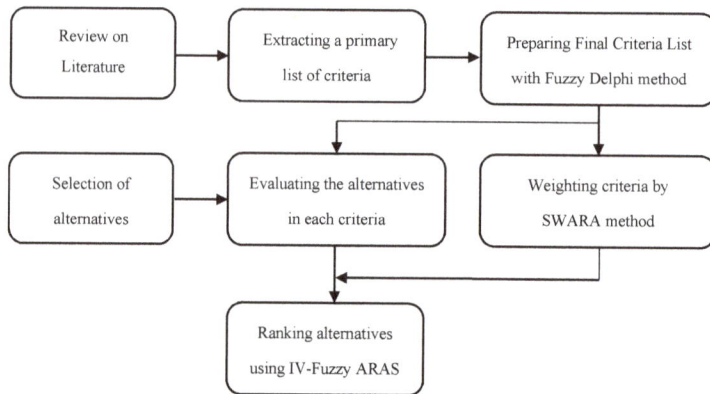

Figure 1. Schematic diagram of research design.

3.1. Fuzzy Delphi Method

The traditional "Project DELPHI" was originally developed by Dalkey and Helmer (1963) to achieve the most reliable consensus among experts on a particular topic [83]; a theoretical consensus obtained through several rounds of extensive consultations through expert interviews [84]. The most important advantage of this method is to avoid direct confrontation of participants [85]. However, the Delphi method has been criticized for higher operational costs, lower degree of convergence and the possibility to remove some key ideas by organizers. Therefore, Murray et al. (1985) proposed the integration of the traditional Delphi method and fuzzy set theory to improve ambiguity and inconsistency in requirements [86,87]. A variety of techniques have been developed to determine metrics by using Delphi method. Here, based on the approach presented by Hsu and Yang (2000) and Kuo and Chen (2008), triangular fuzzy numbers are used to incorporate expert opinions and to implement the fuzzy Delphi technique. Maximum and minimum values obtained based on expert opinions are defined as two endpoints of triangular fuzzy numbers, while the geometric mean is expressed as the membership degree in triangular fuzzy numbers in order to avoid the portion of terminal values [86,88]. Kuo and Chen believed that this method enables decision-makers to achieve a better solution for the selection [86].

Further, the authors highlighted the simplicity and the lack of survey repetition, as well as the use of all expert opinions. Each fuzzy number (T_A) is defined as follows (1):

$$T_A = (L_A, M_A, U_A), \quad L_A = min(X_{Ai}), \quad U_A = max(X_{Ai}), \quad M_A = \sqrt[n]{\prod_{i=1}^{n} X_{Ai}} \tag{1}$$

where, X_{A_i} is the proposed value of the i-th decision-maker in terms of the critical factor A; ($i = 1,2 \dots$). L_A, U_A and M_A represent the values of lower bound, upper bound and geometric mean for the critical factor A, respectively. In the next step, the Center of Area (COA) defuzzification process is performed using the model developed by Zheng and Teng (1993) [89]. The formula is presented as follows (2):

$$DF_k = \frac{(U_k - L_k) + (M_k - L_k)}{3} + L_k \tag{2}$$

where, the index k represents the number of criteria and L_k, U_k and M_k represent the values of lower bound, upper bound and geometric mean for the critical factor k.

The final step is to determine the threshold value for accepting or rejecting the criteria. To this end, the score of 0.7 is determined based on expert opinions. Finally, those criteria with a number less than the threshold value are removed from the list and the final list provides the necessary evaluation criteria.

3.2. SWARA Method

For a large number of multi-attribute decision-making problems, weighting indicators are included among the most important procedures in problem solution [90]. Accordingly, experts play a vital role in criteria evaluation and weighting and form inevitable part of the decision-making process. Th Step-Wise Weight Assessment Ratio Analysis (SWARA) method newly proposed by Krešulienė et al. (2010) allows decision-makers to select, evaluate and weight the criteria [91]. The most important advantage of this approach is its potential for evaluating the accuracy of expert opinions about weights allocated by the process [91]. Further, expert consultations can yield more accurate results than other common methods of multi-criteria decision-making (MCDM) [92]. The main steps of determining the criteria weights based on the SWARA method are described below:

- Step 1: Rank Criteria—First, the criteria determined by decision-makers are selected as the final criteria and then all the criteria are ranked in order of their importance. Accordingly, the most/least important criteria take the highest/lowest position of ranking.
- Step 2: Determine Relative Importance for Criteria (S_j)—Now, the relative importance of each criterion is measured against the most important criterion. This value is represented by S_j.
- Step 3: Calculate Coefficient Value of K_j—As a function of the relative importance for each criterion, the coefficient K_j is determined using Equation (3).

$$K_j = S_j + 1 \tag{3}$$

- Step 4: Calculate Initial Weights for Criteria—In this step, the initial weights of each criterion are calculated by Equation (4). Note that the initial weight for the first—i.e., the most important—criterion is generally considered equal to 1 ($q_1 = 1$).

$$q_j = \frac{q_{j-1}}{K_j} \tag{4}$$

- Step 5: Calculate Final Normalized Weights—As the final step of SWARA, the final weights which is also known as the normalized weights are determined by Equation (5).

$$w_j = \frac{q_j}{\sum q_j} \tag{5}$$

As mentioned before, SWARA is a newly established method for weighting which has been recently used by different studies [93].

3.3. Interval-Valued Fuzzy Additive Ratio Assessment

3.3.1. Generalized Fuzzy Numbers

A generalized fuzzy number $\tilde{A}$ defined as (6),

$$\tilde{A} = (a, b, c, d; \omega), \ 0 \leq a \leq b \leq c \leq d \leq 1, \ 0 \leq \omega \leq 1 \tag{6}$$

It is a fuzzy subset of the real line R with the membership function ($\mu_{\tilde{A}}$) which has the following features [94]:

$\mu_{\tilde{A}}$ is a continuous mapping from R to the closed interval [0, 1].

$$\forall x \in (-\infty, a] \ \rightarrow \ \mu_{\tilde{A}}(x) = 0$$

$\mu_{\tilde{A}}(x)$ is strictly increasing on [a, b].

$\mu_{\tilde{A}}(x) = \omega$ for all $x \in [b, c]$, where ω is a constant on $[0, 1]$, $0 \le \omega \le 1$.

$\mu_{\tilde{A}}(x)$ is strictly decreasing on $[c, d]$.

$\mu_{\tilde{A}}(x) = 0$ for all $x \in [d, +\infty]$.

If $\mu_{\tilde{A}}$ is linear on the intervals $[a, b]$ and $[c, d]$, then a generalized fuzzy number is called a generalized trapezoidal fuzzy number. Figure 2 shows a relationship between the generalized fuzzy number, $\tilde{B}$ and the normalized trapezoidal fuzzy number, $\tilde{A}$.

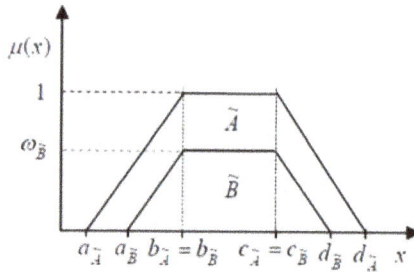

Figure 2. Generalized fuzzy number.

As seen from Figure 2, the normalized trapezoidal fuzzy numbers demonstrate certain cases of generalized fuzzy numbers, where $\omega = 1$. Also, if $b = c$, then the trapezoidal fuzzy number becomes a triangular fuzzy number.

3.3.2. Interval-Valued Fuzzy Numbers

The interval-valued fuzzy numbers are special forms of generalized fuzzy numbers. Similar to generalized fuzzy numbers, these numbers can find a trapezoidal shape. Moreover, interval-valued triangular fuzzy numbers have a triangular shape. Figure 3 shows a graphical representation of an interval-valued triangular fuzzy number.

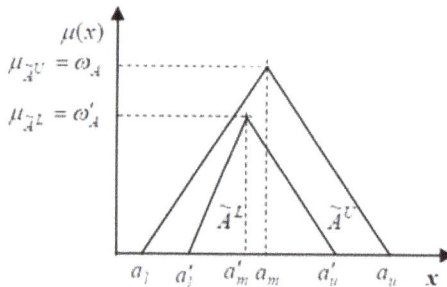

Figure 3. Interval-valued triangular fuzzy number.

An interval-valued triangular fuzzy number can be defined as (7) [95]:

$$\tilde{A} = \left[\tilde{A}^L, \tilde{A}^U\right] = \left[(a'_l, a'_m, a'_u; \omega'_A), (a_l, a_m, a_u; \omega_A)\right] \tag{7}$$

where $\tilde{A}^L$ and $\tilde{A}^U$ are the lower and upper triangular fuzzy numbers, $\tilde{A}^L \subset \tilde{A}^U$ and $\mu_{\tilde{A}}(x)$ are their membership functions. However, $\mu_{\tilde{A}^L}(x) = \omega'_A$ and $\mu_{\tilde{A}^U}(x) = \omega_A$ denote the lower and upper membership functions.

Suppose $\tilde{A} = [\tilde{A}^L, \tilde{A}^U]$ and $\tilde{B} = [\tilde{B}^L, \tilde{B}^U]$ are two interval-valued triangular fuzzy numbers. Then, the basic arithmetic operations on these fuzzy numbers can be represented as (8) to (11):

$$\tilde{A} + \tilde{B} = \left[(a'_l + b'_l, a'_m + b'_m, a'_u + b'_u; min\left(\omega'_A, \omega'_B\right)), (a_l + b_l, a_m + b_m, a_u + b_u; min(\omega_A, \omega_B)) \right] \quad (8)$$

$$\tilde{A} - \tilde{B} = \left[(a'_l - b'_u, a'_m - b'_m, a'_u - b'_l; min\left(\omega'_A, \omega'_B\right)), (a_l - b_u, a_m - b_m, a_u - b_l; min(\omega_A, \omega_B)) \right] \quad (9)$$

$$\tilde{A} \times \tilde{B} = \left[(a'_l \times b'_l, a'_m \times b'_m, a'_u \times b'_u; min\left(\omega'_A, \omega'_B\right)), (a_l \times b_l, a_m \times b_m, a_u \times b_u; min(\omega_A, \omega_B)) \right] \quad (10)$$

$$\tilde{A} \div \tilde{B} = \left[(a'_l \div b'_u, a'_m \div b'_m, a'_u \div b'_l; min\left(\omega'_A, \omega'_B\right)), (a_l \div b_u, a_m \div b_m, a_u \div b_l; min(\omega_A, \omega_B)) \right] \quad (11)$$

Figure 4 shows a certain case of generalized interval-valued fuzzy numbers, normalized with the same mode $(a'_m = a_m)$ and it can be represented as (12)

$$\tilde{A} = \left[\tilde{A}^L, \tilde{A}^U \right] = \left[(a_l, a'_l), a_m, (a'_u, a_u) \right] \quad (12)$$

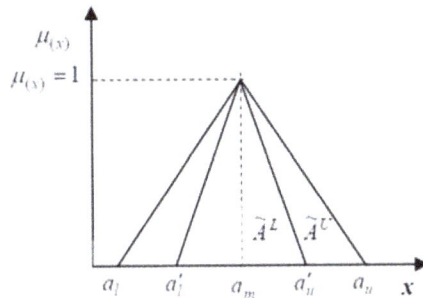

Figure 4. Normalized interval-valued triangular fuzzy number with the same mode.

Suppose $\tilde{A} = \left[\tilde{A}^L, \tilde{A}^U \right] = \left[(a_l, a'_l), a_m, (a'_u, a_u) \right]$ and $\tilde{B} = \left[\tilde{B}^L, \tilde{B}^U \right] = \left[(b_l, b'_l), b_m, (b'_u, b_u) \right]$ denote two normalized interval-valued triangular fuzzy numbers with the same mode. Then, the basic arithmetic operations on these fuzzy numbers can be defined as (13) to (16) [96]:

$$\tilde{A} + \tilde{B} = \left[(a_l + b_l, a'_l + b'_l), a_m + b_m, (a'_u + b'_u, a_u + b_u) \right] \quad (13)$$

$$\tilde{A} - \tilde{B} = \left[(a_l - b_u, a'_l - b'_u), a_m - b_m, (a'_u - b'_l, a_u - b_l) \right] \quad (14)$$

$$\tilde{A} \times \tilde{B} = \left[(a_l \times b_l, a'_l \times b'_l), a_m \times b_m, (a'_u \times b'_u, a_u \times b_u) \right] \quad (15)$$

$$\tilde{A} \div \tilde{B} = \left[(a_l \div b_u, a'_l \div b'_u), a_m \div b_m, (a'_u \div b'_l, a_u \div b_l) \right] \quad (16)$$

In addition, the following unary operation defined on interval-valued triangular fuzzy numbers is of great importance. It is denoted as (17):

$$\frac{1}{k} \times \tilde{A} = \left[\left(\frac{1}{k} \times a_l, \frac{1}{k} \times a'_l \right), \frac{1}{k} \times a_m, \left(\frac{1}{k} \times a'_u, \frac{1}{k} \times a_u \right) \right] \quad (17)$$

3.3.3. Linguistic Variables

The linguistic variables refer to variables whose values correspond to words or sentences in a natural or artificial language. A linguistic variable has a practical potential for dealing with many real-world decision-making problems, which are usually complex and relatively uncertain. A wide range of research studies have reported different linguistic variables with triangular fuzzy numbers [97–99]. Also, the literature provides linguistic variables based on the use of interval-valued fuzzy numbers. Wei and Chen

(2009), for example, developed a scale of nine level linguistic terms using interval-valued trapezoidal fuzzy numbers [100]. Ashtiani et al. (2009) presented a seven-level linguistic terms scale based on interval-valued triangular fuzzy numbers. Tables 2 and 3 show the linguistic variables for the weights of criteria and performance ratings, based on the use of triangular fuzzy numbers and interval-valued triangular fuzzy numbers [57].

Table 2. Linguistic variables for the weights of criteria.

Linguistic Variables	Triangular Fuzzy Number	Interval-Valued Triangular Fuzzy Number
Very low (VL)	(0.0, 0.0, 0.1)	[(0.00, 0.00), 0.0, (0.10, 0.15)]
Low (L)	(0.0, 0.1, 0.3)	[(0.00, 0.50), 0.1, (0.25, 0.35)]
Medium low (ML)	(0.1, 0.3, 0.5)	[(0.00, 0.15), 0.3, (0.45, 0.55)]
Medium (M)	(0.3, 0.5, 0.7)	[(0.25, 0.35), 0.5, (0.65, 0.75)]
Medium high (MH)	(0.5, 0.7, 0.9)	[(0.45, 0.55), 0.7, (0.80, 0.95)]
High (H)	(0.7, 0.7, 1.0)	[(0.55, 0.75), 0.9, (0.95, 1.00)]
Very high (VH)	(0.9, 1.0, 1.0)	[(0.85, 0.95), 1.0, (1.00, 1.00)]

Table 3. Linguistic variables for the performance ratings.

Linguistic Variables	Triangular Fuzzy Number	Interval-Valued Triangular Fuzzy Number
Very poor (VP)	(0.0, 0.0, 0.1)	[(0.00, 0.00), 0.0, (0.10, 0.15)]
Poor (P)	(0.0, 0.1, 0.3)	[(0.00, 0.50), 0.1, (0.25, 0.35)]
Medium poor (MP)	(0.1, 0.3, 0.5)	[(0.00, 0.15), 0.3, (0.45, 0.55)]
Fair (F)	(0.3, 0.5, 0.7)	[(0.25, 0.35), 0.5, (0.65, 0.75)]
Medium good (MG)	(0.5, 0.7, 0.9)	[(0.45, 0.55), 0.7, (0.80, 0.95)]
Good (G)	(0.7, 0.7, 1.0)	[(0.55, 0.75), 0.9, (0.95, 1.00)]
Very good (VG)	(0.9, 1.0, 1.0)	[(0.85, 0.95), 1.0, (1.00, 1.00)]

Since interval-valued fuzzy numbers are more complex than ordinary fuzzy numbers, the transformation of the ordinary fuzzy numbers into the corresponding interval-valued fuzzy numbers can raise some advantages. To transform their weights and performance ratings, the following equations are given:

$$l = \min_{k}\left(l^{k}\right) \tag{18}$$

$$l' = \left(\prod_{k=1}^{K} l^{k}\right)^{\frac{1}{K}} \tag{19}$$

$$m = \left(\prod_{k=1}^{K} m^{k}\right)^{\frac{1}{K}} \tag{20}$$

$$u' = \left(\prod_{k=1}^{K} u^{k}\right)^{\frac{1}{K}} \tag{21}$$

$$u = \max_{k}\left(u^{k}\right) \tag{22}$$

$\tilde{x} = [(l, l'), m, (u', u)]$ is the corresponding interval-valued triangular fuzzy number, while $\tilde{x}^{k} = \left(l^{k}, m^{k}, u^{k}\right)$ is the triangular fuzzy number obtained on the basis of opinion of kth decision maker. The parameters l and u denote the smallest and the greatest performance ratings among all stakeholders, respectively; which reflect the extreme attitudes provided by the experts. Unlike these parameters, other parameters of the interval-valued triangular fuzzy number reflect the expert opinions much more effectively. The reason is that these numbers are obtained as the geometric mean of attitudes from all experts.

3.3.4. Defuzzification of Interval-Valued Triangular Fuzzy Numbers

Since the results of arithmetic operations will be fuzzy numbers, they can be transformed into non-fuzzy numbers in order to rank and compare alternatives. Different procedures have been proposed for ranking fuzzy numbers and for their defuzzification but these procedures concern mainly the trapezoidal or triangular fuzzy numbers. By small changes, however, the same procedures can be used for the defuzzification of interval-valued triangular fuzzy numbers. Equations (23) and (24) are two general defuzzification equations for triangular fuzzy numbers (λ is a coefficient on [0, 1]):

$$gm(\tilde{A}) = \frac{1}{2}[(1 - \lambda)l + m + \lambda u] \tag{23}$$

$$gm(\tilde{A}) = \frac{1 + m + u}{3} \tag{24}$$

Moreover, (25) and (26) are proposed for defuzzification of interval-valued triangular fuzzy numbers:

$$gm(\tilde{B}) = \frac{l + l' + m + u' + u}{5} \tag{25}$$

$$gm(\tilde{B}) = \frac{(1 - \lambda)l + \lambda l' + m + \lambda u' + (1 - \lambda)u}{5} \tag{26}$$

where $\tilde{A}$ represents ordinary triangular fuzzy numbers, whereas $\tilde{B}$ presents interval-valued fuzzy numbers. λ is a coefficient on [0, 1]. Equation (25) is a simple extension of (24), providing an effective way for the defuzzification of known interval-valued fuzzy numbers, represented as the Best Non-Fuzzy Performance (BNP). In contrast, Equation (25) is relatively more complex but it has some advantages. For instance, varying the coefficient λ makes a greater importance to the parameters l' and u' against l and u and vice versa.

3.3.5. Additive Ratio Assessment (ARAS) Method

As a relatively new tool for MCDM, the ARAS method has received significant interested recently, still based on the theory that complex phenomena of the world could be accurately perceived trough simple relative comparisons [101–103]. The ARAS method uses the concept of optimality degree to find a ranking. It is the sum of normalized weighted values of the criteria with respect to each alternative divided by the sum of normalized weighted values of the best alternative.

- Step 1: First, a decision matrix is assembled as $m \times n$, where m denotes alternatives and n denotes criteria.

$$X = \begin{bmatrix} x_{01} & \cdots & x_{0j} & \cdots & x_{0n} \\ \vdots & \ddots & \vdots & \ddots & \vdots \\ x_{i1} & \cdots & x_{ij} & \cdots & x_{in} \\ \vdots & \ddots & \vdots & \ddots & \vdots \\ x_{m1} & \cdots & x_{mj} & \cdots & x_{mn} \end{bmatrix}; i = \overline{0, m}; j = \overline{1, n} \tag{27}$$

x_{ij} is the performance measure of the i-th alternative on the j-th criterion. Also, x_{0j} shows the optimum value for the j-th criterion. If the optimum value of the variable j is undetermined, then it can be determined as follows:

$$\text{when } \max_i x_{ij} \text{ is optimal, } x_{0j} = \max_i x_{ij}$$
$$\text{when } \min_i x_{ij}^* \text{ is optimal, } x_{0j} = \min_i x_{ij}^* \tag{28}$$

In general, the evaluation values of alternatives with respect to criteria (x_{ij}) and the weights for each criterion (w_j) are given as the inputs in the decision matrix. Note that each criterion reflects its certain dimensions; therefore, a comparative analysis and preventing potential consequences from different dimensions require derive dimensionless quantities. To do this, the weighted values are simply divided by optimum obtained as (28). Numerous methods are available for deriving dimensionless useful dimensionless values which will be described below. Through normalization, the values of an original decision matrix are converted into the values on [0, 1] or on [0, ∞].

- Step 2: The primary inputs are normalized for all criteria, represented by $\overline{x}_{ij}$ and formed the matrix elements.

$$\overline{X} = \begin{bmatrix} \overline{x}_{01} & \cdots & \overline{x}_{0j} & \cdots & \overline{x}_{0n} \\ \vdots & \ddots & \vdots & \ddots & \vdots \\ \overline{x}_{i1} & \cdots & \overline{x}_{ij} & \cdots & \overline{x}_{in} \\ \vdots & \ddots & \vdots & \ddots & \vdots \\ \overline{x}_{m1} & \cdots & \overline{x}_{mj} & \cdots & \overline{x}_{mn} \end{bmatrix} ; i = \overline{0,m}; \ j = \overline{1,n} \tag{29}$$

Since there are benefit type and cost type criteria, then the normalization is processed positively or negatively by using (30) and (31), respectively.

$$\overline{x}_{ij} = \frac{x_{ij}}{\sum_{i=0}^{m} x_{ij}} \tag{30}$$

$$x_{ij} = \frac{1}{x_{ij}^*} \overline{x}_{ij} = \frac{x_{ij}}{\sum_{i=0}^{m} x_{ij}} \tag{31}$$

The achievement of dimensionless quantities provides a framework for comparing each criterion against all others.

- Step 3: Here, the weighted normalized decision matrix, $\hat{X}$, is calculated by applying the weight values on the normalized decision matrix $\overline{X}$. The weights are determined by expert panels and should meet the following requirements:

$$0 < w_j < 1, \ \sum_{j=1}^{n} w_j = 1$$

$$\hat{X} = \begin{bmatrix} \hat{x}_{01} & \cdots & \hat{x}_{0j} & \cdots & \hat{x}_{0n} \\ \vdots & \ddots & \vdots & \ddots & \vdots \\ \hat{x}_{i1} & \cdots & \hat{x}_{ij} & \cdots & \hat{x}_{in} \\ \vdots & \ddots & \vdots & \ddots & \vdots \\ \hat{x}_{m1} & \cdots & \hat{x}_{mj} & \cdots & \hat{x}_{mn} \end{bmatrix} ; i = \overline{0,m}; \ j = \overline{1,n} \tag{32}$$

$$\hat{x}_{ij} = \overline{x}_{ij} w_j; \ i = \overline{0,m} \tag{33}$$

Again, w_j denotes the weight value for the *j*th criterion and $\overline{x}_{ij}$ represents the normalized value for the *i*th alternative. Therefore, the value of an optimal function can be calculated as follows:

$$S_i = \sum_{j=1}^{n} \hat{x}_{ij}; \ i = \overline{0,m} \tag{34}$$

According to the logic of ARAS, the best alternative is the only one with the greatest value for an optimal function. Clearly, the worst alternative obtains the value of minimum for the optimal function. To put it differently, the alternative ranking is determined based on the value of S_i.

The degree of utility can be measured by comparing each alternative against the best/optimal one with the best value, represented by S_0. The degree of utility, K_i, of the alternative A_i follows Equation (35):

$$K_i = \frac{S_i}{S_0}; \ i = \overline{0,m} \tag{35}$$

where, S_0 and S_i are derived from Equation (34). Clearly, K_i places on the interval $[0, 1]$ and its value is used for ranking all alternatives.

3.3.6. An Extension of ARAS Method Based on Interval-Valued Triangular Fuzzy Numbers

- Step 1: Determine the optimal performance rating for each criterion

The first point to be considered is that the optimal performance rating for each criterion should be calculated as an interval-valued fuzzy number. Therefore, optimal interval-valued fuzzy performance ratings can be determined as follows:

$$\tilde{x}_{0j} = \left[\left(l_{0j}, l'_{0j} \right), m_{0j}, \left(u'_{0j}, u_{0j} \right) \right] \tag{36}$$

where, $\tilde{x}_{0j}$ represents the optimal interval-valued fuzzy performance rating of the jth criterion. Also, other criteria are defined as follows:

$$l_{0j} = \begin{cases} \max_i l_{ij}; j \in \Omega_{max} \\ \min_i l_{ij}; j \in \Omega_{min} \end{cases} \tag{37}$$

$$l'_{0j} = \begin{cases} \max_i l'_{ij}; j \in \Omega_{max} \\ \min_i l'_{ij}; j \in \Omega_{min} \end{cases} \tag{38}$$

$$m_{0j} = \begin{cases} \max_i m_{ij}; j \in \Omega_{max} \\ \min_i m_{ij}; j \in \Omega_{min} \end{cases} \tag{39}$$

$$u'_{0j} = \begin{cases} \max_i u'_{ij}; j \in \Omega_{max} \\ \min_i u'_{ij}; j \in \Omega_{min} \end{cases} \tag{40}$$

$$u_{0j} = \begin{cases} \max_i u_{ij}; j \in \Omega_{max} \\ \min_i u_{ij}; j \in \Omega_{min} \end{cases} \tag{41}$$

Ω_{max} denotes the benefit criteria, i.e., the higher the values are, the better it is; and Ω_{min} denotes the set of cost criteria, i.e., the lower the values are, the better it is.

- Step 2: Calculate the normalized decision matrix

To enable the use of these interval-valued fuzzy numbers, the normalization process requires some modifications. So, (34) can be replaced by (42):

$$\tilde{r}_{ij} = \begin{cases} \left[\left(\frac{a_{ij}}{c_j^+}, \frac{a'_{ij}}{c_j^+} \right), \frac{b_{ij}}{c_j^+}, \left(\frac{c'_{ij}}{c_j^+}, \frac{c_{ij}}{c_j^+} \right) \right] ; j \in \Omega_{max} \\ \left[\left(\frac{\frac{1}{a_{ij}}}{a_j^-}, \frac{\frac{1}{a'_{ij}}}{a_j^-} \right), \frac{\frac{1}{b_{ij}}}{a_j^-}, \left(\frac{\frac{1}{c'_{ij}}}{a_j^-}, \frac{\frac{1}{c_{ij}}}{a_j^-} \right) \right] ; j \in \Omega_{min} \end{cases} \tag{42}$$

Here, $\tilde{r}_{ij}$ is the optimal interval-valued fuzzy performance rating for the ith alternative on the jth criterion. Further,

$$a_j^- = \sum_{i=0}^{m} \frac{1}{a_{ij}}, \ c_j^+ = \sum_{i=0}^{m} c_{ij}, \ i = 0, 1, \dots, m$$

- Step 3: Calculate the normalized weighted interval-valued decision matrix

This is principally similar to the third step in the original ARAS method. The difference is that fuzzy numbers are to be now multiplied by using the multiplication operation on interval-valued triangular fuzzy numbers. Therefore, this can be expressed as follow:

$$\tilde{v}_{ij} = \tilde{w}_j.\tilde{r}_{ij} \tag{43}$$

where, $\tilde{v}_{ij}$ is the normalized weighted interval-valued fuzzy performance rating for the i-th alternative on the j-th criterion.

- Step 4: Compute the overall interval-valued fuzzy performance ratings

This step can be expressed using (44):

$$\tilde{S}_i = \sum_{j=1}^{n} \tilde{v}_{ij} \tag{44}$$

where, $\tilde{S}_i$ is the overall interval-valued fuzzy performance rating for the ith alternative.

- Step 5: Measure the degree of utility for each alternative

Since the result obtained from the previous step is provided as an interval-valued fuzzy numbers, the calculation process is often more complex with the overall degree of utility. Obviously, it should be transformed into a non-fuzzy number. The degree of utility can be calculated as follow:

$$\tilde{Q}_i = \frac{\tilde{S}_i}{\tilde{S}_0} \tag{45}$$

Again, as the products of (45) are still interval-valued fuzzy numbers, these typically need to be defuzzified. The defuzzification process must be initiated prior of determining the degree of utility. There are a wide range of defuzzification methods with a variety of impacts on resultant outputs. Therefore, it is important how to choose an appropriate defuzzification technique.

- Step 6: Rank alternatives and select the most efficient one

This step follows the similar process as the original Additive Ratio Assessment method.

4. Results: A Case Study of Well-Drilling Projects

The case study of this research includes a company active in the Iranian drilling industry, implementing seven oil and gas well drilling projects. The locations of these projects are Iranian oil reservoirs which have a major portion of the country's oil production. Concerns about environmental issues show other challenges related to such projects. Based on the opinion of the senior management, a group of three experienced professionals in the field of Iranian oil and gas drilling industry was established as the expert team to provide expertized views in different stages. In the first step, the initial questionnaire obtained from Table 1 was distributed among 15 experts in Iran's oil and gas wells projects who were introduced by the expert team. The questionnaire consisted of two main parts: The first part included the importance evaluation of the criteria derived from previous studies (Table 1) based on a Likert scale. In the second part, the expert was asked to propose some measures important in the project evaluation but not included in the first part. As the first round of survey completed, a total of 13 criteria were confirmed by the Delphi method, while 14 new criteria were suggested by experts. Based on these results, the second questionnaire was formulated and its results were also analyzed as in the first stage. Finally, a list of 20 criteria was developed and categorized for evaluating oil and gas well drilling projects using the Fuzzy Delphi method. According to the expert opinions, these criteria were classified in six main criteria (Table 4).

Table 4. Evaluation Criteria in Oil & Gas Well-Drilling Projects using Fuzzy Delphi method.

Criteria	Code	Sub-Criteria	Code
Materials & Equipment	A	Number of drilling rigs used	A1
		Type of drilling rigs	A2
		Quality of materials and goods used	A3
		Quality of drilling and support service	A4
Human Resource	B	Number of operational experts working on the project	B1
		Scientific levels of drilling specialists working in the project	B2
		Experience of drilling specialists working in Project (Ave.)	B3
		Average salaries of employees in the project (Million Rls)	B4
Planning	C	Type of drilled wells in terms of operational risk	C1
		Type of drilled wells in terms of depth	C2
		Type and number of fields under operation	C3
		Cost spent for drilling project (Billion Rls)	C4
		Status of cash flows in project	C5
Quality	D	Employer/Senior Manager Satisfaction	D1
		Waiting time percentage to total well drilling time	D2
		Number of failure reports	D3
		Number of accidents caused by non-compliance with safety regulation, or environmental factors	D4
		Actual cost compliance percentage with planned cost	D5
Number of planned wells	E	number of planned wells in a certain period of time	E
Number of drilled wells	F	number of drilled wells in a certain period of time	F

Next, the SWARA technique was implemented to determine weights of all criteria and sub-criteria based on the opinion of each expert. Also, the obtained values were integrated to calculate the final weights. Table 5 shows the results.

Table 5. Calculation of Criteria Weights for Expert #1.

Code	Criterion	S_j	K_j ($K_j = 1 + S_j$)	Initial Weight	Normalized Final Weights
D	Quality	1	1	1	0.277
A	Materials & Equipment	0.35	1.35	0.741	0.205
E	Number of drilled wells	0.33	1.33	0.557	0.154
B	Human Resource	0.11	1.11	0.502	0.139
C	Planning	0.06	1.06	0.473	0.131
F	Number of planned wells	0.4	1.4	0.338	0.094

As seen in Table 5, based on the first step of the SWARA method, the expert was asked to rank the criteria in a descending order of importance. The results are displayed in the second column of Table 5. Also, the second, third and fourth steps of the SWARA method are presented in the columns 3, 4 and 5 of Table 5, respectively. By implementing the final step of the SWARA method to normalize the weights of criteria, the final weighting values are shown in the column 6 of Table 5. A similar procedure was carried out for the second and third experts. Moreover, for comparative analysis of the weights of all criteria according to expert opinions and the geometric means, the weights and the geometric means obtained from each expert are presented in Figure 5. The same was carried out for the sub-criteria. According to the results, the most important sub-criteria for the materials and equipment, human resource, planning and the quality criteria are types of drilling rigs, experience of drilling specialists working in project, cost spent for drilling project and actual cost compliance percentage with planned cost. Now, to determine the final weights of the sub-criteria, the weights of one sub-criterion obtained from each expert multiplied by the weights of the criteria corresponding to that sub-criterion. In order to aggregate expert opinions, the geometric means for the weights obtained for each sub-criterion from three experts were calculated, shown in the last column of Table 6. Finally, each element of the column was divided by the sum of all its elements to determine the normalized final weight of each sub-criterion. The results are shown in the last column of Table 6.

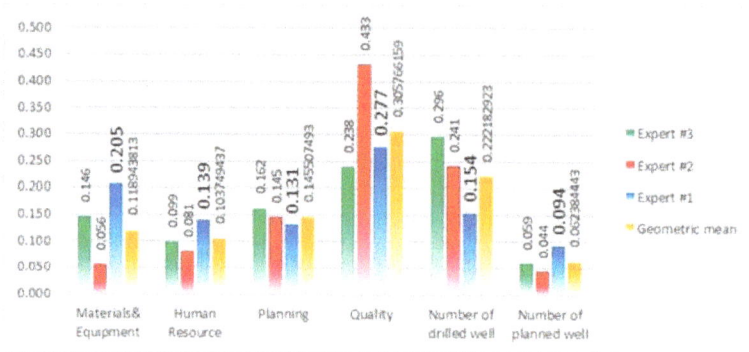

Figure 5. Comparative Analysis of Criteria Weights based on Expert Opinions.

As found, some criteria such as the number of drilled wells and the sub-criteria such as the actual cost compliance percentage with planned costs and the percentage of waiting time to total well drilling time have been identified as the most important sub-criteria. Table 7 shows the ranking of the criteria in terms of importance.

Now, as the weight and importance of the criteria determined, the evaluation of each alternative in terms of all criteria was carried out by the experts using the linguistic variables in Table 2. The results are presented in Table 8.

Then, the qualitative values were converted into quantitative values using Table 2, in order to establish the initial decision-making tables for the three experts. Tables 9–11 provide the quantitative values for the experts.

Next, using Equations (18)–(22), the fuzzy values in the decision matrix of three experts were converted into fuzzy numbers with interval values, as shown in Table 12.

Table 12 presents the ideal alternative using Equations (35)–(40). In the next step using Equation (41), the interval valued fuzzy decision matrix (Table 12) was normalized. The normalized matrix is shown in Table 13.

Here, the weighting values obtained from the SWARA method normalized and aggregated in the final column of Table 6 and also Equation (43) were applied to achieve the normalized decision matrix with fuzzy intervals, as shown in Table 14.

Following the fourth step of the ARAS method and using Equation (44), the values of S, presented as interval-valued fuzzy numbers, were calculated. To perform the fifth step of the ARAS method and compute the Q values, we have to convert the interval-valued fuzzy numbers of S to Crisp numbers by using Equations (23)–(26). The conversion and calculations for each alternative are displayed for different values of λ in Table 15.

As seen, for different values of λ, Project #2 can be selected as the ideal alternative. The analytical discussion confirms its higher score for the criteria such as the quality of materials and goods used, scientific levels of drilling specialists working in the project, as well as the actual cost compliance percentage with planned cost.

Table 6. Calculation of Sub-Criteria Weights for individual Experts and Normalized Final Weights.

	Expert #1				Expert #2					Expert #3					Geometric Mean of Sub-Criteria Weight	Normalized Final Weights
Criterion Code / Criterion Weight	Sub-Criterion Code	Weight in Each Criterion	Final Weight of Sub-Criteria	Criterion Code / Criterion Weight	Sub-Criterion Code	Weight in Each Criterion	Final Weight of Sub-Criteria		Criterion Code / Criterion Weight	Sub-Criterion Code	Weight in Each Criterion	Final Weight of Sub-Criteria				
A, 0.205	A1	0.221	0.045	A, 0.056	A1	0.262	0.015		A, 0.146	A1	0.312	0.046			0.031	0.033
	A2	0.307	0.063		A2	0.445	0.025			A2	0.295	0.043			0.041	0.044
	A3	0.272	0.056		A3	0.114	0.006			A3	0.185	0.027			0.021	0.023
	A4	0.201	0.041		A4	0.178	0.010			A4	0.208	0.030			0.023	0.025
B, 0.139	B1	0.236	0.033	B, 0.081	B1	0.108	0.009		B, 0.099	B1	0.301	0.030			0.020	0.022
	B2	0.278	0.039		B2	0.287	0.023			B2	0.193	0.019			0.026	0.028
	B3	0.286	0.040		B3	0.425	0.035			B3	0.389	0.038			0.038	0.040
	B4	0.200	0.028		B4	0.180	0.015			B4	0.117	0.012			0.017	0.018
C, 0.131	C1	0.186	0.024	C, 0.145	C1	0.348	0.050		C, 0.162	C1	0.180	0.029			0.033	0.035
	C2	0.307	0.040		C2	0.170	0.025			C2	0.249	0.040			0.034	0.037
	C3	0.109	0.014		C3	0.132	0.019			C3	0.127	0.021			0.018	0.019
	C4	0.279	0.037		C4	0.264	0.038			C4	0.3	0.060			0.044	0.047
	C5	0.119	0.016		C5	0.087	0.013			C5	0.073	0.012			0.013	0.014
D, 0.277	D1	0.248	0.069	D, 0.433	D1	0.108	0.047		D, 0.238	D1	0.278	0.066			0.060	0.064
	D2	0.199	0.055		D2	0.423	0.183			D2	0.177	0.042			0.075	0.080
	D3	0.129	0.036		D3	0.159	0.069			D3	0.135	0.032			0.043	0.046
	D4	0.136	0.038		D4	0.083	0.036			D4	0.076	0.018			0.029	0.031
	D5	0.288	0.080		D5	0.226	0.098			D5	0.334	0.080			0.085	0.091
E			0.154	E			0.241		E			0.296			0.222	0.237
F			0.094	F			0.044		F			0.059			0.062	0.067

Table 7. Rankings of Sub-criteria in order of importance.

Code	Sub-Criteria	Final Weight
E	Number of drilled wells	0.237
D5	Actual cost compliance percentage with planned cost	0.091
D2	Waiting time percentage to total well drilling time	0.080
F	Number of planned wells	0.067
D1	Employer/Senior Manager Satisfaction	0.064
C4	Cost spent for drilling project (Billion Rls)	0.047
D3	Number of failure reports	0.046
A2	Type of drilling rigs	0.044
B3	Experience of drilling specialists working in Project	0.040
C2	Type of drilled wells in terms of depth	0.037
C1	Type of drilled wells in terms of operational risk	0.035
A1	Number of drilling rigs used	0.033
D4	Number of accidents caused by non-compliance with safety regulation	0.031
B2	Scientific levels of drilling specialists working in the project	0.028
A4	Quality of drilling and support service	0.025
A3	Quality of materials and goods used	0.023
B1	Number of operational experts working on the project	0.022
C3	Type and number of fields under operation	0.019
B4	Average salaries of employees in the project (Million Rls)	0.018
C5	Status of cash flows in project	0.014

Table 8. Initial Decision Matrix on Linguistic Variables.

Code	A1	A2	A3	A4	B1	B2	B3	B4	C1	C2	C3	C4	C5	D1	D2	D3	D4	D5	E	F
Weight	0.033	0.044	0.023	0.025	0.022	0.028	0.040	0.018	0.035	0.037	0.019	0.047	0.014	0.064	0.080	0.046	0.031	0.091	0.237	0.067
Sign	+	+	+	+	+	+	+	+	−	−	−	−	+	+	−	−	−	+	+	−
Expert #1																				
Project #1	G	G	G	VG	F	F	VG	G	F	VG	VG	F	G	G	F	F	F	G	P	G
Project #2	F	VG	F	G	G	G	G	P	P	G	F	G	F	G	P	P	F	VG	G	G
Project #3	F	G	VG	G	G	F	F	P	F	G	P	F	VG	F	F	G	VP	P	VP	G
Project #4	P	G	G	F	F	G	VG	F	G	F	F	VG	G	VG	P	P	G	G	F	VG
Project #5	G	F	F	G	F	G	G	F	VG	F	F	F	F	G	G	VG	G	VP	VG	F
Project #6	F	VG	P	VG	VG	VG	F	P	F	VG	G	P	F	F	VP	P	P	F	P	F
Project #7	G	G	G	F	G	F	F	P	VG	VG	VG	VG	G	VG	P	G	G	F	VP	G
Expert #2																				
Project #1	VG	G	VG	VG	P	F	VG	F	F	VG	VG	F	VG	G	F	F	F	VG	P	G
Project #2	F	VG	P	F	F	G	G	P	VP	G	F	G	F	VG	P	P	F	VG	G	G
Project #3	F	G	G	VG	G	F	P	P	G	G	P	P	VG	F	G	G	VP	P	VP	G
Project #4	P	P	F	F	F	P	VG	F	G	P	F	VG	G	VG	P	P	G	G	P	VG
Project #5	VG	F	F	F	P	G	G	F	VG	F	F	VG	P	G	VG	VP	VG	VP	VG	F
Project #6	F	VG	P	G	VG	VG	P	P	F	VG	G	P	F	P	VP	P	P	P	P	VG
Project #7	VG	G	G	P	G	F	F	P	VG	VG	VG	VG	G	VG	P	G	VG	P	VP	G
Expert #3																				
Project #1	G	G	F	VG	F	F	VG	G	P	VG	VG	F	G	VG	F	F	G	G	G	P
Project #2	F	VG	F	G	G	G	F	P	P	P	F	G	F	G	P	P	F	VG	G	G
Project #3	F	G	VG	F	F	F	P	P	F	G	P	F	VG	P	G	VG	VP	P	P	G
Project #4	VP	F	P	G	G	P	G	G	VG	VP	G	VG	VG	F	P	P	VG	VG	VG	VG
Project #5	VG	VP	F	G	P	G	G	F	G	VG	F	P	G	F	VG	VP	VG	VP	VG	F
Project #6	P	VG	P	VG	VG	VG	F	P	P	G	G	F	F	VP	VP	P	F	F	VG	VP
Project #7	G	G	G	P	G	F	VP	VP	G	VG	VG	G	G	VG	F	G	VG	G	VP	F

Table 9. Initial Decision Matrix for Expert #1 with Quantitative Values.

Code	A1	A2	A3	A4	B1	B2	B3	B4	C1	C2
Weight	0.033	0.044	0.023	0.025	0.022	0.028	0.040	0.018	0.035	0.037
Sign	+	+	+	+	+	+	+	+	−	−
Project #1	(0.9, 1, 1)	(0.7, 0.9, 1)	(0.9, 1, 1)	(0.9, 1, 1)	(0, 0.1, 0.3)	(0.3, 0.5, 0.7)	(0.9, 1, 1)	(0.3, 0.5, 0.7)	(0.3, 0.5, 0.7)	(0.9, 1, 1)
Project #2	(0.3, 0.5, 0.7)	(0.9, 1, 1)	(0, 0.1, 0.3)	(0.3, 0.5, 0.7)	(0.3, 0.5, 0.7)	(0.7, 0.9, 1)	(0.7, 0.9, 1)	(0, 0.1, 0.3)	(0, 0, 0.1)	(0.7, 0.9, 1)
Project #3	(0.3, 0.5, 0.7)	(0.7, 0.9, 1)	(0.7, 0.9, 1)	(0.9, 1, 1)	(0.7, 0.9, 1)	(0.3, 0.5, 0.7)	(0, 0.1, 0.3)	(0, 0.1, 0.3)	(0.7, 0.9, 1)	(0.7, 0.9, 1)
Project #4	(0, 0.1, 0.3)	(0, 0.1, 0.3)	(0.3, 0.5, 0.7)	(0, 0.1, 0.3)	(0.3, 0.5, 0.7)	(0.7, 0.9, 1)	(0.9, 1, 1)	(0.3, 0.5, 0.7)	(0.7, 0.9, 1)	(0, 0.1, 0.3)
Project #5	(0.9, 1, 1)	(0.3, 0.5, 0.7)	(0.3, 0.5, 0.7)	(0.3, 0.5, 0.7)	(0, 0.1, 0.3)	(0.9, 1, 1)	(0.7, 0.9, 1)	(0.3, 0.5, 0.7)	(0.9, 1, 1)	(0.3, 0.5, 0.7)
Project #6	(0.3, 0.5, 0.7)	(0.9, 1, 1)	(0, 0.1, 0.3)	(0.7, 0.9, 1)	(0.9, 1, 1)	(0.3, 0.5, 0.7)	(0, 0.1, 0.3)	(0, 0.1, 0.3)	(0.3, 0.5, 0.7)	(0.9, 1, 1)
Project #7	(0.9, 1, 1)	(0.7, 0.9, 1)	(0.7, 0.9, 1)	(0, 0.1, 0.3)	(0.7, 0.9, 1)	(0.3, 0.5, 0.7)	(0.3, 0.5, 0.7)	(0, 0.1, 0.3)	(0.9, 1, 1)	(0.9, 1, 1)

Code	C3	C4	C5	D1	D2	D3	D4	D5	E	F
Weight	0.019	0.047	0.014	0.064	0.080	0.046	0.031	0.091	0.237	0.067
Sign	−	−	+	+	−	−	−	+	+	−
Project #1	(0.9, 1, 1)	(0.3, 0.5, 0.7)	(0.9, 1, 1)	(0.7, 0.9, 1)	(0.3, 0.5, 0.7)	(0.3, 0.5, 0.7)	(0.3, 0.5, 0.7)	(0.9, 1, 1)	(0, 0.1, 0.3)	(0.7, 0.9, 1)
Project #2	(0.3, 0.5, 0.7)	(0.7, 0.9, 1)	(0.3, 0.5, 0.7)	(0.9, 1, 1)	(0, 0.1, 0.3)	(0.3, 0.5, 0.7)	(0.3, 0.5, 0.7)	(0.9, 1, 1)	(0.7, 0.9, 1)	(0.7, 0.9, 1)
Project #3	(0, 0.1, 0.3)	(0, 0.1, 0.3)	(0.9, 1, 1)	(0.3, 0.5, 0.7)	(0.7, 0.9, 1)	(0, 0, 0.1)	(0, 0, 0.1)	(0, 0.1, 0.3)	(0, 0, 0.1)	(0.7, 0.9, 1)
Project #4	(0.3, 0.5, 0.7)	(0.9, 1, 1)	(0.7, 0.9, 1)	(0.9, 1, 1)	(0, 0.1, 0.3)	(0.7, 0.9, 1)	(0.7, 0.9, 1)	(0.7, 0.9, 1)	(0, 0.1, 0.3)	(0.9, 1, 1)
Project #5	(0.3, 0.5, 0.7)	(0.9, 1, 1)	(0.7, 0.9, 1)	(0.7, 0.9, 1)	(0.9, 1, 1)	(0, 0, 0.1)	(0.9, 1, 1)	(0, 0, 0.1)	(0.9, 1, 1)	(0.3, 0.5, 0.7)
Project #6	(0.7, 0.9, 1)	(0, 0.1, 0.3)	(0, 0.1, 0.3)	(0, 0.1, 0.3)	(0, 0, 0.1)	(0, 0.1, 0.3)	(0, 0.1, 0.3)	(0, 0.1, 0.3)	(0, 0.1, 0.3)	(0.9, 1, 1)
Project #7	(0.9, 1, 1)	(0.9, 1, 1)	(0.7, 0.9, 1)	(0.9, 1, 1)	(0, 0.1, 0.3)	(0.7, 0.9, 1)	(0.9, 1, 1)	(0, 0.1, 0.3)	(0, 0, 0.1)	(0.7, 0.9, 1)

Table 10. Initial Decision Matrix for Expert #2 with Quantitative Values.

Code	A1	A2	A3	A4	B1	B2	B3	B4	C1	C2
Weight	0.033	0.044	0.023	0.025	0.022	0.028	0.040	0.018	0.035	0.037
Sign	+	+	+	+	+	+	+	+	–	–
Project #1	(0.9, 1, 1)	(0.7, 0.9, 1)	(0.9, 1, 1)	(0.9, 1, 1)	(0, 0.1, 0.3)	(0.3, 0.5, 0.7)	(0.9, 1, 1)	(0.3, 0.5, 0.7)	(0.3, 0.5, 0.7)	(0.9, 1, 1)
Project #2	(0.3, 0.5, 0.7)	(0.9, 1, 1)	(0, 0.1, 0.3)	(0.3, 0.5, 0.7)	(0.3, 0.5, 0.7)	(0.7, 0.9, 1)	(0.7, 0.9, 1)	(0, 0.1, 0.3)	(0, 0, 0.1)	(0.7, 0.9, 1)
Project #3	(0.3, 0.5, 0.7)	(0.7, 0.9, 1)	(0.7, 0.9, 1)	(0.9, 1, 1)	(0.7, 0.9, 1)	(0.3, 0.5, 0.7)	(0, 0.1, 0.3)	(0, 0.1, 0.3)	(0.7, 0.9, 1)	(0.7, 0.9, 1)
Project #4	(0, 0.1, 0.3)	(0, 0.1, 0.3)	(0.3, 0.5, 0.7)	(0, 0.1, 0.3)	(0.3, 0.5, 0.7)	(0.7, 0.9, 1)	(0.9, 1, 1)	(0.3, 0.5, 0.7)	(0.7, 0.9, 1)	(0, 0.1, 0.3)
Project #5	(0.9, 1, 1)	(0.3, 0.5, 0.7)	(0.3, 0.5, 0.7)	(0.3, 0.5, 0.7)	(0, 0.1, 0.3)	(0.7, 0.9, 1)	(0.7, 0.9, 1)	(0.3, 0.5, 0.7)	(0.9, 1, 1)	(0.3, 0.5, 0.7)
Project #6	(0.3, 0.5, 0.7)	(0.9, 1, 1)	(0, 0.1, 0.3)	(0.7, 0.9, 1)	(0.9, 1, 1)	(0.9, 1, 1)	(0, 0.1, 0.3)	(0, 0.1, 0.3)	(0.3, 0.5, 0.7)	(0.9, 1, 1)
Project #7	(0.9, 1, 1)	(0.7, 0.9, 1)	(0.7, 0.9, 1)	(0, 0.1, 0.3)	(0.7, 0.9, 1)	(0.3, 0.5, 0.7)	(0.3, 0.5, 0.7)	(0, 0.1, 0.3)	(0.9, 1, 1)	(0.9, 1, 1)

Code	C3	C4	C5	D1	D2	D3	D4	D5	E	F
Weight	0.019	0.047	0.014	0.064	0.080	0.046	0.031	0.091	0.237	0.067
Sign	–	–	+	+	–	–	–	+	+	–
Project #1	(0.9, 1, 1)	(0.3, 0.5, 0.7)	(0.9, 1, 1)	(0.7, 0.9, 1)	(0, 0.1, 0.3)	(0.3, 0.5, 0.7)	(0.3, 0.5, 0.7)	(0.9, 1, 1)	(0, 0.1, 0.3)	(0.7, 0.9, 1)
Project #2	(0.3, 0.5, 0.7)	(0.7, 0.9, 1)	(0.3, 0.5, 0.7)	(0.9, 1, 1)	(0.3, 0.5, 0.7)	(0, 0.1, 0.3)	(0.3, 0.5, 0.7)	(0.9, 1, 1)	(0.7, 0.9, 1)	(0.7, 0.9, 1)
Project #3	(0, 0.1, 0.3)	(0, 0.1, 0.3)	(0.9, 1, 1)	(0.3, 0.5, 0.7)	(0.7, 0.9, 1)	(0.7, 0.9, 1)	(0, 0, 0.1)	(0, 0.1, 0.3)	(0, 0, 0.1)	(0.7, 0.9, 1)
Project #4	(0.3, 0.5, 0.7)	(0.9, 1, 1)	(0.7, 0.9, 1)	(0.9, 1, 1)	(0.9, 1, 1)	(0, 0.1, 0.3)	(0.7, 0.9, 1)	(0.7, 0.9, 1)	(0, 0.1, 0.3)	(0.9, 1, 1)
Project #5	(0.3, 0.5, 0.7)	(0.9, 1, 1)	(0.7, 0.9, 1)	(0.7, 0.9, 1)	(0, 0, 0.1)	(0, 0, 0.1)	(0.9, 1, 1)	(0, 0, 0.1)	(0.9, 1, 1)	(0.3, 0.5, 0.7)
Project #6	(0.7, 0.9, 1)	(0, 0.1, 0.3)	(0, 0.1, 0.3)	(0, 0.1, 0.3)	(0, 0, 0.1)	(0, 0.1, 0.3)	(0, 0.1, 0.3)	(0, 0.1, 0.3)	(0, 0.1, 0.3)	(0.9, 1, 1)
Project #7	(0.9, 1, 1)	(0.9, 1, 1)	(0.7, 0.9, 1)	(0.9, 1, 1)	(0.7, 0.9, 1)	(0.7, 0.9, 1)	(0.9, 1, 1)	(0, 0.1, 0.3)	(0, 0, 0.1)	(0.7, 0.9, 1)

Table 11. Initial Decision Matrix for Expert #3 with Quantitative Values.

Code	A1	A2	A3	A4	B1	B2	B3	B4	C1	C2
Weight	0.033	0.044	0.023	0.025	0.022	0.028	0.040	0.018	0.035	0.037
Sign	+	+	+	+	+	+	+	+	–	–
Project #1	(0.7, 0.9, 1)	(0.7, 0.9, 1)	(0.3, 0.5, 0.7)	(0.9, 1, 1)	(0.3, 0.5, 0.7)	(0.3, 0.5, 0.7)	(0.9, 1, 1)	(0.7, 0.9, 1)	(0, 0.1, 0.3)	(0.9, 1, 1)
Project #2	(0.3, 0.5, 0.7)	(0.9, 1, 1)	(0.3, 0.5, 0.7)	(0.7, 0.9, 1)	(0.7, 0.9, 1)	(0.7, 0.9, 1)	(0.3, 0.5, 0.7)	(0, 0.1, 0.3)	(0, 0.1, 0.3)	(0, 0.1, 0.3)
Project #3	(0.3, 0.5, 0.7)	(0.7, 0.9, 1)	(0.9, 1, 1)	(0.3, 0.5, 0.7)	(0.3, 0.5, 0.7)	(0.3, 0.5, 0.7)	(0, 0.1, 0.3)	(0, 0.1, 0.3)	(0.3, 0.5, 0.7)	(0.7, 0.9, 1)
Project #4	(0, 0, 0.1)	(0.3, 0.5, 0.7)	(0, 0.1, 0.3)	(0.7, 0.9, 1)	(0.7, 0.9, 1)	(0, 0.1, 0.3)	(0.7, 0.9, 1)	(0.7, 0.9, 1)	(0.9, 1, 1)	(0, 0, 0.1)
Project #5	(0.9, 1, 1)	(0, 0, 0.1)	(0.3, 0.5, 0.7)	(0.7, 0.9, 1)	(0, 0.1, 0.3)	(0.7, 0.9, 1)	(0.7, 0.9, 1)	(0.3, 0.5, 0.7)	(0.7, 0.9, 1)	(0.9, 1, 1)
Project #6	(0, 0.1, 0.3)	(0.9, 1, 1)	(0, 0.1, 0.3)	(0.9, 1, 1)	(0.9, 1, 1)	(0.9, 1, 1)	(0.3, 0.5, 0.7)	(0, 0.1, 0.3)	(0, 0.1, 0.3)	(0.7, 0.9, 1)
Project #7	(0.7, 0.9, 1)	(0.7, 0.9, 1)	(0.7, 0.9, 1)	(0, 0.1, 0.3)	(0.7, 0.9, 1)	(0.3, 0.5, 0.7)	(0, 0, 0.1)	(0, 0, 0.1)	(0.7, 0.9, 1)	(0.9, 1, 1)

Code	C3	C4	C5	D1	D2	D3	D4	D5	E	F
Weight	0.019	0.047	0.014	0.064	0.080	0.046	0.031	0.091	0.237	0.067
Sign	–	–	+	+	–	–	–	+	+	–
Project #1	(0.9, 1, 1)	(0.3, 0.5, 0.7)	(0.7, 0.9, 1)	(0.9, 1, 1)	(0.3, 0.5, 0.7)	(0.3, 0.5, 0.7)	(0.7, 0.9, 1)	(0.7, 0.9, 1)	(0.7, 0.9, 1)	(0, 0.1, 0.3)
Project #2	(0.3, 0.5, 0.7)	(0.7, 0.9, 1)	(0.3, 0.5, 0.7)	(0.7, 0.9, 1)	(0, 0.1, 0.3)	(0.7, 0.9, 1)	(0.3, 0.5, 0.7)	(0.9, 1, 1)	(0.7, 0.9, 1)	(0.7, 0.9, 1)
Project #3	(0, 0.1, 0.3)	(0.3, 0.5, 0.7)	(0.9, 1, 1)	(0, 0.1, 0.3)	(0.7, 0.9, 1)	(0.9, 1, 1)	(0, 0, 0.1)	(0, 0.1, 0.3)	(0, 0.1, 0.3)	(0.7, 0.9, 1)
Project #4	(0.7, 0.9, 1)	(0.9, 1, 1)	(0.7, 0.9, 1)	(0.7, 0.9, 1)	(0, 0.1, 0.3)	(0, 0.1, 0.3)	(0.9, 1, 1)	(0.9, 1, 1)	(0.7, 0.9, 1)	(0.9, 1, 1)
Project #5	(0.3, 0.5, 0.7)	(0, 0.1, 0.3)	(0.7, 0.9, 1)	(0.3, 0.5, 0.7)	(0.9, 1, 1)	(0.9, 1, 1)	(0.9, 1, 1)	(0, 0, 0.1)	(0.9, 1, 1)	(0.3, 0.5, 0.7)
Project #6	(0.7, 0.9, 1)	(0.3, 0.5, 0.7)	(0.3, 0.5, 0.7)	(0, 0, 0.1)	(0, 0, 0.1)	(0, 0.1, 0.3)	(0.3, 0.5, 0.7)	(0.3, 0.5, 0.7)	(0.9, 1, 1)	(0, 0, 0.1)
Project #7	(0.9, 1, 1)	(0.7, 0.9, 1)	(0.7, 0.9, 1)	(0.9, 1, 1)	(0.3, 0.5, 0.7)	(0.7, 0.9, 1)	(0.9, 1, 1)	(0.7, 0.9, 1)	(0, 0, 0.1)	(0.3, 0.5, 0.7)

Table 12. Aggregated Decision Matrix as Fuzzy Numbers with Interval Values.

Code	A1	A2	A3	A4	B1	B2	B3	B4	C1	C2
Weight	0.033	0.044	0.023	0.025	0.022	0.028	0.040	0.018	0.035	0.037
Sign	+	+	+	+	+	+	+	+	+	–
Ideal Alternative (X0)	[0.7, 0.8277), 0.9655, (1, 1)]	[0.9, 0.9), 1, (1, 1)]	[0.7, 0.8277), 0.9655, (1, 1)]	[0.9, 0.9), 1, (1, 1)]	[0.9, 0.9), 1, (1, 1)]	[0.9, 0.9), 1, (1, 1)]	[0.9, 0.9), 1, (1, 1)]	[0.3, 0.5278), 0.7399, (0.8879, 1)]	[(0, 0), 0, (0.2080, 0.3)]	[(0, 0), 0, (0.2759, 0.7)]
Project #1	[(0.7, 0.76), 0.93, (1, 1)]	[(0.7, 0.7), 0.90, (1, 1)]	[(0.30, 0.57), 0.77, (0.89, 1)]	[(0.90, 0.90), 1, (1, 1)]	[(0, 0), 0.29, (0.53, 0.7)]	[(0.30, 0.30), 0.50, (0.7, 0.7)]	[(0.90, 0.90), 1, (1, 1)]	[(0.30, 0.53), 0.74, (0.89, 1)]	[(0, 0), 0.29, (0.53, 0.7)]	[(0.90, 0.90), 1, (1, 1)]
Project #2	[(0.30, 0.30), 0.50, (0.7, 0.7)]	[(0.90, 0.90), 1, (1, 1)]	[(0, 0), 0.29, (0.53, 0.7)]	[(0.30, 0.53), 0.74, (0.89, 1)]	[(0.30, 0.53), 0.74, (0.89, 1)]	[(0.7, 0.7), 0.90, (1, 1)]	[(0.30, 0.53), 0.74, (0.89, 1)]	[(0, 0), 0.10, (0.30, 0.30)]	[(0, 0), 0, (0.21, 0.30)]	[(0, 0), 0.43, (0.67, 1)]
Project #3	[(0, 0), 0.10, (0.30, 0.30)]	[(0.7, 0.7), 0.90, (1, 1)]	[(0.7, 0.83), 0.97, (1, 1)]	[(0.30, 0.57), 0.77, (0.89, 1)]	[(0.30, 0.53), 0.74, (0.89, 1)]	[(0.30, 0.30), 0.50, (0.7, 0.7)]	[(0, 0), 0.17, (0.40, 0.7)]	[(0, 0), 0.10, (0.30, 0.30)]	[(0.30, 0.40), 0.61, (0.79, 1)]	[(0.7, 0.7), 0.90, (1, 1)]
Project #4	[(0.30, 0.40), 0.61, (0.79, 1)]	[(0, 0), 0.36, (0.59, 1)]	[(0, 0), 0.36, (0.59, 1)]	[(0, 0), 0.36, (0.59, 1)]	[(0, 0), 0.10, (0.30, 0.30)]	[(0, 0), 0.10, (0.30, 0.30)]	[(0.7, 0.83), 0.97, (1, 1)]	[(0.30, 0.40), 0.61, (0.79, 1)]	[(0.7, 0.76), 0.93, (1, 1)]	[(0, 0), 0, (0.28, 0.7)]
Project #5	[(0.7, 0.83), 0.97, (1, 1)]	[(0, 0), 0, (0.37, 0.7)]	[(0.30, 0.30), 0.50, (0.7, 0.7)]	[(0.30, 0.53), 0.74, (0.89, 1)]	[(0, 0), 0.17, (0.40, 0.7)]	[(0.7, 0.7), 0.90, (1, 1)]	[(0.7, 0.7), 0.90, (1, 1)]	[(0.30, 0.30), 0.50, (0.7, 0.7)]	[(0.7, 0.83), 0.97, (1, 1)]	[(0.30, 0.43), 0.63, (0.79, 1)]
Project #6	[(0, 0), 0.29, (0.53, 0.7)]	[(0.90, 0.90), 1, (1, 1)]	[(0, 0), 0.10, (0.30, 0.30)]	[(0, 0), 0, (0.28, 0.7)]	[(0.90, 0.90), 1, (1, 1)]	[(0.90, 0.90), 1, (1, 1)]	[(0, 0), 0.29, (0.53, 0.7)]	[(0, 0), 0.10, (0.30, 0.30)]	[(0, 0), 0.29, (0.53, 0.7)]	[(0.7, 0.83), 0.97, (1, 1)]
Project #7	[(0.7, 0.76), 0.93, (1, 1)]	[(0.7, 0.7), 0.90, (1, 1)]	[(0.7, 0.7), 0.90, (1, 1)]	[(0, 0), 0.17, (0.40, 0.7)]	[(0.7, 0.7), 0.90, (1, 1)]	[(0.7, 0.7), 0.90, (1, 1)]	[(0, 0), 0, (0.37, 0.7)]	[(0, 0), 0, (0.21, 0.30)]	[(0.7, 0.83), 0.97, (1, 1)]	[(0.90, 0.90), 1, (1, 1)]

Code	C3	C4	C5	D1	D2	D3	D4	D5	E	F
Weight	0.019	0.047	0.014	0.064	0.080	0.046	0.031	0.091	0.237	0.067
Sign	–	–	+	+	–	–	–	+	+	–
Ideal Alternative (X0)	[(0, 0), 0.1, (0.3, 0.3)]	[(0, 0), 0.1710, (0.3979, 0.7)]	[(0.9, 0.9), 1, (1, 1)]	[(0.9, 0.9), 1, (1, 1)]	[(0, 0), 0, (0.2154, 0.3)]	[(0, 0), 0, (0.1, 0.1)]	[(0, 0), 0, (0.1, 0.1)]	[(0.9, 0.9), 1, (1, 1)]	[(0.9, 0.9), 1, (1, 1)]	[(0, 0), 0, (0.4121, 0.7)]
Project #1	[(0.90, 0.90), 1, (1, 1)]	[(0.30, 0.30), 0.50, (0.7, 0.7)]	[(0.7, 0.76), 0.93, (1, 1)]	[(0.7, 0.76), 0.93, (1, 1)]	[(0.30, 0.30), 0.50, (0.7, 0.7)]	[(0.30, 0.40), 0.61, (0.79, 1)]	[(0.30, 0.30), 0.50, (0.7, 0.7)]	[(0.7, 0.76), 0.93, (1, 1)]	[(0, 0), 0.21, (0.45, 1)]	[(0, 0), 0.43, (0.67, 1)]
Project #2	[(0.30, 0.30), 0.50, (0.7, 0.7)]	[(0.7, 0.7), 0.90, (1, 1)]	[(0.30, 0.30), 0.50, (0.7, 0.7)]	[(0.7, 0.76), 0.93, (1, 1)]	[(0, 0), 0.10, (0.30, 0.30)]	[(0, 0), 0.10, (0.30, 0.30)]	[(0, 0), 0, (0.10, 0.10)]	[(0.90, 0.90), 1, (1, 1)]	[(0.7, 0.7), 0.90, (1, 1)]	[(0.7, 0.7), 0.90, (1, 1)]
Project #3	[(0, 0), 0.10, (0.30, 0.30)]	[(0, 0), 0.29, (0.53, 0.7)]	[(0.90, 0.90), 1, (1, 1)]	[(0, 0), 0.29, (0.53, 0.7)]	[(0.30, 0.53), 0.74, (0.89, 1)]	[(0.7, 0.76), 0.93, (1, 1)]	[(0, 0), 0, (0.10, 0.10)]	[(0.90, 0.90), 1, (1, 1)]	[(0, 0), 0, (0.14, 0.30)]	[(0.7, 0.7), 0.90, (1, 1)]
Project #4	[(0.30, 0.40), 0.61, (0.79, 1)]	[(0.90, 0.90), 1, (1, 1)]	[(0.7, 0.7), 0.90, (1, 1)]	[(0, 0), 0.29, (0.53, 0.7)]	[(0, 0), 0.10, (0.30, 0.30)]	[(0, 0), 0.10, (0.30, 0.30)]	[(0.7, 0.76), 0.93, (1, 1)]	[(0.7, 0.76), 0.93, (1, 1)]	[(0, 0), 0.36, (0.59, 1)]	[(0.90, 0.90), 1, (1, 1)]
Project #5	[(0.30, 0.30), 0.50, (0.7, 0.7)]	[(0, 0), 0.37, (0.59, 1)]	[(0.7, 0.7), 0.90, (1, 1)]	[(0.30, 0.53), 0.74, (0.89, 1)]	[(0, 0), 0.10, (0.30, 0.30)]	[(0.7, 0.83), 0.97, (1, 1)]	[(0.7, 0.83), 0.97, (1, 1)]	[(0, 0), 0, (0.10, 0.10)]	[(0.90, 0.90), 1, (1, 1)]	[(0.30, 0.30), 0.50, (0.7, 0.7)]
Project #6	[(0.7, 0.7), 0.90, (1, 1)]	[(0, 0), 0.17, (0.40, 0.7)]	[(0, 0), 0.29, (0.53, 0.7)]	[(0, 0), 0, (0.28, 0.7)]	[(0, 0), 0, (0.10, 0.10)]	[(0, 0), 0.17, (0.40, 0.7)]	[(0, 0), 0.10, (0.30, 0.30)]	[(0, 0), 0.29, (0.53, 0.7)]	[(0, 0), 0.22, (0.45, 1)]	[(0, 0), 0, (0.41, 1)]
Project #7	[(0.90, 0.90), 1, (1, 1)]	[(0.7, 0.83), 0.97, (1, 1)]	[(0.7, 0.7), 0.90, (1, 1)]	[(0.90, 0.90), 1, (1, 1)]	[(0.7, 0.7), 0.90, (1, 1)]	[(0.7, 0.83), 0.97, (1, 1)]	[(0, 0), 0.17, (0.40, 0.7)]	[(0.7, 0.83), 0.97, (1, 1)]	[(0, 0), 0, (0.10, 0.10)]	[(0.30, 0.53), 0.74, (0.89, 1)]

Table 13. Normalized Decision Matrix as Fuzzy Numbers with Interval Values.

Code	A1	A2	A3	A4	B1	B2	B3	B4	C1	C2
Weight	0.033	0.044	0.023	0.025	0.022	0.028	0.040	0.018	0.035	0.037
Sign	+	+	+	+	+	+	+	+	+	—
Ideal Alternative (X0)	[(0.1169, 0.1169), 0.1299, (0.1299, 0.1299)]	[(0.1045, 0.1235), 0.1441, (0.1493, 0.1493)]	[(0.1169, 0.1169), 0.1299, (0.1299, 0.1299)]	[(0.1216, 0.1216), 0.1351, (0.1351, 0.1351)]	[(0.1265, 0.12668), 0.1408, (0.1408, 0.1408)]	[(0.1265, 0.12668), 0.1408, (0.1408, 0.1408)]	[(0.0612, 0.1077), 0.1510, (0.1812, 0.2041)]	[(0.2007, 0.20007), 0.2007, (0, 0)]	[(0.2687, 0.2687), 0.2687, (0, 0)]	[(0.5132, 0.51132), 0, (0, 0)]
Project #1	[(0.0909, 0.0909), 0.1169, (0.1299, 0.1299)]	[(0.0448, 0.0857), 0.1144, (0.1325, 0.1493)]	[(0.1169, 0.1169), 0.1299, (0.1299, 0.1299)]	[(0, 0), 0.0395, (0.0713, 0.0946)]	[(0.0423, 0.0423), 0.0704, (0.0986, 0.0986)]	[(0.0423, 0.0423), 0.0704, (0.1408, 0.1408)]	[(0.0612, 0.1077), 0.1510, (0.1812, 0.2041)]	[(0.1911, 0.1911), 0, (0, 0)]	[(0.4326, 0.4326), 0, (0, 0)]	[(0, 0), 0, (0, 0)]
Project #2	[(0, 0), 0.0436, (0.0788, 0.1045)]	[(0, 0), 0.0436, (0.0788, 0.1045)]	[(0.0390, 0.0685), 0.0961, (0.1153, 0.1299)]	[(0.0405, 0.0713), 0.1000, (0.1200, 0.1351)]	[(0.0986, 0.0986), 0.1268, (0.1408, 0.1408)]	[(0.0423, 0.0743), 0.1042, (0.1251, 0.1408)]	[(0, 0), 0.0204, (0.0612, 0.0612)]	[(0.3232, 0.3232), 0.3232, (0, 0)]	[(0.4326, 0.4326), 0, (0, 0)]	[(0, 0), 0, (0, 0)]
Project #3	[(0.0909, 0.0909), 0.1169, (0.1299, 0.1299)]	[(0.1045, 0.1235), 0.1441, (0.1493, 0.1493)]	[(0.0390, 0.0745), 0.0995, (0.1153, 0.1299)]	[(0.0405, 0.0713), 0.1000, (0.1200, 0.1351)]	[(0.0423, 0.0423), 0.0704, (0.0986, 0.0986)]	[(0, 0), 0.0241, (0.0560, 0.0986)]	[(0, 0), 0.0204, (0.0612, 0.0612)]	[(0, 0), 0, (0, 0)]	[(0, 0), 0, (0, 0)]	[(0.4868, 0.4868), 0, (0, 0)]
Project #4	[(0, 0), 0.0462, (0.0772, 0.1299)]	[(0, 0), 0.0531, (0.0887, 0.1493)]	[(0, 0), 0.0462, (0.0772, 0.1299)]	[(0, 0), 0.0231, (0.0538, 0.1065, 0.1351)]	[(0, 0), 0.0609, (0.0943, 0.1408)]	[(0.0986, 0.1166), 0.1360, (0.1408, 0.1408)]	[(0.0612, 0.0812), 0.1241, (0.1609, 0.2041)]	[(0, 0), 0, (0, 0)]	[(0.2987, 0.2987), 0.2987, (0, 0)]	[(0, 0), 0, (0, 0)]
Project #5	[(0.1169, 0.1169), 0.1299, (0.1299, 0.1299)]	[(0.0448, 0.0448), 0.0746, (0.1045, 0.1045)]	[(0.0909, 0.1075), 0.1254, (0.1299, 0.1299)]	[(0.1216, 0.1216), 0.1351, (0.1351, 0.1351)]	[(0.1268, 0.12668), 0.1408, (0.1408, 0.1408)]	[(0.0986, 0.0986), 0.1268, (0.1408, 0.1408)]	[(0.0612, 0.0612), 0.1020, (0.1429, 0.1429)]	[(0.2849, 0.2849), 0, (0, 0)]	[(0, 0), 0, (0, 0)]	[(0, 0), 0, (0, 0)]
Project #6	[(0.0909, 0.0909), 0.1169, (0.1299, 0.1299)]	[(0, 0), 0.0149, (0.0448, 0.0446)]	[(0.0909, 0.1075), 0.1254, (0.1299, 0.1299)]	[(0.0946, 0.0946), 0.1216, (0.1351, 0.1351)]	[(0.1268, 0.12668), 0.1408, (0.1408, 0.1408)]	[(0, 0), 0.0412, (0.0743, 0.0986)]	[(0, 0), 0.0204, (0.0612, 0.0612)]	[(0.2849, 0.2849), 0, (0, 0)]	[(0, 0), 0, (0, 0)]	[(0, 0), 0, (0, 0)]
Project #7	[(0.0909, 0.0909), 0.1169, (0.1299, 0.1299)]	[(0.1045, 0.1045), 0.1343, (0.1493, 0.1493)]	[(0, 0), 0.0222, (0.0517, 0.0909)]	[(0.0946, 0.0946), 0.1216, (0.1351, 0.1351)]	[(0.0423, 0.0423), 0.0704, (0.0986, 0.0986)]	[(0, 0), 0, (0.0515, 0.0986)]	[(0, 0), 0, (0.0425, 0.0612)]	[(0, 0), 0, (0, 0)]	[(0, 0), 0, (0, 0)]	[(0, 0), 0, (0, 0)]

Code	C3	C4	C5	D1	D2	D3	D4	D5	E	F
Weight	0.019	0.047	0.014	0.064	0.080	0.046	0.031	0.091	0.237	0.067
Sign	—	—	+	+	—	—	—	+	+	+
Ideal Alternative (X0)	[(0.1992, 0.1992), 0, (0, 0)]	[(0.1216, 0.1216), 0.1351, (0.1351, 0.1351)]	[(0.1216, 0.1216), 0.1351, (0.1351, 0.1351)]	[(0.1584, 0.1584), 0.1584, (0, 0)]	[(0.1472, 0.1472), 0.1472, (0, 0)]	[(0.1475, 0.1475), 0.1639, (0.1639, 0.1639)]	[(0.1406, 0.1406), 0.1563, (0.1563, 0.1563)]	[(0.1406, 0.1406), 0.1563, (0.1563, 0.1563)]	[(0.2966, 0.2966), 0.2966, (0, 0)]	—
Project #1	[(0, 0), 0, (0, 0)]	[(0.0946, 0.1029), 0.1260, (0.1351, 0.1351)]	[(0.0946, 0.1029), 0.1260, (0.1351, 0.1351)]	[(0, 0), 0, (0, 0)]	[(0, 0), 0, (0, 0)]	[(0.1148, 0.1248), 0.1528, (0.1639, 0.1639)]	[(0, 0), 0.0325, (0.0700, 0.1563)]	[(0.1094, 0.1094), 0.1406, (0.1563, 0.1563)]	[(0.2824, 0.2824), 0, (0, 0)]	[(0, 0), 0, (0, 0)]
Project #2	[(0, 0), 0, (0, 0)]	[(0.0405, 0.0405), 0.0676, (0.0946, 0.0946)]	[(0.0946, 0.1029), 0.1260, (0.1351, 0.1351)]	[(0.2550, 0.2550), 0, (0, 0)]	[(0.2370, 0.2370), 0, (0, 0)]	[(0.1475, 0.1475), 0.1639, (0.1639, 0.1639)]	[(0, 0), 0.0325, (0.0700, 0.1563)]	[(0.1406, 0.1406), 0.1563, (0.1563, 0.1563)]	[(0, 0), 0, (0, 0)]	[(0, 0), 0, (0, 0)]
Project #3	[(0.1889, 0.1889), 0, (0, 0)]	[(0.1216, 0.1216), 0.1351, (0.1351, 0.1351)]	[(0, 0), 0.0395, (0.0713, 0.0946)]	[(0, 0), 0, (0, 0)]	[(0, 0), 0, (0, 0)]	[(0, 0), 0.0164, (0.0492, 0.0492)]	[(0, 0), 0, (0.0225, 0.0469)]	[(0, 0), 0, (0.0225, 0.0469)]	[(0, 0), 0, (0, 0)]	[(0, 0), 0, (0, 0)]
Project #4	[(0, 0), 0, (0, 0)]	[(0.0946, 0.0946), 0.1216, (0.1351, 0.1351)]	[(0.0946, 0.1118), 0.1305, (0.1351, 0.1351)]	[(0.1636, 0.1636), 0, (0, 0)]	[(0, 0), 0, (0, 0)]	[(0.1148, 0.1248), 0.1528, (0.1639, 0.1639)]	[(0, 0), 0.0556, (0.0929, 0.1563)]	[(0, 0), 0.0556, (0.0929, 0.1563)]	[(0, 0), 0, (0, 0)]	[(0, 0), 0, (0, 0)]
Project #5	[(0.3293, 0.3293), 0, (0, 0)]	[(0.0946, 0.0946), 0.1216, (0.1351, 0.1351)]	[(0.0405, 0.0713), 0.1000, (0.1200, 0.1351)]	[(0.2433, 0.2433), 0.2433, (0, 0)]	[(0.2433, 0.2433), 0.2433, (0, 0)]	[(0, 0), 0.0164, (0.0164)]	[(0.1406, 0.1406), 0.1563, (0.1563, 0.1563)]	[(0.1406, 0.1406), 0.1563, (0.1563, 0.1563)]	[(0, 0), 0, (0, 0)]	[(0, 0), 0, (0, 0)]
Project #6	[(0.2827, 0.2827), 0, (0, 0)]	[(0, 0), 0.0395, (0.0713, 0.0946)]	[(0, 0), 0, (0.0373, 0.0946)]	[(0.2089, 0.2089), 0, (0, 0)]	[(0.4215, 0.4215), 0, (0, 0)]	[(0, 0), 0.0479, (0.0865, 0.1148)]	[(0, 0), 0.0337, (0.0700, 0.1563)]	[(0, 0), 0.0337, (0.0700, 0.1563)]	[(0.4210, 0.4210), 0.4210, (0, 0)]	[(0, 0), 0, (0, 0)]
Project #7	[(0, 0), 0, (0, 0)]	[(0.0946, 0.0946), 0.1216, (0.1351, 0.1351)]	[(0.1216, 0.1216), 0.1351, (0.1351, 0.1351)]	[(0.1858, 0.1858), 0, (0, 0)]	[(0, 0), 0, (0, 0)]	[(0, 0), 0.0583, (0.0974, 0.1639)]	[(0, 0), 0, (0.0156, 0.0156)]	[(0, 0), 0, (0.0156, 0.0156)]	[(0, 0), 0, (0, 0)]	[(0, 0), 0, (0, 0)]

Table 14. Normalized weighted Decision Matrix with Interval-valued Fuzzy Numbers.

Code	A1	A2	A3	A4	B1	B2	B3	B4	C1	C2
Weight	0.033	0.044	0.023	0.025	0.022	0.028	0.040	0.018	0.035	0.037
Sign	+	+	+	+	+	+	+	+	−	−
Ideal Alternative (X0)	[(0.0036, 0.0043), 0.0050, (0.0052, 0.0052)]	[(0.0051, 0.0051), 0.0057, (0.0057, 0.0057)]	[(0.0024, 0.0028), 0.0033, (0.0034, 0.0034)]	[(0.0029, 0.0029), 0.0032, (0.0032, 0.0032)]	[(0.0027, 0.0027), 0.0030, (0.0030, 0.0030)]	[(0.0035, 0.0035), 0.0039, (0.0039, 0.0039)]	[(0.0051, 0.0051), 0.0056, (0.0056, 0.0056)]	[(0.0011, 0.0019), 0.0027, (0.0032, 0.0036)]	[(0.0071, 0.0071), 0.0071, (0, 0)]	[(0.0098, 0.0098), 0.0098, (0, 0)]
Project #1	[(0.0036, 0.0040), 0.0049, (0.0052, 0.0052)]	[(0.0040, 0.0040), 0.0051, (0.0057, 0.0057)]	[(0.0010, 0.0019), 0.0026, (0.0030, 0.0034)]	[(0.0029, 0.0029), 0.0032, (0.0032, 0.0032)]	[(0, 0), 0.0005, (0.0016, 0.0021)]	[(0.0012, 0.0012), 0.0019, (0.0027, 0.0027)]	[(0.0051, 0.0051), 0.0056, (0.0056, 0.0056)]	[(0.0011, 0.0019), 0.0027, (0.0032, 0.0036)]	[(0.0067, 0.0067), 0, (0, 0)]	[(0, 0), 0, (0, 0)]
Project #2	[(0.0016, 0.0016), 0.0026, (0.0036, 0.0036)]	[(0.0051, 0.0051), 0.0057, (0.0057, 0.0057)]	[(0, 0), 0.0010, (0.0018, 0.0024)]	[(0.0010, 0.0017), 0.0024, (0.0029, 0.0032)]	[(0.0009, 0.0016), 0.0022, (0.0026, 0.0030)]	[(0.0027, 0.0027), 0.0035, (0.0039, 0.0039)]	[(0.0017, 0.0030), 0.0042, (0.0050, 0.0056)]	[(0, 0), 0.0004, (0.0011, 0.0011)]	[(0.0114, 0.0114), 0.0114, (0, 0)]	[(0.0158, 0.0158), 0, (0, 0)]
Project #3	[(0.0016, 0.0016), 0.0026, (0.0036, 0.0036)]	[(0.0040, 0.0040), 0.0051, (0.0057, 0.0057)]	[(0.0024, 0.0028), 0.0033, (0.0034, 0.0034)]	[(0.0010, 0.0018), 0.0025, (0.0029, 0.0032)]	[(0.0009, 0.0016), 0.0022, (0.0026, 0.0030)]	[(0.0012, 0.0012), 0.0019, (0.0027, 0.0027)]	[(0, 0), 0.0010, (0.0022, 0.0040)]	[(0, 0), 0.0004, (0.0011, 0.0011)]	[(0, 0), 0, (0, 0)]	[(0, 0), 0, (0, 0)]
Project #4	[(0, 0), 0, (0.0011, 0.0016)]	[(0, 0), 0.0020, (0.0034, 0.0057)]	[(0, 0), 0.0012, (0.0020, 0.0034)]	[(0, 0), 0.0011, (0.0019, 0.0022)]	[(0.0009, 0.0012), 0.0018, (0.0023, 0.0030)]	[(0, 0), 0.0017, (0.0026, 0.0039)]	[(0.0040, 0.0047), 0.0055, (0.0056, 0.0056)]	[(0, 0), 0.0004, (0.0011, 0.0011)]	[(0, 0), 0, (0, 0)]	[(0.0109, 0.0109), 0.0109, (0, 0)]
Project #5	[(0.0036, 0.0043), 0.0050, (0.0052, 0.0052)]	[(0, 0), 0, (0.0021, 0.0040)]	[(0.0010, 0.0010), 0.0017, (0.0024, 0.0024)]	[(0.0010, 0.0017), 0.0024, (0.0032, 0.0032)]	[(0, 0), 0.0005, (0.0012, 0.0021)]	[(0.0027, 0.0027), 0.0035, (0.0039, 0.0039)]	[(0.0040, 0.0040), 0.0051, (0.0056, 0.0056)]	[(0.0011, 0.0011), 0.0018, (0.0026, 0.0026)]	[(0, 0), 0, (0, 0)]	[(0, 0), 0, (0, 0)]
Project #6	[(0, 0), 0.0015, (0.0027, 0.0036)]	[(0, 0), 0, (0.0021, 0.0040)]	[(0, 0), 0.0003, (0.0010, 0.0010)]	[(0, 0), 0.0006, (0.0013, 0.0013)]	[(0, 0), 0.0005, (0.0012, 0.0021)]	[(0.0035, 0.0035), 0.0035, (0.0039, 0.0039)]	[(0, 0), 0.0017, (0.0030, 0.0040)]	[(0, 0), 0.0004, (0.0011, 0.0011)]	[(0.0100, 0.0100), 0, (0, 0)]	[(0, 0), 0, (0, 0)]
Project #7	[(0.0036, 0.0040), 0.0049, (0.0052, 0.0052)]	[(0.0040, 0.0040), 0.0051, (0.0057, 0.0057)]	[(0.0024, 0.0024), 0.0031, (0.0034, 0.0034)]	[(0, 0), 0.0006, (0.0013, 0.0023)]	[(0.0021, 0.0021), 0.0027, (0.0030, 0.0030)]	[(0.0012, 0.0012), 0.0019, (0.0027, 0.0027)]	[(0, 0), 0, (0.0021, 0.0040)]	[(0, 0), 0, (0.0008, 0.0011)]	[(0, 0), 0, (0, 0)]	[(0, 0), 0, (0, 0)]

Code	C3	C4	C5	D1	D2	D3	D4	D5	E	F
Weight	0.019	0.047	0.014	0.064	0.080	0.046	0.031	0.091	0.237	0.067
Sign	−	+	+	+	+	+	+	+	+	−
Ideal Alternative (X0)	[(0.0097, 0.0097), 0, (0, 0)]	[(0.0093, 0.0093), 0, (0, 0)]	[(0.0017, 0.0017), 0.0019, (0.0019, 0.0019)]	[(0.0078, 0.0078), 0.0086, (0.0086, 0.0086)]	[(0.0127, 0.0127), 0.0127, (0, 0)]	[(0.0068, 0.0068), 0.0068, (0, 0)]	[(0.0092, 0.0092), 0.0092, (0, 0)]	[(0.0134, 0.0134), 0.0149, (0.0149, 0.0149)]	[(0.0334, 0.0334), 0.0371, (0.0371, 0.0371)]	[(0.0198, 0.0198), 0.0198, (0, 0)]
Project #1	[(0, 0), 0, (0, 0)]	[(0, 0), 0, (0, 0)]	[(0.0013, 0.0015), 0.0018, (0.0019, 0.0019)]	[(0.0064, 0.0066), 0.0080, (0.0086, 0.0086)]	[(0, 0), 0, (0, 0)]	[(0, 0), 0, (0, 0)]	[(0, 0), 0, (0, 0)]	[(0.0105, 0.0114), 0.0139, (0.0149, 0.0149)]	[(0, 0), 0.0077, (0.0166, 0.0371)]	[(0.0188, 0.0188), 0, (0, 0)]
Project #2	[(0, 0), 0, (0, 0)]	[(0, 0), 0, (0, 0)]	[(0.0006, 0.0006), 0.0010, (0.0013, 0.0013)]	[(0.0060, 0.0066), 0.0080, (0.0086, 0.0086)]	[(0.0205, 0.0205), 0, (0, 0)]	[(0.0109, 0.0109), 0, (0, 0)]	[(0, 0), 0, (0, 0)]	[(0.0134, 0.0134), 0.0149, (0.0149, 0.0149)]	[(0.0260, 0.0260), 0.0334, (0.0371, 0.0371)]	[(0, 0), 0, (0, 0)]
Project #3	[(0.0092, 0.0092), 0, (0, 0)]	[(0.0088, 0.0088), 0, (0, 0)]	[(0.0017, 0.0017), 0.0019, (0.0019, 0.0019)]	[(0, 0), 0.0025, (0.0046, 0.0060)]	[(0, 0), 0, (0, 0)]	[(0, 0), 0, (0, 0)]	[(0.0088, 0.0088), 0.0088, (0, 0)]	[(0, 0), 0.0015, (0.0045, 0.0045)]	[(0, 0), 0.0053, 0.0111]	[(0, 0), 0, (0, 0)]
Project #4	[(0, 0), 0, (0, 0)]	[(0, 0), 0, (0, 0)]	[(0.0013, 0.0013), 0.0017, (0.0019, 0.0019)]	[(0.0060, 0.0071), 0.0083, (0.0086, 0.0086)]	[(0.0141, 0.0141), 0, (0, 0)]	[(0.0075, 0.0075), 0, (0, 0)]	[(0, 0), 0, (0, 0)]	[(0.0105, 0.0114), 0.0139, (0.0149, 0.0149)]	[(0, 0), 0.0132, (0.0220, 0.0371)]	[(0, 0), 0, (0, 0)]
Project #5	[(0, 0), 0, (0, 0)]	[(0.0154, 0.0154), 0, (0, 0)]	[(0.0013, 0.0013), 0.0017, (0.0019, 0.0019)]	[(0.0026, 0.0046), 0.0064, (0.0077, 0.0086)]	[(0.0112, 0.0112), 0.0112, (0, 0)]	[(0.0112, 0.0112), 0.0112, (0, 0)]	[(0, 0), 0, (0, 0)]	[(0, 0), 0, (0.0015, 0.0015)]	[(0.0334, 0.0334), 0.0371, (0.0371, 0.0371)]	[(0, 0), 0, (0, 0)]
Project #6	[(0, 0), 0, (0, 0)]	[(0.0132, 0.0132), 0, (0, 0)]	[(0, 0), 0.0006, (0.0010, 0.0013)]	[(0, 0), 0.0024, 0.0060)]	[(0.0096, 0.0096), 0, (0, 0)]	[(0.0096, 0.0096), 0, (0, 0)]	[(0.0133, 0.0131), 0, (0, 0)]	[(0, 0), 0.0044, (0.0079, 0.0105)]	[(0, 0), 0.0080, (0.0166, 0.0371)]	[(0.0281, 0.0281), 0.0281, (0, 0)]
Project #7	[(0, 0), 0, (0, 0)]	[(0, 0), 0, (0, 0)]	[(0.0013, 0.0013), 0.0017, (0.0019, 0.0019)]	[(0.0078, 0.0078), 0.0086, (0.0086, 0.0086)]	[(0.0149, 0.0149), 0, (0, 0)]	[(0, 0), 0, (0, 0)]	[(0, 0), 0, (0, 0)]	[(0, 0), 0.0053, (0.0089, 0.0149)]	[(0, 0), 0, (0.0037, 0.0037)]	[(0, 0), 0, (0, 0)]

Table 15. Final ARAS calculations with interval values and Alternative rankings.

Alternatives	S	λ = 0			λ = 0.5			λ = 1		
		BNP	Q	Rank	BNP	Q	Rank	BNP	Q	Rank
Ideal Alternative	[(0.167, 0.169), 0.160, (0.096, 0.096)]	0.144	1	0	0.142	1	0	0.161	1	0
Project #1	[(0.062, 0.066),0.058, (0.072, 0.094)]	0.073	0.505	4	0.069	0.485	4	0.066	0.411	4
Project #2	[(0.118, 0.121), 0.091, (0.089, 0.090)]	0.101	0.704	1	0.010	0.705	1	0.094	0.581	1
Project #3	[(0.039, 0.041),0.034, (0.041, 0.050)]	0.043	0.2959	7	0.040	0.281	7	0.038	0.238	7
Project #4	[(0.056, 0.060), 0.064, (0.069, 0.093)]	0.070	0.483	5	0.068	0.477	5	0.066	0.409	5
Project #5	[(0.077, 0.081), 0.076, (0.074, 0.078)]	0.076	0.528	3	0.077	0.545	3	0.079	0.489	2
Project #6	[(0.106, 0.106), 0.079, (0.051, 0.080)]	0.080	0.553	2	0.083	0.590	2	0.068	0.420	3
Project #7	[(0.037, 0.038), 0.034, (0.047, 0.056)]	0.045	0.310	6	0.041	0.290	6	0.039	0.244	6

5. Conclusions and Recommendations

The necessity of project performance evaluation is crucial and undeniable which helps projects identify weaknesses and strengths as well as recognize and improve inefficiencies. Therefore several project performance evaluation approach has been used and advanced over the years that MCDM models is one of them. As a result, due to the complexity of the projects, and in order to achieve a better imagination of environmental ambiguities and overcome the inherent uncertainty of the projects, in this study, an Interval-Valued Fuzzy Additive Ratio Assessment (ARAS) Method proposed as a novel MCDM approach for evaluating the projects. On the other hand, despite the importance and certain position of oil and gas well drilling projects as well as its impact on the economies of countries, no structured assessment methodology has been presented for these types of projects and research works show the lack of literature in this field. So, the oil and gas well drilling projects selected as a case study. Given the limited research on performance evaluation in the context of such operations, the initial list of criteria and sub-criteria of evaluation obtained from available literature, was improved using the Fuzzy Delphi method. In order to determine criteria weights, the importance value of each final criterion was calculated through SWARA method based on expert opinions. As the performance scores of projects were determined with respect to defined criteria, a set of seven Iranian oil and gas projects were ranked by using interval-valued fuzzy ARAS method. The results revealed six general assessment criteria for oil and gas well-drilling projects including: number of planned wells, number of drilled wells, materials & equipment, Human resource, quality and planning. However, the third to sixth criteria consist of several sub-criteria among which the number of drilled wells, actual cost compliance percentage with planned costs and percentage of waiting time to total well drilling time were defined as the most important sub-criteria. From the projects being studied, Project #2 was defined as the best alternative due to its distinguished performance in terms of sub-criteria of type of drilling rigs, number of operational experts working on the project, scientific levels of drilling specialists working in the project, type of drilled wells in terms of operational risk and actual cost compliance percentage with planned cost. Rather, Project #3 was considered as the weakest alternative. The managerial investigation of Project #2 showed that management stability, along with the use of new methods for employee assessment, as well as the implementation of knowledge management methods had significant role in the success of that project, while Project #3 presented frequently-changed management. The innovations of this study include the development of a comprehensive list of criteria and sub-criteria in performance evaluation of oil and gas well-drilling projects by using Fuzzy Delphi, SWARA and interval-valued fuzzy ARAS methods.

As mentioned before, there are unpredictable issues in the process of conducting research that are one of the sources of uncertainties having been emphasized in the literature. It seems that stochastic MADM methods which have been considered in the recent year's literature can be used in future researches. On the other hand, performing drilling projects is time-consuming and the evaluation of options and the criteria weights may change over time due to changes in both the macro environment and the preferences and the priorities of the experts. Hence, the use of the prospective MADM method is recommended to better cover possible changes. Also, in future researches, we can use aggregate operators to integrate experts' opinions and hesitant fuzzy sets to further investigate the sensitive analysis of decisions which are taken.

Finally, we assumed that identified criteria are independent from each other. Future research can examine the existence of this dependency and in case of confirmation of the relationship between criteria; they can use appropriate methods (Like ANP, DEMATEL and ISM).

Author Contributions: Each author has participated and contributed sufficiently to take public responsibility for appropriate portions of the content. (Jalil Heidary Dahooie and Edmundas Kazimieras Zavadskas developed the methodology; Amirsalar Vanaki analyzed the results of case study; Mahdi Abolhasani wrote the paper; Zenonas Turskis revised and improved the paper.)

Conflicts of Interest: The authors declare no conflict of interest.

References

1. Salazar-Aramayo, J.L.; Rodrigues-da-Silveira, R.; Rodrigues-de-Almeida, M.; de Castro-Dantas, T.N. A conceptual model for project management of exploration and production in the oil and gas industry: The case of a Brazilian company. *Int. J. Proj. Manag.* **2013**, *31*, 589–601. [CrossRef]
2. Lewis, J.P. *The Project Manager's Desk Reference: A Comprehensive Guide to Project Planning, Scheduling, Evaluation, Control and Systems*; McGraw-Hill Inc.: Chicago, IL, USA, 1999.
3. Institute, P.M.; Staff, P.M.I. *A Guide to The Project Management Body of Knowledge (PMBOK Guide)*, 3rd ed.; An American National Standard, ANSI/PMI 99-001-2004; Project Management Institute: Sydney, Australia, 2004.
4. Ahari, R.M.; Niaki, S.T.A. A hybrid approach based on locally linear neuro-fuzzy modeling and TOPSIS to determine the quality grade of gas well-drilling projects. *J. Pet. Sci. Eng.* **2014**, *114*, 99–106. [CrossRef]
5. Butt, A.; Naaranoja, M.; Savolainen, J. Project change stakeholder communication. *Int. J. Proj. Manag.* **2016**, *34*, 1579–1595. [CrossRef]
6. Xu, Y.; Yeh, C.-H. A performance-based approach to project assignment and performance evaluation. *Int. J. Proj. Manag.* **2014**, *32*, 218–228. [CrossRef]
7. Cao, Q.; Hoffman, J.J. A case study approach for developing a project performance evaluation system. *Int. J. Proj. Manag.* **2011**, *29*, 155–164. [CrossRef]
8. Büyüközkan, G.; Karabulut, Y. Energy project performance evaluation with sustainability perspective. *Energy* **2017**, *119*, 549–560. [CrossRef]
9. Haralambopoulos, D.; Polatidis, H. Renewable energy projects: Structuring a multi-criteria group decision-making framework. *Renew. Energy* **2003**, *28*, 961–973. [CrossRef]
10. Barfod, M.B. An MCDA approach for the selection of bike projects based on structuring and appraising activities. *Eur. J. Oper. Res.* **2012**, *218*, 810–818. [CrossRef]
11. Wei, G.; Zhao, X. Induced hesitant interval-valued fuzzy Einstein aggregation operators and their application to multiple attribute decision making. *J. Intell. Fuzzy Syst.* **2013**, *24*, 789–803.
12. Zadeh, L.A. The concept of a linguistic variable and its application to approximate reasoning—I. *Inf. Sci.* **1975**, *8*, 199–249. [CrossRef]
13. Gorzałczany, M.B. A method of inference in approximate reasoning based on interval-valued fuzzy sets. *Fuzzy Sets Syst.* **1987**, *21*, 1–17. [CrossRef]
14. Atanassov, K.T. Intuitionistic fuzzy sets. *Fuzzy Sets Syst.* **1986**, *20*, 87–96. [CrossRef]
15. Baležentis, T.; Zeng, S. Group multi-criteria decision making based upon interval-valued fuzzy numbers: An extension of the MULTIMOORA method. *Expert Syst. Appl.* **2013**, *40*, 543–550. [CrossRef]
16. Zavadskas, E.K.; Antucheviciene, J.; Hajiagha, S.H.R.; Hashemi, S.S. Extension of weighted aggregated sum product assessment with interval-valued intuitionistic fuzzy numbers (WASPAS-IVIF). *Appl. Soft Comput.* **2014**, *24*, 1013–1021. [CrossRef]
17. Grimsey, D.; Lewis, M.K. Evaluating the risks of public private partnerships for infrastructure projects. *Int. J. Proj. Manag.* **2002**, *20*, 107–118. [CrossRef]
18. Castillo, L.; Dorao, C.A. Decision-making in the oil and gas projects based on game theory: Conceptual process design. *Energy Convers. Manag.* **2013**, *66*, 48–55. [CrossRef]
19. Luo, D.; Zhao, X. Modeling the operating costs for petroleum exploration and development projects. *Energy* **2012**, *40*, 189–195. [CrossRef]
20. Osmundsen, P.; Toft, A.; Agnar Dragvik, K. Design of drilling contracts—Economic incentives and safety issues. *Energy Policy* **2006**, *34*, 2324–2329. [CrossRef]
21. Ahari, R.M.; Niaki, S.T.A. Contractor selection in gas well-drilling projects with quality evaluation using Neuro-fuzzy networks. *IERI Procedia* **2014**, *10*, 274–279. [CrossRef]
22. Dachyar, M.; Pratama, N.R. Performance evaluation of a drilling project in oil and gas service company in Indonesia by MACBETH method. *J. Phys. Conf. Ser.* **2014**, *495*, 012012. [CrossRef]
23. Martinsuo, M.; Korhonen, T.; Laine, T. Identifying, framing and managing uncertainties in project portfolios. *Int. J. Proj. Manag.* **2014**, *32*, 732–746. [CrossRef]
24. Marques, G.; Gourc, D.; Lauras, M. Multi-criteria performance analysis for decision making in project management. *Int. J. Proj. Manag.* **2011**, *29*, 1057–1069. [CrossRef]

25. Kähkönen, K. Level of complexity in projects and its impacts on managerial solutions. *Int. Proj. Manag. Assoc.* **2008**, *XXIX*, 3.

26. Turner, J.R.; Müller, R. On the nature of the project as a temporary organization. *Int. J. Proj. Manag.* **2003**, *21*, 1–8. [CrossRef]

27. Cheng, C.-H.; Liou, J.J.; Chiu, C.-Y. A Consistent Fuzzy Preference Relations Based ANP Model for R&D Project Selection. *Sustainability* **2017**, *9*, 1352.

28. Strojny, J.; Szulc, J.; Baran, M. Applying the AHP Method into the Assessment of Project Attitudes. In *Eurasian Business Perspectives*; Springer: New York, NY, USA, 2018; pp. 183–197.

29. Aragonés-Beltrán, P.; Chaparro-González, F.; Pastor-Ferrando, J.-P.; Pla-Rubio, A. An AHP (Analytic Hierarchy Process)/ANP (Analytic Network Process)-based multi-criteria decision approach for the selection of solar-thermal power plant investment projects. *Energy* **2014**, *66*, 222–238. [CrossRef]

30. He, Q.; Luo, L.; Hu, Y.; Chan, A.P. Measuring the complexity of mega construction projects in China—A fuzzy analytic network process analysis. *Int. J. Proj. Manag.* **2015**, *33*, 549–563. [CrossRef]

31. Ozcan, U.; Dogan, A.; Soylemez, I. Evaluation of Research Projects of Undergraduate Students in an Engineering Department Using Topsis Method. *Eurasia Proc. Educ. Soc. Sci.* **2016**, *5*, 420–424.

32. Altuntas, S.; Dereli, T. A novel approach based on DEMATEL method and patent citation analysis for prioritizing a portfolio of investment projects. *Expert Syst. Appl.* **2015**, *42*, 1003–1012. [CrossRef]

33. Beccali, M.; Cellura, M.; Mistretta, M. Decision-making in energy planning. Application of the Electre method at regional level for the diffusion of renewable energy technology. *Renew. Energy* **2003**, *28*, 2063–2087. [CrossRef]

34. Çolak, M.; Kaya, İ. Prioritization of renewable energy alternatives by using an integrated fuzzy MCDM model: A real case application for Turkey. *Renew. Sustain. Energy Rev.* **2017**, *80*, 840–853. [CrossRef]

35. Anagnostopoulos, K.; Doukas, H.; Psarras, J. A linguistic multicriteria analysis system combining fuzzy sets theory, ideal and anti-ideal points for location site selection. *Expert Syst. Appl.* **2008**, *35*, 2041–2048. [CrossRef]

36. Wang, J.; Wang, J.-Q.; Zhang, H.-Y.; Chen, X.-H. Multi-criteria decision-making based on hesitant fuzzy linguistic term sets: An outranking approach. *Knowl. Based Syst.* **2015**, *86*, 224–236. [CrossRef]

37. Riera, J.V.; Massanet, S.; Herrera-Viedma, E.; Torrens, J. Some interesting properties of the fuzzy linguistic model based on discrete fuzzy numbers to manage hesitant fuzzy linguistic information. *Appl. Soft Comput.* **2015**, *36*, 383–391. [CrossRef]

38. Yavuz, M.; Oztaysi, B.; Onar, S.C.; Kahraman, C. Multi-criteria evaluation of alternative-fuel vehicles via a hierarchical hesitant fuzzy linguistic model. *Expert Syst. Appl.* **2015**, *42*, 2835–2848. [CrossRef]

39. Sanchez, M.A.; Castro, J.R.; Castillo, O.; Mendoza, O.; Rodriguez-Diaz, A.; Melin, P. Fuzzy higher type information granules from an uncertainty measurement. *Granul. Comput.* **2017**, *2*, 95–103. [CrossRef]

40. Das, S.; Kar, S.; Pal, T. Robust decision making using intuitionistic fuzzy numbers. *Granul. Comput.* **2017**, *2*, 41–54. [CrossRef]

41. Liu, P.; Mahmood, T.; Khan, Q. Multi-Attribute Decision-Making Based on Prioritized Aggregation Operator under Hesitant Intuitionistic Fuzzy Linguistic Environment. *Symmetry* **2017**, *9*, 270. [CrossRef]

42. Dubois, D.J. *Fuzzy Sets and Systems: Theory and Applications*; Academic Press: Cambridge, MA, USA, 1980; Volume 144.

43. Feng, F.; Li, Y.; Leoreanu-Fotea, V. Application of level soft sets in decision making based on interval-valued fuzzy soft sets. *Comput. Math. Appl.* **2010**, *60*, 1756–1767. [CrossRef]

44. Xu, Z.; Gou, X. An overview of interval-valued intuitionistic fuzzy information aggregations and applications. *Granul. Comput.* **2017**, *2*, 13–39. [CrossRef]

45. Jiang, Y.; Xu, Z.; Shu, Y. Interval-valued intuitionistic multiplicative aggregation in group decision making. *Granul. Comput.* **2017**, *2*, 387–407. [CrossRef]

46. Meng, S.; Liu, N.; He, Y. GIFIHIA operator and its application to the selection of cold chain logistics enterprises. *Granul. Comput.* **2017**, *2*, 187–197. [CrossRef]

47. Wang, C.; Fu, X.; Meng, S.; He, Y. Multi-attribute decision-making based on the SPIFGIA operators. *Granul. Comput.* **2017**, *2*, 321–331. [CrossRef]

48. Liang, D.; Xu, Z. The new extension of TOPSIS method for multiple criteria decision making with hesitant Pythagorean fuzzy sets. *Appl. Soft Comput.* **2017**, *60*, 167–179. [CrossRef]

49. Karnik, N.N.; Mendel, J.M. Operations on type-2 fuzzy sets. *Fuzzy Sets Syst.* **2001**, *122*, 327–348. [CrossRef]

50. Liu, P. A weighted aggregation operators multi-attribute group decision-making method based on interval-valued trapezoidal fuzzy numbers. *Expert Syst. Appl.* **2011**, *38*, 1053–1060. [CrossRef]

51. Xu, Z.; Wang, H. Managing multi-granularity linguistic information in qualitative group decision making: An overview. *Granul. Comput.* **2016**, *1*, 21–35. [CrossRef]

52. Zhang, H. Some interval-valued 2-tuple linguistic aggregation operators and application in multiattribute group decision making. *Appl. Math. Model.* **2013**, *37*, 4269–4282. [CrossRef]

53. Mendel, J.M. A comparison of three approaches for estimating (synthesizing) an interval type-2 fuzzy set model of a linguistic term for computing with words. *Granul. Comput.* **2016**, *1*, 59–69. [CrossRef]

54. Wu, J.; Chiclana, F. A social network analysis trust—Consensus based approach to group decision-making problems with interval-valued fuzzy reciprocal preference relations. *Knowl. Based Syst.* **2014**, *59*, 97–107. [CrossRef]

55. Reiser, R.H.S.; Bedregal, B.; Dos Reis, G. Interval-valued fuzzy coimplications and related dual interval-valued conjugate functions. *J. Comput. Syst. Sci.* **2014**, *80*, 410–425. [CrossRef]

56. Qin, J. Interval type-2 fuzzy Hamy mean operators and their application in multiple criteria decision making. *Granul. Comput.* **2017**, *2*, 249–269. [CrossRef]

57. Vahdani, B.; Hadipour, H.; Sadaghiani, J.S.; Amiri, M. Extension of VIKOR method based on interval-valued fuzzy sets. *Int. J. Adv. Manuf. Technol.* **2010**, *47*, 1231–1239. [CrossRef]

58. Chatterjee, K.; Kar, S. Unified Granular-number-based AHP-VIKOR multi-criteria decision framework. *Granul. Comput.* **2017**, *2*, 199–221. [CrossRef]

59. Ashtiani, B.; Haghighirad, F.; Makui, A.; ali Montazer, G. Extension of fuzzy TOPSIS method based on interval-valued fuzzy sets. *Appl. Soft Comput.* **2009**, *9*, 457–461. [CrossRef]

60. Liu, P.; You, X. Probabilistic linguistic TODIM approach for multiple attribute decision-making. *Granul. Comput.* **2017**, *2*, 333–342. [CrossRef]

61. Campeol, G.; Carollo, S.; Masotto, N. Infrastructural projects and territorial development in Veneto Dolomites: Evaluation of performances through AHP. *Procedia Soc. Behav. Sci.* **2016**, *223*, 468–474. [CrossRef]

62. Osmundsen, P.; Sørenes, T.; Toft, A. Offshore oil service contracts new incentive schemes to promote drilling efficiency. *J. Pet. Sci. Eng.* **2010**, *72*, 220–228. [CrossRef]

63. Liu, J.; Li, Q.; Wang, Y. Risk analysis in ultra deep scientific drilling project—A fuzzy synthetic evaluation approach. *Int. J. Proj. Manag.* **2013**, *31*, 449–458. [CrossRef]

64. Ngacho, C.; Das, D. A performance evaluation framework of development projects: An empirical study of constituency development fund (CDF) construction projects in Kenya. *Int. J. Proj. Manag.* **2014**, *32*, 492–507. [CrossRef]

65. Bassioni, H.A.; Price, A.D.F.; Hassan, T.M. Performance measurement in construction. *J. Manag. Eng.* **2004**, *20*, 42–50. [CrossRef]

66. Hare, B.; Cameron, I.; Roy Duff, A. Exploring the integration of health and safety with pre-construction planning. *Eng. Constr. Archit. Manag.* **2006**, *13*, 438–450. [CrossRef]

67. Haslam, R.A.; Hide, S.A.; Gibb, A.G.F.; Gyi, D.E.; Pavitt, T.; Atkinson, S.; Duff, A.R. Contributing factors in construction accidents. *Appl. Ergon.* **2005**, *36*, 401–415. [CrossRef] [PubMed]

68. Ortega, I. Systematic prevention of construction failures: An overview. *Technol. Law Insur.* **2000**, *5*, 15–22. [CrossRef]

69. Tabish, S.Z.S.; Jha, K.N. Analyses and evaluation of irregularities in public procurement in India. *Constr. Manag. Econ.* **2011**, *29*, 261–274. [CrossRef]

70. Eriksson, P.E.; Westerberg, M. Effects of cooperative procurement procedures on construction project performance: A conceptual framework. *Int. J. Proj. Manag.* **2011**, *29*, 197–208. [CrossRef]

71. Ali, A.S.; Rahmat, I. The performance measurement of construction projects managed by ISO-certified contractors in Malaysia. *J. Retail Leis. Prop.* **2010**, *9*, 25–35. [CrossRef]

72. Chan, A.P.C. Key performance indicators for measuring construction success. *Benchmarking* **2004**, *11*, 203–221. [CrossRef]

73. Fuentes, R.; Fuster, B.; Lillo-Bañuls, A. A three-stage DEA model to evaluate learning-teaching technical efficiency: Key performance indicators and contextual variables. *Expert Syst. Appl.* **2016**, *48*, 89–99. [CrossRef]

74. Tone, K.; Tsutsui, M. Network DEA: A slacks-based measure approach. *Eur. J. Oper. Res.* **2009**, *197*, 243–252. [CrossRef]

75. Li, K.; Lin, B. Impact of energy conservation policies on the green productivity in China's manufacturing sector: Evidence from a three-stage DEA model. *Appl. Energy* **2016**, *168*, 351–363. [CrossRef]
76. Oral, M.; Oukil, A.; Malouin, J.-L.; Kettani, O. The appreciative democratic voice of DEA: A case of faculty academic performance evaluation. *Socio Econ. Plan. Sci.* **2014**, *48*, 20–28. [CrossRef]
77. Tongzon, J. Efficiency measurement of selected Australian and other international ports using data envelopment analysis. *Transp. Res. Part A Policy Pract.* **2001**, *35*, 107–122. [CrossRef]
78. Menkhoff, L.; Schmidt, U.; Brozynski, T. The impact of experience on risk taking, overconfidence, and herding of fund managers: Complementary survey evidence. *Eur. Econ. Rev.* **2006**, *50*, 1753–1766. [CrossRef]
79. Schmidt, F.L.; Hunter, J.E.; Outerbridge, A.N. Impact of job experience and ability on job knowledge, work sample performance, and supervisory ratings of job performance. *J. Appl. Psychol.* **1986**, *71*, 432–439. [CrossRef]
80. Paquin, J.-P.; Gauthier, C.; Morin, P.-P. The downside risk of project portfolios: The impact of capital investment projects and the value of project efficiency and project risk management programmes. *Int. J. Proj. Manag.* **2016**, *34*, 1460–1470. [CrossRef]
81. Jankowski, J.E. Do we need a price index for industrial R&D? *Res. Policy* **1993**, *22*, 195–205.
82. McGrath, M. The R&D effectiveness index: A metric for product development performance. *J. Prod. Innov. Manag.* **1994**, *11*, 213–220.
83. Meredith, J. Alternative research paradigms in operations. *J. Oper. Manag.* **1989**, *8*, 297–326. [CrossRef]
84. Kardaras, D.K.; Karakostas, B.; Mamakou, X.J. Content presentation personalisation and media adaptation in tourism web sites using fuzzy Delphi method and fuzzy cognitive maps. *Expert Syst. Appl.* **2013**, *40*, 2331–2342. [CrossRef]
85. Okoli, C.; Pawlowski, S.D. The Delphi method as a research tool: An example, design considerations and applications. *Inf. Manag.* **2004**, *42*, 15–29. [CrossRef]
86. Kuo, Y.-F.; Chen, P.-C. Constructing performance appraisal indicators for mobility of the service industries using fuzzy Delphi method. *Expert Syst. Appl.* **2008**, *35*, 1930–1939. [CrossRef]
87. Murray, T.J.; Pipino, L.L.; van Gigch, J.P. A pilot study of fuzzy set modification of Delphi. *Hum. Syst. Manag.* **1985**, *5*, 76–80.
88. Hsu, T.; Yang, T. Application of fuzzy analytic hierarchy process in the selection of advertising media. *J. Manag. Syst.* **2000**, *7*, 19–39.
89. Tzeng, G.H.; Teng, J.Y. Transportation investment project selection with fuzzy multiobjectives. *Transp. Plan. Technol.* **1993**, *17*, 91–112. [CrossRef]
90. Hashemkhani Zolfani, S.; Aghdaie, M.H.; Derakhti, A.; Zavadskas, E.K.; Morshed Varzandeh, M.H. Decision making on business issues with foresight perspective; an application of new hybrid MCDM model in shopping mall locating. *Expert Syst. Appl.* **2013**, *40*, 7111–7121. [CrossRef]
91. Keršuliene, V.; Zavadskas, E.K.; Turskis, Z. Selection of rational dispute resolution method by applying new step-wise weight assessment ratio analysis (Swara). *J. Bus. Econ. Manag.* **2010**, *11*, 243–258. [CrossRef]
92. Dehnavi, A.; Aghdam, I.N.; Pradhan, B.; Morshed Varzandeh, M.H. A new hybrid model using Step-Wise Weight Assessment Ratio Analysis (SWARA) technique and Adaptive Neuro-Fuzzy Inference System (ANFIS) for regional landslide hazard assessment in Iran. *CATENA* **2015**, *135*, 122–148. [CrossRef]
93. Hashemkhani Zolfani, S.; Bahrami, M. Investment prioritizing in high tech industries based on SWARA-COPRAS approach. *Technol. Econ. Dev. Econ.* **2014**, *20*, 534–553. [CrossRef]
94. Chen, S.-J.; Chen, S.-M. Fuzzy risk analysis based on similarity measures of generalized fuzzy numbers. *IEEE Trans. Fuzzy Syst.* **2003**, *11*, 45–56. [CrossRef]
95. Yao, J.-S.; Lin, F.-T. Constructing a fuzzy flow-shop sequencing model based on statistical data. *Int. J. Approx. Reason.* **2002**, *29*, 215–234. [CrossRef]
96. Chen, S.-J.; Chen, S.-M. Fuzzy risk analysis based on measures of similarity between interval-valued fuzzy numbers. *Comput. Math. Appl.* **2008**, *55*, 1670–1685. [CrossRef]
97. Chen, C.-T. Extensions of the TOPSIS for group decision-making under fuzzy environment. *Fuzzy Sets Syst.* **2000**, *114*, 1–9. [CrossRef]
98. Mahdavi, I.; Mahdavi-Amiri, N.; Heidarzade, A.; Nourifar, R. Designing a model of fuzzy TOPSIS in multiple criteria decision making. *Appl. Math. Comput.* **2008**, *206*, 607–617. [CrossRef]
99. Wang, Y.-M.; Elhag, T.M. Fuzzy TOPSIS method based on alpha level sets with an application to bridge risk assessment. *Expert Syst. Appl.* **2006**, *31*, 309–319. [CrossRef]

100. Wei, S.-H.; Chen, S.-M. Fuzzy risk analysis based on interval-valued fuzzy numbers. *Expert Syst. Appl.* **2009**, *36*, 2285–2299. [CrossRef]

101. Tupenaite, L.; Zavadskas, E.K.; Kaklauskas, A.; Turskis, Z.; Seniut, M. Multiple criteria assessment of alternatives for built and human environment renovation. *J. Civ. Eng. Manag.* **2010**, *16*, 257–266. [CrossRef]

102. Zavadskas, E.K.; Turskis, Z. A new additive ratio assessment (ARAS) method in multicriteria decision-making. *Technol. Econ. Dev. Econ.* **2010**, *16*, 159–172. [CrossRef]

103. Zavadskas, E.; Turskis, Z.; Vilutiene, T. Multiple criteria analysis of foundation instalment alternatives by applying Additive Ratio Assessment (ARAS) method. *Arch. Civ. Mech. Eng.* **2010**, *10*, 123–141. [CrossRef]

symmetry

MDPI

Article

A Model for Shovel Capital Cost Estimation, Using a Hybrid Model of Multivariate Regression and Neural Networks

Abdolreza Yazdani-Chamzini [1] (ID), Edmundas Kazimieras Zavadskas [2,3] (ID), Jurgita Antucheviciene [2,*] (ID) and Romualdas Bausys [4]

[1] Young Researchers and Elite Club, South Tehran Branch, Islamic Azad University, 14115/344 Tehran, Iran; abdalrezaych@gmail.com
[2] Department of Construction Management and Real Estate, Vilnius Gediminas Technical University, Sauletekio al. 11, 10223 Vilnius, Lithuania; edmundas.zavadskas@vgtu.lt
[3] Institute of Sustainable Construction, Vilnius Gediminas Technical University, Sauletekio al. 11, 10223 Vilnius, Lithuania
[4] Department of Graphical Systems, Vilnius Gediminas Technical University, Sauletekio al. 11, 10223 Vilnius, Lithuania; romualdas.bausys@vgtu.lt
* Correspondence: jurgita.antucheviciene@vgtu.lt; Tel.: +370-5-274-5232

Received: 27 October 2017; Accepted: 29 November 2017; Published: 1 December 2017

Abstract: Cost estimation is an essential issue in feasibility studies in civil engineering. Many different methods can be applied to modelling costs. These methods can be divided into several main groups: (1) artificial intelligence, (2) statistical methods, and (3) analytical methods. In this paper, the multivariate regression (MVR) method, which is one of the most popular linear models, and the artificial neural network (ANN) method, which is widely applied to solving different prediction problems with a high degree of accuracy, have been combined to provide a cost estimate model for a shovel machine. This hybrid methodology is proposed, taking the advantages of MVR and ANN models in linear and nonlinear modelling, respectively. In the proposed model, the unique advantages of the MVR model in linear modelling are used first to recognize the existing linear structure in data, and, then, the ANN for determining nonlinear patterns in preprocessed data is applied. The results with three indices indicate that the proposed model is efficient and capable of increasing the prediction accuracy.

Keywords: cost estimation; shovel machine; neural network; multivariate regression; hybrid model

1. Introduction

Earthmoving and loading/unloading works are essential parts of construction processes. They require the use of heavy equipment. In spite of the fact that energy use and emissions are critical when selecting the equipment [1], the cost is usually a crucial factor in the industry. In mining, loading is one of the most important operations, influencing mine preparation, design, and production. This activity has a significant share of the capital and operational expenditures. In the process of mine planning, the loading system mainly determines what other pieces of equipment and what mode of operation will be used, and it is a key to low-cost production [2]. A shovel machine, which is a tool for digging, lifting, and moving bulk materials, is the most common piece of equipment that is used to load rock material in surface mines. A shovel is a machine capable of handling hard, dense, abrasive, as well as highly fragmented ground, which can accurately spot for loading into dump trucks, rail cars, loading hoppers, etc. [2]. Shovels are usually grouped into four classes [3]:

- small shovels (0.5–2 m^3 bucket size);
- medium shovels (2–5 m^3 bucket size);
- large-size shovels (5–25 m^3 bucket size);
- very large-size shovels (with a bucket larger than 25 m^3).

The selection of the best shovel among a pool of the existing alternatives based on the considered criteria can be performed applying the feasibility studies, which should be conducted to analyse the involved different technical, economic, and operational aspects. The most frequent technological parameters of shovels operations are analysed in the literature. As an example that is related to mathematical simulation a study presenting automated designing of swing circuit for a hydraulic shovel [4] can be mentioned. The effectiveness of mining equipment, namely trucks and shovel, regarding its useful employment without time losses were analysed [5]. For the considered equipment selection problem, the mentioned numerous aspects of different nature can be effectively investigated by using a multiple criteria decision making (MCDM) tool. MCDM methodology allows for combining several important issues (criteria), also estimate the relative importance of the analysed criteria, as well as to compare potential alternative equipment and to select the best suited in the analysed situation [6].

Besides equipment selection that is based on technical requirements, the production rate and cost-benefit analysis make the main parts of the feasibility studies. Cost estimation with various purposes takes a significant place at different stages and processes in mining and related industries. During the planning stage, machine specifications, like technical and operating features, and, accordingly, the costs are not available in in the mineral fields, unlike other areas [7,8]. Therefore, developing an up-to-date model with sufficient accuracy is essential. There is a number of models for estimating mineral industry costs, such as the cost estimation to small underground mines [9], the evaluation of mining equipment, as well as mineral processing equipment costs and other capital expenditures that are related to mining and processing operations [10,11], as well as cost estimation that is adapted for the peculiarities of the Australian mining industry [12], the system of cost estimation in mining of metallic, as well as nonmetallic minerals in two countries, comprising Canada and the United States [13], a cost model for preliminary feasibility study and a different cost model for a detailed feasibility study, including exponential regression and multivariable linear regression, implemented in Iranian deposit of copper [14], a case study of South Africa for calculation of capital costs for setting up a coal mine [15], and Hard-rock LHD cost estimation [7]. A methodology of evaluation of ore bodies and a guide for practical application was suggested [16–19].

An individual question of feasibility evaluations applying simplified cost models was analysed [20]. A guide for the mining sector, including mining and energy valuation, and being focused on Australian investors and managers [21], was prepared. Also, energy costs of equipment as a significant part of costs in mining activity were evaluated, including not only processing costs but also transportation and exploitation costs of machinery [22].

The models mentioned above often take into account only one parameter as an independent variable, while the constructed models are merely based on the regression analysis. Therefore, it is essential to develop a more accurate and efficient model for cost estimation of equipment.

The different techniques can be applied for the optimization problems. During the last decade, significant research has been performed on fuzzy logic control of the nonlinear systems [23,24]. Another direction of the optimization techniques is artificial neural networks.

The Artificial Neural Network (ANN), which is one of the artificial intelligence methods, as well as multivariate regression (as one of the statistical models), are powerful tools for pattern recognition and modelling. These two methods are used by various researchers.

An ANN-based model for estimating distillation process, using the Levenberg–Marquardt approach is developed by Singh et al. [25]. Yamamura [26] predicted pharmacokinetic parameters by an ANN modelling. An ANN prediction model for determining the failure depth of coal seam floors was developed by Lian-guo et al. [27]. They compared calculation results with the real case study and stated that the predicted results by applying the suggested model agreed well with practical measurements.

An ANN model for an industrial gas turbine was developed by Fast et al. [28]. They used the operational data with a multilayer feed-forward network to construct an ANN model. Analysing the results, they made a conclusion that some of the functional and performance parameters of the gas turbine, including a critical parameter of identification of the anti-icing mode, can be accurately predicted in a changing modelling environment. Lee et al. [29] suggested a way to improve the reliability of the Bridge Management System, using the ANN-based Backwards Prediction Model.

Jalali-Heravi et al. [30] developed the shuffling multivariate adaptive regression splines and the adaptive neuro-fuzzy inference system as tools for studying a quantitative structure-activity relationship (QSAR) of severe acute respiratory syndrome (SARS) inhibitors.

Verlinden et al. [31] presented a case study and estimated the cost of sheet metal parts, using a combination of multiple regression and artificial neural networks. Mesroghli et al. [32] used the regression and artificial neural networks for estimating gross calorific value based on coal analysis. Sahoo et al. [33] developed the models for predicting stream water temperature, using three techniques, namely the regression analysis, an artificial neural network, and also combining them with chaotic non-linear dynamic models. Therefore, it is clear that the ANN and regression have demonstrated their capabilities of modelling engineering practice problems.

An application of neural networks for improving the weighting precision, having the aim to optimise loading of trucks, as well as production efficiency of electric shovels, was presented by Gu et al. [34]. The combination with fuzzy logic was suggested to decrease uncertainties in the process.

During the recent years, the different aspects of civil engineering problems have been considered when applying artificial neural networks (ANNs). Suspended-dome model updating was performed by back propagation network approach to evaluate the discrepancy between actual structure and the corresponding numerical approximation [35]. The ANN approach was applied for the estimation of the axial bearing capacity of the rectangular concrete-filled tubular columns [36]. A stochastic conceptual cost evaluation of the highway projects is performed by generating an empirical distribution of the estimated cost range, without additional initial assumptions [37]. This empirical distribution is constructed applying ANN techniques and bootstrap sampling. The application of ANN techniques was implemented to study construction labour productivity [38]. Additionally, the different ANN activation and transfer functions are applied to estimate the most influencing factors to model construction labour productivity. The parameter sensitivity analysis for civil engineering problems was performed implementing ANN algorithms [39]. This paper dealt with parameter sensitivity analysis paradigm, in which the essential element is neural network ensemble.

Usually, structural health monitoring of the engineering systems is governed by fixed or hand-crafted features, and this fact significantly reduces the reliability of such monitoring systems. The proposed structural damage detection system, which was constructed by implementing one-dimensional (1D) convolutional neural networks, allows for extracting damage-sensitive features from raw acceleration information [40].

The prediction problem of the capacity characteristics of the pile structures has been modelled implementing ANN and principal component analysis [41]. The issue of the claim management in the construction projects was solved applying neural network approach [42]. The proposed approach allows for not only classifying and ranking emerging claims, but also to predict the claim frequency in the construction projects.

The integration of two different information flows: spatial planning of buildings and territorial planning system, was performed while applying ANN technique [43]. The evaluation of the uncertainty influence to the estimated cost of the construction projects was carried out implementing different models for the training and evaluating phases [44]. For training phase, fuzzy adaptive learning control network and the fast, messy genetic algorithm were applied. For evaluating phase, component ratios, regression, and multi-factor evaluation sub-models are accomplished.

In this paper, multivariate regression and ANN models are employed to construct a hybrid model to yield a more accurate and precise model than traditional multivariate regression and ANN models.

In the proposed approach, a model is considered as a function of linear and nonlinear components, so that the multivariate regression model is first employed to recognise the existing linear pattern in data. Then, the ANN is applied as a nonlinear function to model the preprocessed data. Finally, the main performance criteria of the model are calculated, including the coefficient of determination (R^2), Normalized Mean Square Error (NMSE), and Mean Absolute Percentage Error (MAPE), and the best model is identified according to calculation results.

2. A Multivariate Regression Model

The method of multivariate regression (MVR) is used to determine the relationships between independent and dependent variables. It might also be used for analysing the data or generating a model [45]. In statistical analysis, the mathematical form of an MVR model is determined by the following equation:

$$Y = X\beta + \varepsilon \tag{1}$$

where ε addresses the $N \times 1$ vector of observations for the disturbance term, X represents the $N \times k$ matrix of observations of the k-independent variables, β expresses the $k \times 1$ vector of parameters, and Y denotes the $N \times 1$ vector of observations of the dependent variable.

The best prediction of Y (the variable Y regressed on X), based on basic assumptions of the linear model, is denoted as $\widehat{Y}$, and can be obtained by the following equation:

$$\widehat{Y} = X(X^T X)^{-1} X^T, \tag{2}$$

where Y is the shovel capital cost (CC), and $X = (1, BC, BL, W, HP)$ is a set of independent variables defined, respectively, as a constant term, bucket capacity (BC), boom length (BL), weight (W), and horsepower (HP). Thus, the model is expressed by

$$CC = b_1 + b_2 BC + b_3 BL + b_4 W + b_5 HP + e, \tag{3}$$

where b_1 is a constant term; b_2, b_3, b_4, and b_5 are the regression coefficients; and, e is an error term.

3. An Artificial Neural Network

The ANN technique, a branch of artificial intelligence methods, is a reliable and useful tool for the formulation of linear and non-linear patterns. Bourquin et al. [46] revealed that an ANN methodology shows a clear superiority as a modelling technique, in comparison to classical methods, for data sets showing non-linear relationships, and this is for both data fitting and prediction abilities. This technique is widely used for many scientific and engineering problems, such as data processing, classification, and pattern recognition. The ANN technique has some unique features, distinguishing it from other data processing systems. This technique, even when partly damaged, can work successfully. This method can also be used for parallel processing, generalisation, and demonstrates low vulnerability to errors in the dataset [47]. An ANN model employs the mechanism that is applied in the human brain to extract the patterns and behaviours of data [48].

The ANN technique has been developed based on the structure of biological neural networks, where neurons are the backbone of the structure. The inspiration for an artificial neuron arises from a biological neuron; so that, an artificial neuron can send signals to other neurons. Then, it collects these signals, and when fired, it transmits a signal to all of the connected neurons [49]. Figure 1 graphically shows a typical artificial neuron.

From Figure 1, the transfer function is shown by f; the activation threshold of the neuron j is determined by θ_j; the connection weight between the ith and jth neurons is assigned by w_{ij}; the input signal of n other neurons to a neuron j is determined by x_i (I = 1, 2, ... , n); and, the output of the neuron j is assigned by y_j, which can be mathematically computed by Equation (4):

$$y_i = f\left(\sum_{i=1}^{n} w_{ij}x_i - \theta_j\right),$$

(4)

where f usually presents hyperbolic tangent sigmoid and linear transfer functions.

A typical neural network comprises three primary layers, including input, intermediate, and output layers. Based on the basic concepts of machine learning, the number of hidden (intermediate) layers does not have a theoretical limit [50]. A typical ANN structure is depicted in Figure 2.

A multilayer feedforward perceptron (MLP), based on a backpropagation learning function, is applied for the estimation of the shovel capital cost. In the process of formulation, the model output with the actual output is compared to adjust the coefficients. An ANN model follows a five-step procedure to obtain a relationship between input(s) and output(s). Firstly, the training dataset, a vector of input–output, is randomly selected. Secondly, the network structure is constructed. Next, the model output vector is calculated for the input vector. Then, by using the model performance measure, connection weights are adjusted. Finally, the process of improving the weights is continued in order to satisfy the model performance.

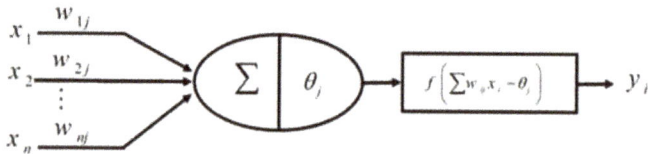

Figure 1. A model for an artificial neuron.

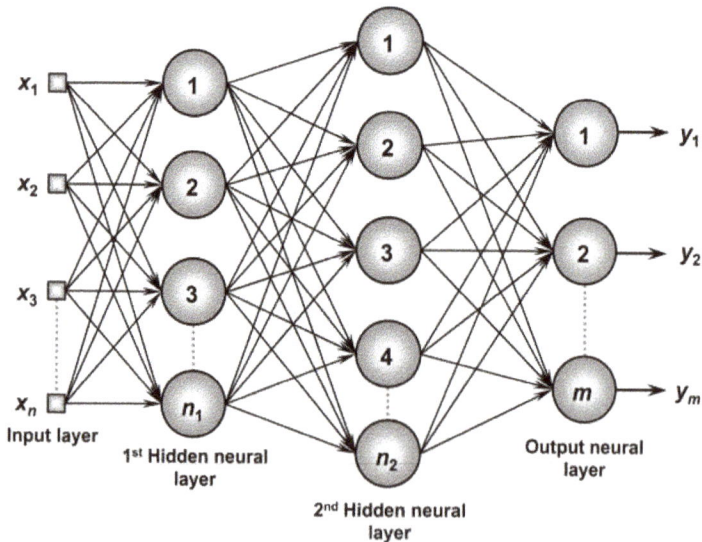

Figure 2. Architecture of a typical artificial neural networks (ANN).

4. The Hybrid Methodology

Both ANN and MVR models have earned success in modelling nonlinear and linear patterns, respectively. However, none of them is a universal model, which is appropriate to all situations. The MVR model ignores the complex nonlinear patterns that are involved in data. On the other hand, using an ANN model for the formulation of a linear problem can lead to mixed results [51]. Since it is impossible to recognise the behaviour of the data correctly, a combined approach, based on both nonlinear and linear formulating proficiency, is a robust plan in modelling a complicated issue [52].

In this study, a new hybrid model is proposed, in which the MVR model and ANNs are integrated to acquire a robust and efficient method for cost estimation of shovel machines. Since the MVR is a linear approach that is unable to recognise nonlinear patterns in data, the ANN is applied to model the residuals and to acquire the nonlinear behaviour, while the result of the ANN is added to the final output. Therefore, the MVR model will be responsible for the linear model, while the ANN will be responsible for the nonlinear part. A diagram of the model can be schematically seen in Figure 3.

Having this in mind, we assume that the shovel costs are classed as the nonlinear and linear parts:

$$Y_t = L_t + NL_t \tag{5}$$

where NL_t and L_t indicate the nonlinear and linear components, respectively. Therefore, the main idea is to employ, in the first place, the MVR model, and, next, to apply the ANN to formulate the residuals of the linear structure. A schematic diagram of the proposed model is shown in Figure 3.

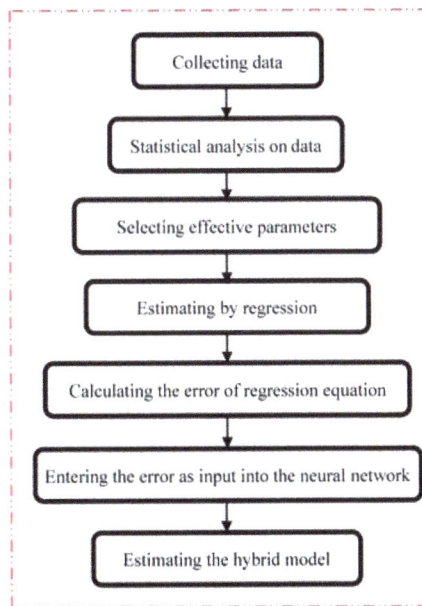

Figure 3. The approach used in the research.

Since the MVR model cannot approximate the nonlinear patterns that are involved in the dataset, then the residuals of the linear pattern may comprise nonlinear behaviour, which cannot be captured by a linear pattern. The combination model includes the exceptional advantages of both MVR and ANN models, allowing it to recognise various patterns. Therefore, it is profitable to separately formulate nonlinear and linear patterns by employing multiple patterns and, then, to integrate the results for improving the outcomes and model efficiency.

5. Model Performance

The model output can be analysed with comparing the assessed values and the real ones for generalising the developed formula to the unidentified output. This paper uses three performance measures, including Normalized Mean Square Error (NMSE), the coefficient of determination (R^2), and Mean Absolute Percentage Error (MAPE), to analyse the performance of each equation. These indicators can be calculated by as follows:

$$R^2 = 1 - \frac{\sum_{i=1}^{N} (A_i - P_i)^2}{\sum_{i=1}^{N} (A_i - \overline{A_i})^2}, \tag{6}$$

$$\text{NMSE} = \frac{\sum_{i=1}^{N} (A_i - P_i)^2}{\sum_{i=1}^{N} (A_i - \overline{A_i})^2}, \tag{7}$$

$$\text{MAPE} = \frac{1}{n} \sum_{i=1}^{n} \left| \frac{P_i - A_i}{A_i} \right| \times 100, \tag{8}$$

where P_i denotes the predicted values, A_i denotes the observed values, N is the number of the datasets, and $\overline{A_i}$ is the average of the observed set.

R^2 shows to what extent the independent variable(s) can explain the variability in the dependent variable. R^2 belongs to the closed interval zero and one and can only take a positive value [53]. The value of close to zero for R^2 shows a poor fit of the predictive model, while the value of close one to presents a good fit. The R^2 value of 0.9 and higher is considered to be very satisfactory, while the values of 0.8–0.9 represent a relatively acceptable formula, and those below 0.8 are taken into account to be unacceptable [54].

Mean absolute percentage error is an unbiased statistic that is used for calculating the differentiation between the predicted and observed values. The mean absolute percentage error contains positive numbers without having an upper bound [55,56]. The normalized mean square error presents non-negative numbers, which can be any numbers between zero and one, while the value of 0 means that the model is ideal and the value of one denotes that the constructed model is non-ideal.

6. The Data

Thirty different sizes of various shovel types (14 cable and 16 hydraulic machines) have been considered, and a data set of technical and economic information that is obtained from the USA equipment manufacturers, mining companies and projects has been formed [57]. The costs were based on the US dollar in 2007. The data in the set are classified by the machine types (cable or hydraulic). The capital cost items include bucket capacity (cubic yard), boom length (ft), weight (Ibs), and horsepower (HP). Table 1 shows some statistical information on the data applied to each variable.

Table 1. Data distribution.

Machine Type	Operator	Bucket Capacity (BC)	Boom Length (BL)	Weight (W)	Horsepower (HP)	Capital Cost (CC)
	Min	7	35	556,000	685	2,800,000
Cable	Max	80	64	4,500,000	7000	17,900,000
	Mean	31.29	50.57	1,517,071	2923.85	7,871,429
	Standard deviation	21.27	9.04	1,084,127.97	1989.71	4,443,392.42
	Min	3	15.4	95,900	250	605,000
Hydraulic	Max	44	39.3	1,399,000	3000	6,900,000
	Mean	12.74	28.74	385,318.8	873.56	2,068,313
	Standard deviation	9.97	6.96	334,189.25	700.14	1,781,125.17

7. The Implementation of the Proposed Model

The 22 data items were randomly used to formulate the model, and, then, other items of the data set were selected to evaluate the performance of the model. First, by using the MVR model, the linear pattern in the data was extracted as

$$CC = 1{,}287{,}582.06 \times D + 129{,}953.24 \times BC + 44{,}841.31 \times BL + 0.12 \times W + 438.241 \times HP - 1324{,}659.59 \quad (9)$$

Then, the residuals of the MVR model were entered into the neural network as an input. Training minimises the error between the network output and the target output by repeatedly changing the values of the ANN's connection weights according to a predetermined algorithm [58].

Different algorithms are developed to model a complex structure that is involved in the dataset. Feed forward back propagation method is the most efficient technique for obtaining good results [59]. In this paper, the selected network was based on feedforward backpropagation (as seen in Figure 2). A backpropagation network is a multilinear perceptron that is constructed by three main layers (i) one layer for connecting independent variables to the formulating system by using nodes, namely input layer; (ii) one or more layers for capturing the patterns involved in data by using nodes, namely intermediate layer(s); and (iii) one layer for connecting dependent variables to the formulating system with nodes, namely output layer [60].

Moreover, the training process of the network is fulfilled by utilizing the triangular function, employing the backpropagation algorithm by adjusting the value of weights and bias concerning gradient descent with the momentum, under Matlab software environment.

The optimum network architecture was constructed while comparing various network architectures concerning the ANNs performance. The best-fitted network, which presents the best prediction accuracy with the test data, is composed of one input, four hidden, and one output neurons (N^{1-4-1}). The correlation between the actual values and the output of the proposed model for testing data is presented in Figure 4.

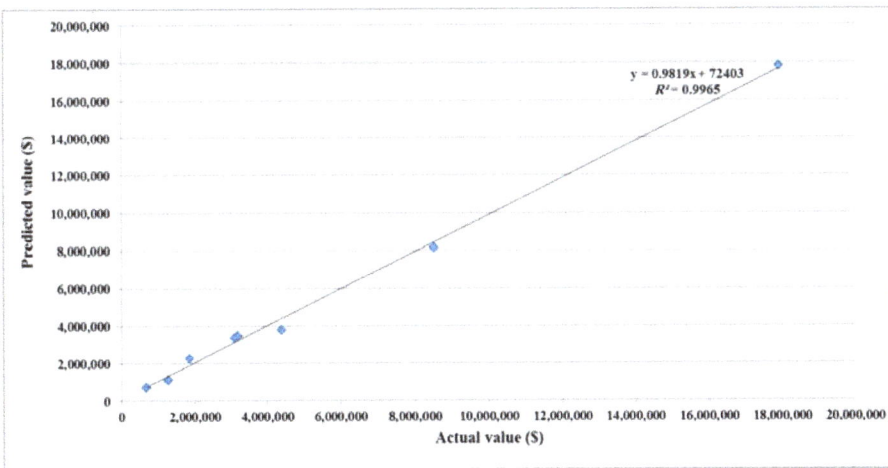

Figure 4. The correlation for shovel cost based on the proposed model.

8. The Comparison of the Developed Model with Other Models

In this section, the predictive capabilities of the proposed model are compared with those of the MVR and ANN models. The ANN model was built with the selected variables that were found through the MVR model and trained with the Levenberg-Marquardt algorithm. The neural network model used is composed of five input, eight hidden, and one output neurons (N^{5-8-1}). The results of different models used for testing data are presented in Table 2.

Table 2. The comparison of different models.

Case	Actual	MVR	ANN	Proposed Model
1	680,000	120,755.6	815,732.3	716,660.3
2	1,263,000	120,2621	1,299,821	1,115,828
3	1,850,000	2,338,051	2,145,405	2,293,947
4	4,400,000	3,903,419	5,124,051	3,802,985
5	3,200,000	3,528,523	3,719,688	3,470,780
6	8,500,000	7,865,014	9,298,305	8,150,003
7	3,100,000	3,431,278	4,803,671	3,378,503
8	17,900,000	16,661,424	18,551,802	17,801,455

The performance measures of the proposed and other models for testing a data set are presented in Table 3. It can be seen that the NMSE value for the proposed model is 0.0035, which is smaller than those obtained by using MVR and ANN, making 0.0059 and 0.0076, respectively.

Table 3. Performance measures of the proposed model and other models for testing data.

Model	R^2	NMSE	MAPE
Proposed model	0.9965	0.0035	9.59%
ANN	0.9924	0.0076	17.44%
MVR	0.9941	0.0059	20%

The MAPE value for the proposed model is 9.59%, which is also dramatically smaller than those obtained by ANN and MVR, and making 17.44% and 20%, respectively. The R^2 value for the proposed model is 0.9965, which is bigger than those that are yielded by MVR and ANN and making 0.9941 and 0.9924, respectively. The experimental results presented in Table 2 show that the hybrid models are more accurate. This conclusion can be derived because the hybrid models integrate linear and nonlinear information for predicting, while the individual model uses only linear or nonlinear information for modeling.

The comparison of the actual values and the values that are predicted by using ANN, MVR, and the hybrid models is presented in Figure 5. It can be seen from the graphs that the estimates that are yielded by the ANN, MVR, and the hybrid models closely follow the actual values. It can also be seen that the predicting ability of ANFIS outperforms that of other models.

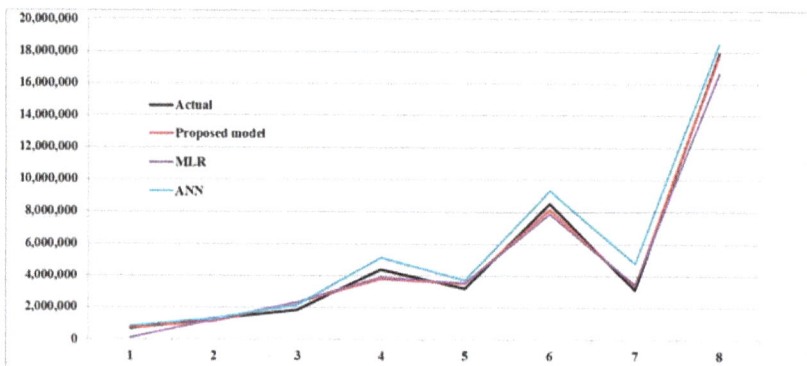

Figure 5. The comparison of cost estimation models.

The obtained result is also in good agreement with the previous studies [8], which state that, in most cases, the accuracy of the estimation by hybrid procedures would be better than that for pure statistical methods due to the essential property of costs, which are cumulative. To facilitate the cost estimation, it is also possible to apply convex optimization algorithms [61,62] and to compare the results in future works.

9. Sensitivity Analysis

Sensitivity analysis is a useful tool for determining the relationship between the considered parameters [63]. The most sensitive factors affecting the shovel capital cost is analysed by the cosine amplitude method (CAM). This method is a useful tool for performing sensitivity analysis.

Based on the concepts of the CAM method, the sensitivity for each independent component can be determined through establishing the degree of the relationship (r_{ij}) between the shovel capital cost and the considered independent component [56]. The larger the value of CAM, the higher its impact on the capital cost. If the shovel capital cost is not related to the independent variable, then, the CAM value is zero. The independent variable plays a positive role the shovel capital cost where the CAM value is non-negative and plays a negative role in the shovel capital cost where the CAM value is non-positive.

Let n be the number of independent variables represented as an array $X = \{x_1, x_2, \ldots, x_n\}$, while each of its elements, x_i, in the data array X, is itself a vector of length m, and can be expressed as:

$$X_i = \{x_{i1}, x_{i2}, x_{i3}, \ldots, x_{im}\}, \tag{10}$$

Thus, each of the data pairs can be viewed as a point in m dimensional space, where each point requires m coordinates for a complete description [64]. Each element of the relation, r_{ij}, results from a pairwise comparison of two data samples. The strength of the relationship between the data samples, x_i and x_j, is given by the membership value, expressing this strength:

$$r_{ij} = \frac{\sum_{k=1}^{m} x_{ik} x_{jk}}{\sqrt{\sum_{k=1}^{m} X_{ik}^2 \sum_{k=1}^{m} X_{jk}^2}}, \ 0 \le r_{ij} \le 1. \tag{11}$$

The strength of the relations (r_{ij} values) between the shovel capital cost and input parameters is shown in Figure 6. As shown in Figure 6, the most effective parameters of the capital cost of hydraulic and cable shovels are horsepower, bucket capacity, and weight, respectively.

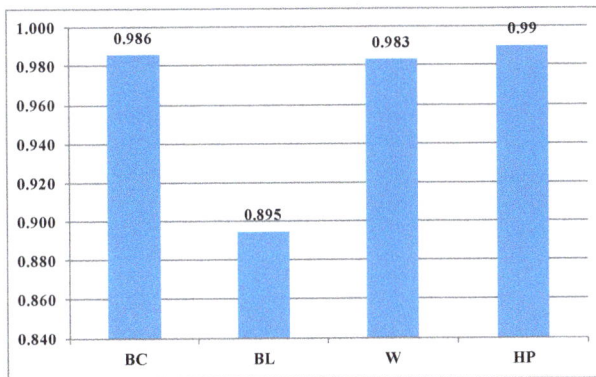

Figure 6. Sensitivity analysis of input parameters.

10. Conclusions

Despite the fact that there are many different prediction models, the improvement of prediction accuracy is still an acute problem that is facing decision makers in many areas. Multivariable regression (MVR) models are among the most popular linear models in predicting. Although various techniques have been widely used with the aim of constructing more accurate models, they cannot recognise nonlinear patterns in the existing data. On the other hand, artificial neural networks (ANNs) are well-known as the useful tools for pattern recognition, clustering, and, mainly, prediction with a high degree of accuracy, but it is hardly reasonable to use the ANNs blindly to model linear problems. The hybrid methods, which decompose a problem into its linear and nonlinear constituent parts, refer to the most efficient models.

In this paper, the hybridisation of the MVR and ANN models is proposed to overcome their limitations mentioned above and to yield a more accurate predictive model that is generated by individual methods. In the proposed model, the unique capability of the MVR model has been utilised in linear modelling to recognise the existing linear structure in data, and, then, an ANN is applied to model the nonlinear forms, using the preprocessed data. The results obtained demonstrate that the proposed model is superior to the individual models regarding three indices, and can yield more accurate data. Moreover, MVR provides the excellent initial approximation of the error due to nonlinear part as the initial data of the ANN. Therefore, this fact enables the reduction computation time of the ANN.

It should be noted that the proposed methodology could be only employed for complex systems. Therefore, it is appropriate for a dataset with at least one nonlinear pattern that is involved in information.

Author Contributions: The individual contribution and responsibilities of the authors were as follows: Abdolreza Yazdani-Chamzini and Edmundas Kazimieras Zavadskas designed the research, Abdolreza Yazdani-Chamzini collected and analyzed the data and the obtained results, performed the development of the paper. Edmundas Kazimieras Zavadskas, Jurgita Antucheviciene and Romualdas Bausys provided extensive advice throughout the study, regarding the research design, methodology, findings and revised the manuscript. All the authors have read and approved the final manuscript.

Conflicts of Interest: The authors declare no conflict of interest.

References

1. Devi, L.P.; Palaniappan, S. A study on energy use for excavation and transport of soil during building construction. *J. Clean. Prod.* **2017**, *164*, 543–556. [CrossRef]
2. Atkinson, T. Selection and sizing of excavating equipment. In *SME Mining Engineering Handbook*; Society for Mining, Metallurgy, and Exploration: Littleton, CO, USA, 1992; pp. 1311–1333.
3. Tatiya, R.R. *Surface and Underground Excavations—Methods, Techniques and Equipment*; Taylor & Francis Group plc: London, UK, 2005.
4. Huang, J.; Dong, Z.; Quan, L.; Jin, Z.; Lan, Y.; Wang, Y. Development of a dual displacement controlled circuit for hydraulic shovel swing motion. *Autom. Constr.* **2015**, *57*, 166–174. [CrossRef]
5. Elevli, S.; Elevli, B. Performance measurement of mining equipments by utilizing OEE. *Acta Montan. Slovaca* **2010**, *15*, 95–101.
6. Lashgari, A.; Yazdani-Chamzini, A.; Fouladgar, M.M.; Zavadskas, E.K.; Shafiee, S.; Abbate, N. Equipment selection using fuzzy multi criteria decision making model: Key study of Gole Gohar iron mine. *Inz. Econ.* **2012**, *23*, 125–136. [CrossRef]
7. Sayadi, A.R.; Lashgari, A.; Fouladgar, M.M.; Skibniewski, M. Estimating capital and operational costs of backhoe shovels. *J. Civ. Eng. Manag.* **2012**, *18*, 378–385. [CrossRef]
8. Mutmansky, J.M.; Suboleski, S.C.; O'Hara, T.A.; Prasad, K.V.K. Cost comparisons. In *SME Mining Engineering Handbook*, 2nd ed.; Hartman, H.L., Ed.; Society for Mining, Metallurgy, and Exploration: Littleton, CO, USA, 1992; pp. 2070–2089.
9. Anon. *Estimating Preproduction and Operating Costs of Small Underground Deposits (CANMET)*; Canadian Government Publishing Centre: Ottawa, ON, Canada, 1986.

10. Mular, A.L.; Poulin, R. *CAPCOSTS: A Handbook for Estimating Mining and Mineral Processing Equipment Costs and Capital Expenditures and Aiding Mineral Project Evaluations*; CIM Bulletin Special, 47; Canadian Institute of Mining and Metallurgy: Montréal, QC, Canada, 1998.

11. Mular, A.L. *Mining and Mineral Processing Equipment Costs and Preliminary Capital Cost Estimation*; CIM Bulletin Special, 25; Canadian Institute of Mining and Metallurgy: Montreal, QC, Canada, 1982.

12. Lanz, T.; Noakes, M. *Cost Estimation Handbook for the Australian Mining Industry*; Australasian Institute of Mining and Metallurgy (Aus IMM): Carlton South, Australia, 1993.

13. USBM. *US Bureau of Mines Cost Estimating System Handbook, Mining and Beneficiation of Metallic and Nonmetallic Minerals Expected Fossil Fuels in the United States and Canada*; United States Bureau of Mines: Denver, CO, USA, 1987; pp. 10–87.

14. Mohutsiwa, M.; Musingwini, C. Parametric estimation of capital costs for establishing a coal mine: South Africa case study. *J. South. Afr. Inst. Min. Metall.* **2015**, *115*, 789–797. [CrossRef]

15. Sayadi, A.R.; Lashgari, A.; Paraszczak, J. Hard-rock LHD cost estimation using single and multiple regressions based on principal component analysis. *Tunn. Undergr. Space Technol.* **2012**, *27*, 133–141. [CrossRef]

16. O'Hara, T.A. Quick guide to the evaluation of ore bodies. *CIM Bull.* **1998**, *88*, 34–43.

17. O'Hara, T.A. *Mine Evaluation, Mineral Industry Costs*; Hoskins, J.R., Ed.; Northwest Mining Association: Spokane, WA, USA, 1982; pp. 89–99.

18. O'Hara, T.A. *Quick Guide to Mine Operating Costs and Revenue*; Paper No. 186; CIM Annual General Meeting: Toronto, ON, Canada, 1987.

19. O'Hara, T.A.; Suboleski, C.S. Costs and Cost Estimation. In *SME Mining Engineering Handbook*; Society for Mining, Metallurgy, and Exploration: Littleton, CO, USA, 1992; Volume 1, pp. 405–424.

20. Camm, T.W. Simplified cost models for pre-feasibility mineral evaluations. *Min. Eng.* **1994**, *46*, 559–562.

21. Rudenno, V. *The Mining Valuation Handbook: Australian Mining and Energy Valuation for Investors and Management*; Australian Print Group: Victoria, Australia, 1998.

22. Sayadi, A.R.; Khademi, J.; Rahimi, M.A. Estimating the energy costs of mine equipment using an information system. In Proceedings of the 24th International Mining Congress and Exhibition of Turkey, Antalya, Turkey, 14–17 April 2015; pp. 9–18.

23. Wei, Y.; Qiu, J.; Shi, P.; Chadli, M. Fixed-order piecewise-affine output feedback controller for fuzzy-affine-model-based nonlinear systems with time-varying delay. *IEEE Trans. Circuits Syst. I Regul. Pap.* **2017**, *64*, 945–958. [CrossRef]

24. Wei, Y.; Qiu, J.; Karimi, H.R. Fuzzy-affine-model-based memory filter design of nonlinear systems with time-varying delay. *IEEE Trans. Fuzzy Syst.* **2017**, *PP*, 1. [CrossRef]

25. Singh, V.; Gupta, I.; Gupta, H.O. ANN-based estimator for distillation using Levenberg-Marquardt approach. *Eng. Appl. Artif. Intell.* **2007**, *20*, 249–259. [CrossRef]

26. Yamamura, S. Clinical application of artificial neural network (ANN) modeling to predict pharmacokinetic parameters of severely ill patients. *Adv. Drug Deliv. Rev.* **2003**, *55*, 1233–1251. [CrossRef]

27. Lian-Guo, W.; Zhi-Kang, Z.; Yin-Long, L.; Hong-Bo, Y.; Sheng-Qiang, Y.; Jian, S.; Jin-Yao, Z. Combined ANN prediction model for failure depth of coal seam floors. *Min. Sci. Technol.* **2009**, *19*, 684–688.

28. Fast, M.; Assadi, M.; De, S. Development and multi-utility of an ANN model for an industrial gas turbine. *Appl. Energy* **2009**, *86*, 9–17. [CrossRef]

29. Lee, J.; Sanmugarasa, K.; Blumenstein, M.; Loo, Y.C. Improving the reliability of a Bridge Management System (BMS) using an ANN-based Backward Prediction Model (BPM). *Autom. Constr.* **2008**, *17*, 758–772. [CrossRef]

30. Jalali-Heravi, M.; Asadollahi-Baboli, M.; Mani-Varnosfaderani, A. Shuffling multivariate adaptive regression splines and adaptive neuro-fuzzy inference system as tools for QSAR study of SARS inhibitors. *J. Pharm. Biomed. Anal.* **2009**, *50*, 853–860. [CrossRef] [PubMed]

31. Verlinden, B.; Duflou, J.R.; Collin, P.; Cattrysse, D. Cost estimation for sheet metal parts using multiple regression and artificial neural networks: A case study. *Int. J. Prod. Econ.* **2008**, *111*, 484–492. [CrossRef]

32. Mesroghli, S.; Jorjani, E.; Chelgani, S.C. Estimation of gross calorific value based on coal analysis using regression and artificial neural networks. *Int. J. Coal Geol.* **2009**, *79*, 49–54. [CrossRef]

33. Sahoo, G.B.; Schladow, S.G.; Reuter, J.E. Forecasting stream water temperature using regression analysis, artificial neural network, and chaotic non-linear dynamic models. *J. Hydrol.* **2009**, *378*, 325–342. [CrossRef]

34. Gu, Y.; Wu, L.; Tang, S. A fuzzy neutral-network-driven weighting system for electric shovel. In *Advances in Neural Networks—ISNN 2008*; Sun, F., Zhang, J., Tan, Y., Cao, J., Yu, W., Eds.; Lecture Notes in Computer Science; Springer: Berlin/Heidelberg, Germany, 2008; Volume 5264, pp. 526–532.

35. Guo, J.; Zhao, X.; Guo, J.; Yuan, X.; Dong, S.; Xiong, Z. Model updating of suspended-dome using artificial neural networks. *Adv. Struct. Eng.* **2017**, *20*, 1727–1743. [CrossRef]

36. Du, Y.; Chen, Z.; Zhang, C.; Cao, X. Research on axial bearing capacity of rectangular concrete-filled steel tubular columns based on artificial neural networks. *Front. Comput. Sci.* **2017**, *11*, 863–873. [CrossRef]

37. Gardner, B.J.; Gransberg, D.D.; Rueda, J.A. Stochastic conceptual cost estimating of highway projects to communicate uncertainty using bootstrap sampling. *ASCE-ASME J. Risk Uncertain. Eng. Part A Civ. Eng.* **2017**, *3*. [CrossRef]

38. El-Gohary, K.M.; Aziz, R.F.; Abdel-Khalek, H.A. Engineering approach using ANN to improve and predict construction labor productivity under different influences. *J. Constr. Eng. Manag.* **2017**, *143*. [CrossRef]

39. Cao, M.S.; Pan, L.X.; Gao, Y.F.; Novak, D.; Ding, Z.C.; Lehky, D.; Li, X.L. Neural network ensemble-based parameter sensitivity analysis in civil engineering systems. *Neural Comput. Appl.* **2017**, *28*, 1583–1590. [CrossRef]

40. Abdeljaber, O.; Avci, O.; Kiranyaz, S.; Gabbouj, M.; Inman, D.J. Real-time vibration-based structural damage detection using one-dimensional convolutional neural networks. *J. Sound Vib.* **2017**, *388*, 154–170. [CrossRef]

41. Benali, A.; Boukhatem, B.; Hussien, M.N.; Nechnech, A.; Karray, M. Prediction of axial capacity of piles driven in non-cohesive soils based on neural networks approach. *J. Civ. Eng. Manag.* **2017**, *23*, 393–408. [CrossRef]

42. Yousefi, V.; Yakhchali, S.H.; Khanzadi, M.; Mehrabanfar, E.; Šaparauskas, J. Proposing a neural network model to predict time and cost claims in construction projects. *J. Civ. Eng. Manag.* **2016**, *22*, 967–978. [CrossRef]

43. Ustinovichius, L.; Peckienė, A.; Popov, V. A model for spatial planning of site and building using BIM methodology. *J. Civ. Eng. Manag.* **2017**, *23*, 173–182. [CrossRef]

44. Wang, W.C.; Bilozerov, T.; Dzeng, R.J.; Hsiao, F.Y.; Wang, K.C. Conceptual cost estimations using neuro-fuzzy and multi-factor evaluation methods for building projects. *J. Civ. Eng. Manag.* **2017**, *23*, 1–14. [CrossRef]

45. Monjezi, M.; Rezaei, M.; Varjani, A.Y. Prediction of rock fragmentation due to blasting in Gol-E-Gohar iron mine using fuzzy logic. *Int. J. Rock Mech. Min. Sci.* **2009**, *46*, 1273–1280. [CrossRef]

46. Bourquin, J.; Schmidli, H.; Hoogevest, P.; Leuenberger, H. Advantages of Artificial Neural Networks (ANNs) as alternative modelling technique for data sets showing non-linear relationships using data from a galenical study on a solid dosage form. *Eur. J. Pharm. Sci.* **1998**, *7*, 5–16. [CrossRef]

47. Malinowski, P.; Ziembicki, P. Analysis of district heating network monitoring by neural networks classification. *J. Civ. Eng. Manag.* **2006**, *12*, 21–28.

48. He, X.; Xu, S. *Process Neural Networks Theory and Applications*; Springer: Berlin/Heidelberg, Germany, 2007.

49. Engelbrecht, A.P. *Computational Intelligence: An Introduction*; John Wiley & Sons, Ltd.: Hoboken, NJ, USA, 2002.

50. Sumathi, S.; Paneerselvam, S. *Computational Intelligence Paradigms: Theory & Applications Using MATLAB*; Taylor and Francis Group, LLC: Oxford, UK, 2010.

51. Khashei, M.; Bijari, M. A novel hybridization of artificial neural networks and ARIMA models for time series forecasting. *Appl. Soft Comput.* **2011**, *11*, 2664–2675. [CrossRef]

52. Zhang, G.P. Time series forecasting using a hybrid ARIMA and neural network model. *Neurocomputing* **2003**, *50*, 159–175. [CrossRef]

53. Farid, M.; Hossein Abadi, M.M.; Yazdani-Chamzini, A.; Yakhchali, S.H.; Basiri, M.H. Developing a new model based on neuro-fuzzy system for predicting roof fall in coal mines. *Neural Comput. Appl.* **2013**, *23*, 129–137. [CrossRef]

54. Nash, J.E.; Sutcliffe, J.V. River flow forecasting through conceptual models, I. A discussion of principles. *J. Hydrol.* **1970**, *10*, 282–290. [CrossRef]

55. Razani, M.; Yazdani-Chamzini, A.; Yakhchali, S.H. A novel fuzzy inference system for predicting roof fall rate in underground coal mines. *Saf. Sci.* **2013**, *55*, 26–33. [CrossRef]

56. Yazdani-Chamzini, A.; Yakhchali, S.H.; Volungevičienė, D.; Zavadskas, E.K. Forecasting gold price changes by using adaptive network fuzzy inference system. *J. Bus. Econ. Manag.* **2012**, *13*, 994–1010. [CrossRef]

57. Info Mine: Mining Cost Service Indexes. Available online: http://www.infomine.com (accessed on 26 June 2015).

58. Wang, W.; Van Gelder, P.H.A.J.M.; Vrijling, J.K.; Ma, J. Forecasting daily streamflow using hybrid ANN models. *J. Hydrol.* **2006**, *324*, 383–399. [CrossRef]
59. Srinivas, Y.; Raj, A.S.; Oliver, D.H.; Muthuraj, D.; Chandrasekar, N. A robust behavior of Feed forward back propagation algorithm of Artificial Neural Networks in the application of vertical electrical sounding data inversion. *Geosci. Front.* **2012**, *3*, 729–736. [CrossRef]
60. Basheer, I.A.; Hajmeer, M. Artificial neural networks: Fundamentals, computing, design, and application. *J. Microbiol. Methods* **2000**, *43*, 3–31. [CrossRef]
61. Wei, Y.; Qiu, J.; Karimi, H.R.; Ji, W. A novel memory filtering design for semi-Markovian jump time-delay systems. *IEEE Trans. Syst. Man Cybern.* **2017**, *PP*, 1–13. [CrossRef]
62. Qiu, J.; Wei, Y.; Karimi, H.R.; Gao, H. Reliable Control of Discrete-Time Piecewise-Affine Time-Delay Systems via Output Feedback. *IEEE Trans. Reliab.* **2017**, *57*, 1491–1550. [CrossRef]
63. Ross, T.J. *Fuzzy Logic with Engineering Applications*, 3rd ed.; John Wiley & Sons Ltd.: Hoboken, NJ, USA, 2010.
64. Yazdani-Chamzini, A.; Razani, M.; Yakhchali, S.H.; Zavadskas, E.K.; Turskis, Z. Developing a fuzzy model based on subtractive clustering for road header performance prediction. *Autom. Constr.* **2013**, *35*, 111–120. [CrossRef]

![symmetry logo] *symmetry*

MDPI

Article

Tool-Wear Analysis Using Image Processing of the Tool Flank

Ovidiu Gheorghe Moldovan [1,†,‡], Simona Dzitac [2,3,‡], Ioan Moga [1,‡], Tiberiu Vesselenyi [1,‡] and Ioan Dzitac [3,4,*,‡]

1   Department of Mechatronics, University of Oradea, 410087 Oradea, Romania;
     omoldovan@uoradea.ro (O.G.M.); imoga@uoradea.ro (I.M.); tvesselenyi@yahoo.co.uk (T.V.)
2   Department of Energy Engineering, University of Oradea, 410087 Oradea, Romania; simona@dzitac.ro
3   Department of Social Sciences, Agora University, 410526 Oradea, Romania
4   Department of Mathematics—Computer Science, Aurel Vlaicu University of Arad, 310025 Arad, Romania
*   Correspondence: ioan.dzitac@ieee.org; Tel.: +40-359-101-032
†   Current address: Department of Mathematics—Computer Science, Aurel Vlaicu University of Arad,
     310025 Arad, Romania.
‡   These authors contributed equally to this work.

Received: 5 November 2017; Accepted: 28 November 2017; Published: 30 November 2017

Abstract: Flexibility of manufacturing systems is an essential factor in maintaining the competitiveness of industrial production. Flexibility can be defined in several ways and according to several factors, but in order to obtain adequate results in implementing a flexible manufacturing system able to compete on the market, a high level of autonomy (free of human intervention) of the manufacturing system must be achieved. There are many factors that can disturb the production process and reduce the autonomy of the system, because of the need of human intervention to overcome these disturbances. One of these factors is tool wear. The aim of this paper is to present an experimental study on the possibility to determine the state of tool wear in a flexible manufacturing cell environment, using image acquisition and processing methods.

Keywords: image processing; flexible manufacturing; tool-flank-wear monitoring; artificial neural networks

---

1. Introduction

The assessment of tool wear is of major importance in a manufacturing system that aims for higher automation and flexibility. The automatic tool readjustment (ATR) function implemented in flexible manufacturing systems (FMSs) prepares a new set of tools in the storage unit of the automatic tool changer (ATC) of the machine. The basic implementation of the ATR function in a FMS is based on a tool list. Each individual machine in the FMS transfers the tool list of its ATC magazine to a central control system; the system also has a list of workpieces to be manufactured that includes a tool list needed for each workpiece to be manufactured. The goal of the ATR is to transfer the required tools to the ATC in "hidden time", meaning that the machine is still working while the tools for the new task are transferred, so the machine will have all the tools needed for each workpiece. Although the implementation of the ATR function significantly decreases the down time of the machines as a result of tool replacement in the ATC magazine, it has no effect in case of tool-life management (TLM). In order to manage the tool life with the ATR function, the system must monitor the tool wear and, on the basis of the standard life-time of the tools, include the tool determined to be close to its usage life in the ATR list of tools that need replacement. This system can improve the autonomy of the manufacturing system and the quality of the pieces. This kind of TLM, although bringing a significant improvement, has its own disadvantages. The efficiency of the system strongly depends on the precision of the tool-life calculation. Depending on the true situation of cutting conditions (tool-material quality and its accordance with the standard, exact composition and homogeneity of the part material, and efficiency of the cooling system), a tool can present early signs of wear, causing a decrease in the quality of manufactured workpieces (the true life-time is shorter than the standard), or a tool can be removed on the basis of a standard life-time even if the quality of the cutting process is satisfactory and the tool is performing with acceptable parameters (the true life-time is longer than the standard). There are considerably many studies regarding different approaches to tool management. A tool-management approach enabling an autonomous cooperation of tools and machine tools within a batch production system is presented in [1]. Considering the type of parameter of the cutting process that is monitored, these range from surface roughness and cutting force to vibration and chip shape. In [2], the authors analyzed whether cutting parameters (feed rate, and spindle speed) have an effect on tool wear and surface roughness. Surface roughness of processed parts is also studied as a parameter that can predict the state of the tool wear. Prediction of the surface roughness of a workpiece by using adaptive neuro-fuzzy inference system (ANFIS) modeling for the monitoring of unmanned production systems with tool-life management is presented in [3]. In other studies, machined surface images are analyzed on the basis of a support vector machine using as input the features extracted from the gray-level co-occurrence matrix [4], and in-process surface-roughness monitoring system for an end-milling operation is analyzed using neural-fuzzy methods in [5]. Another parameter monitored in order to identify tool wear is the cutting force. In [6], force-based tool-condition monitoring for a turning process using support vector regression analysis is used to establish the flank wear of the cutting tool. Cutting force signals are also used in [7] to estimate the tool wear and the surface quality, and in [8], a partial least-square regression method is presented to make the tool-wear prediction also on the basis of the force signal. Related to cutting-force measurements are measurements of the current amplitude of the main drive of the machine tool, presented in [9,10], where it is shown that this parameter can also be linked to tool-wear development. Vibrations and machine tool dynamics are studied in order to find their relation with tool wear [11]. Some studies combine signals from different sensors, such as, for example, in [12], where tool-wear prediction in milling is analyzed using the simultaneous detection of acceleration and spindle drive current. Additionally, chip morphologies are analyzed in order to evaluate tool-flank wear and its effects on surface roughness [13]. More direct methods are focused on measuring spatial tool wear using a three-dimensional (3D) laser profile sensor [14]. Regarding processing and analysis techniques used to identify tool wear, there are also a large number of methods employed: machine learning and computer vision techniques [15], wavelet extreme learning machine models [16] support vector regression [17], analytical mathematical models [18], empirical models [19] and co-evolutionary particle swarm optimization-based selective neural network ensembles [20]. Tool wear is also studied experimentally; researchers aim to develop different models to predict tool wear from experimental data [21–23]. In this paper, we analyzed a system on the basis of image acquisition and processing, which can provide useful information regarding the cutting tool usage on the basis of the cutting edge wear. Our system is intended to be used in conjunction with methods to detect tool breakage, such as, for example, measuring the main drive current or torque, which are already built into the computer numerical control (CNC) of modern machine tools. Regarding the image processing methods, we describe one artificial neural network (ANN) applied to classify the image features obtained by processing the images with classical image processing methods (filtering, edge detection and morphological operations) and two ANNs applied directly on the images without their preprocessing (one single-hidden-layer ANN and one two-hidden-layer autoencoder). In order to compare the results obtained for each ANN, we describe in detail the training and testing of the ANNs. We also studied the Training Success Rate (TSR) for these ANNs for a large range of nodes in each layer. The integration of such a system with the ATR function of a FMS could increase the autonomy of the system and the quality of manufactured parts, as well as ensure a rational tool usage.

2. Tool-Flank-Wear Monitoring System

This section may be divided by subheadings. It should provide a concise and precise description of the experimental results and their interpretation, as well as the experimental conclusions that can be drawn.

This area is relatively protected from chips and cooling liquid used during the processing, which could substantially impede the acquisition of the images. Thus, after every instance that the tool has been used and placed in the magazine, a tool-flank image can be acquired and processed. In this approach, it is important that one of the tool flanks is in the area covered by the camera. We assume that all the teeth of the tool are more or less equally affected by wear, so that it is enough to acquire the image of the flank of only one tooth. One of the conditions for image acquisition is to have one tooth of the tool oriented toward the camera, so that the tool has to be positioned in the tool holder accordingly. In order to acquire more accurate images, the camera is placed on an positioning device, which moves the camera in the appropriate position for each tool (Figure 1). The coordinates for each tool are stored in the tool database on the FMS controller. This controller also transmits the coordinates to the camera positioning controller. The acquisitioned images are processed on a separate computer (tool wear identification in Figure 1), which decides whether the tool is worn or not. If the system decides that the tool-flank wear exceeds the acceptable limit, the information is transmitted to the FMS, which in turn replaces the tool in the magazine or updates the information in the tool list so that a replacement tool is used from that time on, if such a replacement tool exists in the magazine. Flank wear of tools can be detected as a result of changes in the angle of the worn surface relative to the unworn flank surface. The worn surface will have a different orientation, causing the light to be reflected at a different angle and then observed on the acquired image (Figure 2).

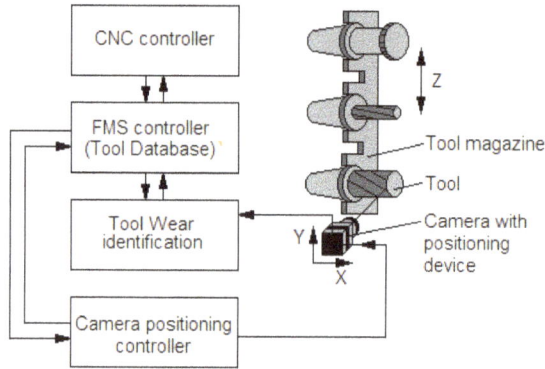

Figure 1. Tool-wear monitoring system.

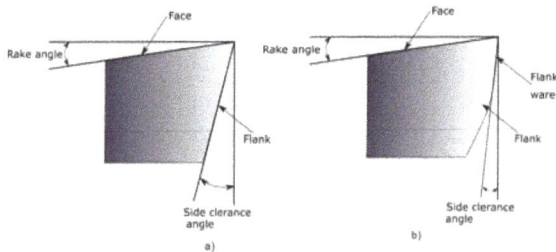

Figure 2. Tool-wear monitoring system. (**a**) unworn flank; (**b**) worn flank.

In practice, these surfaces are not ideal planes but are packed with irregularities. The new tool flank has marks as a consequence of sharpening, and the worn surface is much more irregular because of particle displacement by friction with the surface of the workpiece and chips. Thus, in the resulting images, the two surfaces (worn and unworn) are not simple to separate.

3. Experimental Setup for Tool-Flank-Wear Detection

The experimental system was developed with the goal to obtain consecutive images of the tool flank, which then can be used to test and analyze different image processing methods. Extensive study of flank wear for different tools, tool and workpiece materials or different processing parameters is beyond the scope of our study.

The system presented in Figure 3 is composed of a computer system (1), National Instruments CCD camera (2), optical microscope with lighting system (3), cutting tool (4), tool holder (5), and National Instruments camera source (6).

Figure 3. Image acquisition system for the automatic determination of tool wear.

For the experimental tests, a high-speed steel end mill HS18-0-1 (AISI T1) was used, with two helical teeth and with a diameter of 14 mm (the tool is presented in Figure 4).

Figure 4. Tool used for the experimental test.

The workpiece was made of C45 carbon steel (AISI 1045) and is shown in Figure 5.

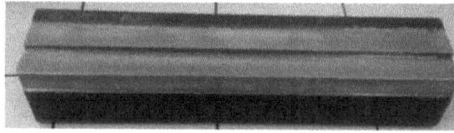

Figure 5. Workpiece used for the experimental test.

Using the experimental setup, images of the cutting tool were acquired. During the experiments, a total of eight complete steps of the cutting tool were made (the tool cut a total of eight entire lengths of the workpiece). Images of the tool flank were acquired after cutting every 200 mm in the workpiece. The processing parameters were set to a feed rate of 50 mm/min, a spindle speed of 500 rpm, a width of cut of 6 mm (representing 43% of the tool diameter) and a depth of cut of 2 mm. During the experiments, 21 images of the tool flank were acquired showing consecutive stages of flank-wear development. A larger number of images obtained for successive stages of wear would be generally similar for larger numbers of tools of the same type, dimension and material and under the same cutting conditions, having little effect on the results of an image processing algorithm. The reason for this is that the number and complexity of the features contained in the images of a tool flank is fairly low. The images have been analyzed and it was concluded that the wear begins to be visible at the 14th image and progresses forward from this, as can be seen in Figure 6. After the images had been acquired, three different algorithms were tested in order to find out which of these, if presented with an unlabeled image, could decide whether the tool had reached an unacceptable degree of wear.

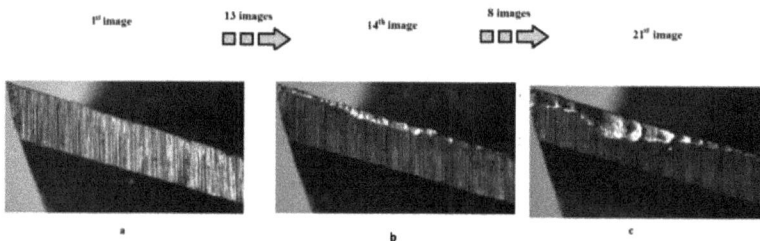

Figure 6. Images taken at successive times during processing showing the stages of tool wear: (**a**) new tool; (**b**) first stage when the wear became visible; and (**c**) last stage with massive wear.

4. Image Processing for Tool-Flank-Wear Detection

4.1. Image Classification Using One-Hidden-Layer ANN on Features Extracted from Image Data

On the basis of a step-by-step analysis of the acquired images, we developed an algorithm in order to extract significant features that can help to distinguish worn and unworn tools. In the following, we describe the steps of the developed algorithm. First, the filtering of the original gray image is done with a range filter, for which each output pixel contains the maximum value minus the minimum value of a 3×3 neighborhood of the filtered pixel. This is followed by transforming the gray-level image into a black and white (b/w) image using Otsu's method. This method establishes the threshold used to transform the image automatically from the image's gray-level histogram, so that no manual adjustment of the threshold is needed. The next step was to find edges in the b/w image using the Sobel edge detection method. Using this method, we obtained a new b/w image of the frontiers of the objects from the previous b/w image. Here, the "object" has to be understood as any white region separated from other similar regions, so that the word object has no physical meaning. The result of the above described steps is shown in Figure 7.

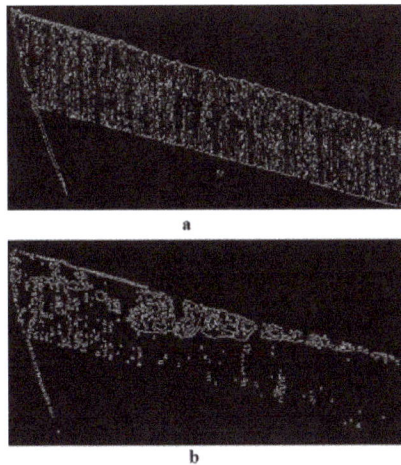

Figure 7. Binary black and white (b/w) image of (**a**) the new and (**b**) last stage of wear of the tool flank.

Starting from the observation that feature extraction can be more computationally efficient if the objects in the image are aligned to lines and columns of the internal computational representation (matrices) of the images, we computed the Hough transform of the image to find the angle of rotation of the tool edge; then we rotated the image with the angle identified by the Hough transform in order to position the edge of the tool horizontally in the image. This was followed by performing an image-opening morphological operation in order to enhance the objects in the image. The last step of the preprocessing was to apply a third-order one-dimensional median filter to the image. The result of these operations is presented in Figure 8.

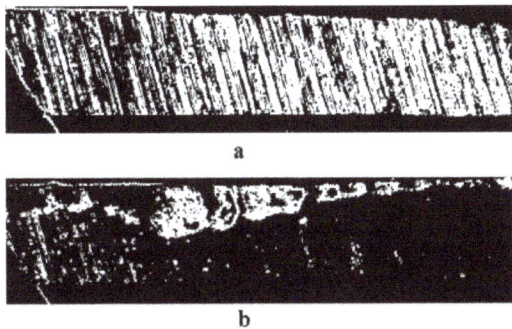

Figure 8. Enhanced and "rotated-to-horizontal" images of (**a**) the flank of the new tool and (**b**) tool with wear.

Studying the structure of the enhanced and rotated images, we tried to use different morphological parameters to extract features that could distinguish between worn and unworn tool-flank images. One of these parameters was the Euler number (EN). The EN is the total number of objects (as defined above) in the image minus the total number of holes (dark regions surrounded by white regions).

Computing the ENs for the whole set of images, we obtained the diagram in Figure 9. As can be seen in the diagram in Figure 9, the EN has a significant drop from the 14th image, which corresponds to the first image that has visible signs of wear. If we establish a threshold (in this case at about EN = 3500), we can fairly discriminate between ENs corresponding to worn and unworn tool-flank images.

Although we obtained good results in this case, in order to consider the EN as a reliable feature, a large number of experiments with different tools should be made and thresholds for the EN diagram have to be established manually (by a human agent) for each tool, which impedes the practicality of the method. Another feature had been found by observing differences between the worn and unworn flank images. We computed the normalized sum of white pixels (denoted NSP), having the value of 1, on each horizontal line of the image. The NSP on each line is computed as the sum of pixels on that line divided by the number of pixels on the line with the maximum number of pixels. Clearly, the sum of the pixels in a line will be equal to the number of white pixels in that line, given that the black pixels have a value of 0. The NSP is defined by the expressions:

$$w_i = \sum_{j=1}^{n} v_{i,j} \tag{1}$$

$$NSP_i = \frac{w_i}{max(w_i)_{i=1...m}} \tag{2}$$

where i is the current line with n as the number of lines in the image; j is the current column with n as the number of columns in the image; $w_{i,j}$ is the value of the pixel intensity on line i and column j, which in this case is 0 for dark and 1 for white pixels; w_i is the sum of pixels on line i; and NSP_i is the value of the NSP on line i.

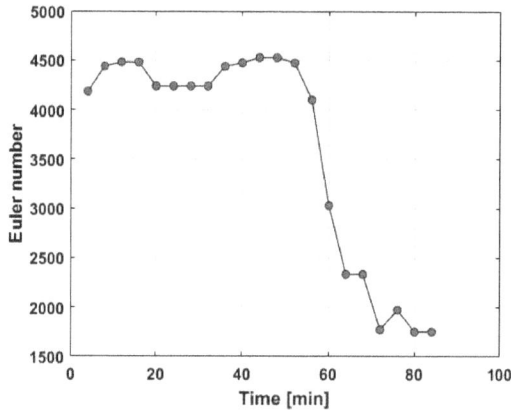

Figure 9. Euler number diagram for the experimental images set (points corresponding to worn tool flank are marked with red).

Computing the NSP for each image, we obtained the diagrams in Figures 10 and 11. It can be seen that the shape of the NSP diagram of worn and unworn tool flanks are quite different.

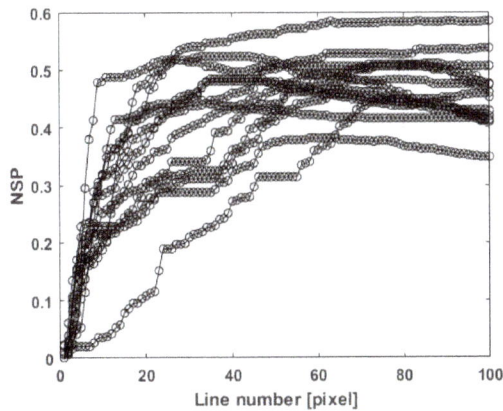

Figure 10. Normalized sum of white pixels (NSP) diagrams for unworn tool flank.

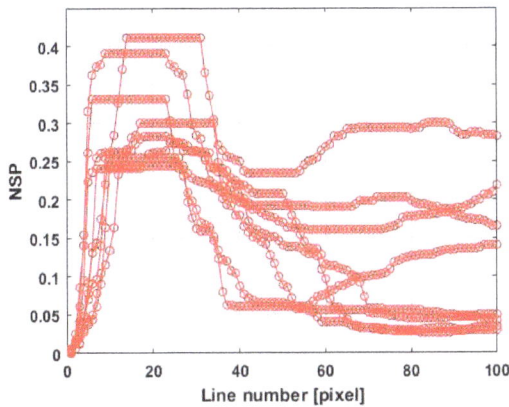

Figure 11. Normalized sum of white pixels (NSP) diagrams for worn tool flank.

Analyzing these diagrams, we can conclude that the NSP can be regarded as a fairly reliable feature to be used as a flank-wear detection parameter. From this point on, the main goal is to find a method to discriminate between the shapes (patterns) of NSP curves representing unworn flank images and those representing worn flank images. In order to do so, we analyzed two methods: one based on approximations with second-degree polynomials and the other based on ANN pattern recognition. The approximation of the NSP curves is shown in Figures 12 and 13. Differentiating the two types of curves are the locations (abscissa or line number *i*) of their maxima: point M.

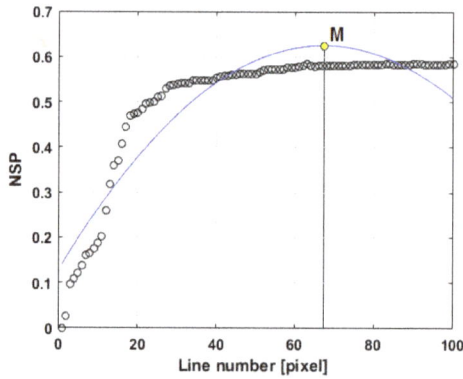

Figure 12. Normalized sum of white pixels (NSP) diagram for unworn tool flank and the second-order polynomial approximation (blue curve) with maxima at point M.

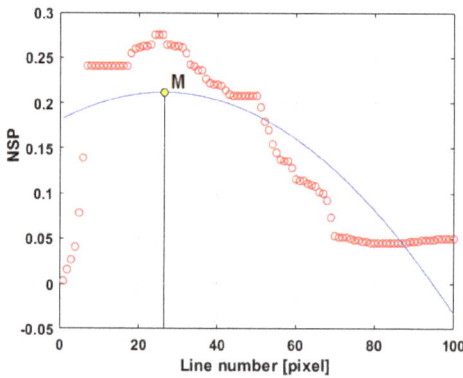

Figure 13. Normalized sum of white pixels (NSP) diagram for worn tool flank and the second-order polynomial approximation (blue curve) with maxima at point M.

Plotting the location of maxima versus the processing time until the image was acquired gives us the diagram in Figure 14.

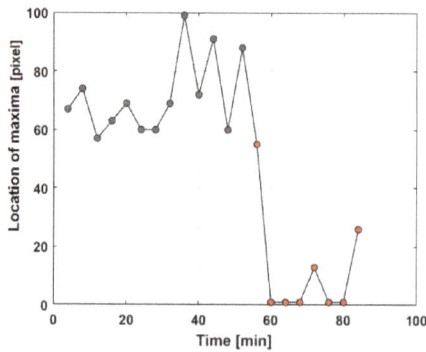

Figure 14. Normalized sum of white pixels (NSP) diagram for worn tool flank and the second-order polynomial approximation (blue curve) with maxima at point M.

As in the case of the EN parameter, the location of maxima of the second-degree polynomial approximation of the NSP curve shows a good separation of worn and unworn flank images, but the manual adjustment of the threshold for the locations of maxima still remains an issue.

Another approach to classify the NSP curves is to use ANN pattern-recognition methods. For this purpose, we used the MATLAB nprtool module of the Neural Network Toolbox. The toolbox employs a two-layer feedforward network, with a sigmoid transfer function in the hidden layer and a softmax transfer function in the output layer [24]. The learning process is based on a scaled conjugate gradient backpropagation algorithm. The input data for the ANN consisted of 21 samples of 100 data points on each NSP curve, from which 13 were extracted from unworn flank images and 8 from worn flank images. These had been further divided into two groups: 13 samples (representing 8 unworn and 5 worn flanks) had been used for training and 8 samples had been used for testing (representing 5 unworn and 3 worn flanks). The input layer consisted of 100 neurons corresponding to the 100 data points of the NSP curve (Figures 10 and 11). The number of output neurons must be equal to the number of elements in the target vector, which is the number of categories of the classification process. In our case, there were two categories: worn and unworn tool-flank images. Essentially, the target vector represented the labeling of the NSP dataset with the dimension of 2×13 for training and 2×8 for testing.

For successful training, the training performance of the network is represented in Figure 15, which shows that, in this case, a very small error had been achieved in a short run of just 23 epochs. We can see the result of a successful testing session presented in Figure 16 in the form of a confusion matrix. In this representation, each column of the matrix represents a predicted class, while each row represents a true class. The green squares represent the correctly classified and the red squares represent the incorrectly classified samples. As can be seen in the confusion matrix, each sample, for this case, was correctly classified.

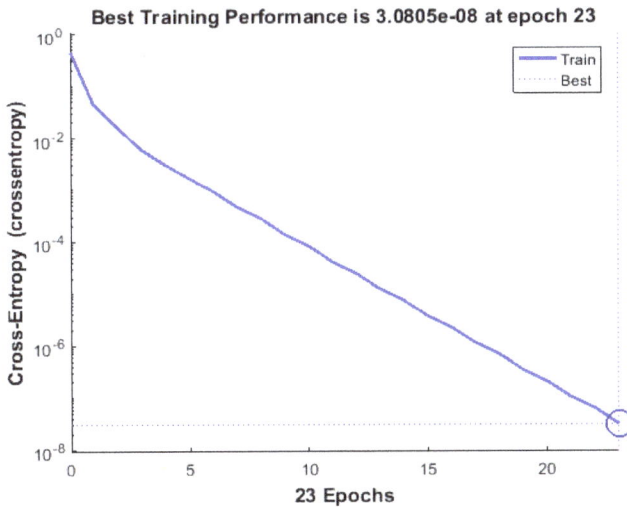

Figure 15. Training performance diagram for the artificial neural network (ANN).

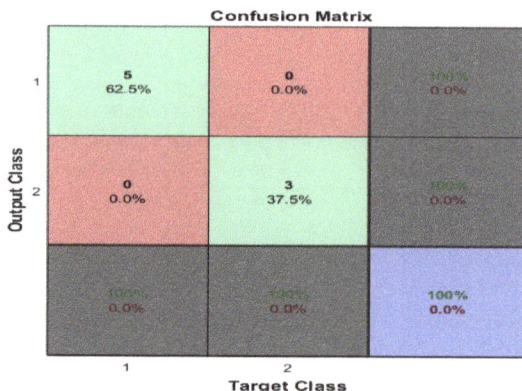

Figure 16. Confusion matrix for the test set.

Although working with ANN software—in our case using the MATLAB Neural Network Toolbox, but generally with other ANN software products as well—is quite straightforward, there are some specific issues which have to be dealt with:

1. The number of neurons in the hidden layer must be established on a somehow empirical basis. The number of neurons in the hidden layers may be important in order to extract the meaningful features of the image.
2. Every training session can produce different results because of the fact that the initial weights and biases of each neuron are set randomly. Training the network with the same number of neurons on the same input datasets can produce different results when tested with unlabeled data.
3. The number of training epochs has to be well established in order to avoid overfitting. If overfitting occurs, the network will be less successful in classifying unlabeled data.

In order to find the influence of these parameters on successfully training the networks, we ran training sessions multiple times with different number of neurons in the hidden layer. Every trained network was tested on the test sample set of five unworn and three worn flank images that were not used in the training sessions. For the network type described in this paragraph, we trained networks with a number of neurons in the hidden layer in the range of 10 to 200 with a step of 10 (10, 20, 30, …, 200 neurons). A way to reduce the influence of the randomly set initial weights is to run training sessions a large number of times for the same number of neurons. For each network with a specific neuron number, we trained the network 100 times. After each training, the network was tested, and we counted the number of times the testing was successful (successfully classified every test image from the test set). We defined the TSR as the percentage of trainings that produced successful testing results from the whole number of trainings (in this case, 100). The TSR is a good indicator of the influence of neuron numbers, as the influence of initial weights is diminished by the high number of trainings. The results are presented in the diagram in Figure 17.

As can be seen, the number of successful trainings decreases with the increase of the number of neurons in the hidden layer. This led us to the conclusion that for this type of classification, it is better to use a small number of neurons (from 10 to 60). Using a small number of neurons is also recommended because the network will use less memory, which is important if we have a large number of types of tools which have to be classified.

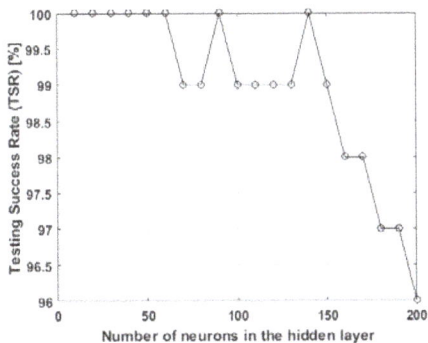

Figure 17. Training success rate (TSR) for networks with 10 to 200 neurons in the hidden layer (100 training sessions for each network; each point in the diagram).

4.2. Image Classification Using One-Hidden-Layer ANN on Image Data

In the above paragraph, we described the application of the ANN pattern-recognition method to discriminate between NSP features extracted from worn and unworn flank images. To avoid lengthy computations to extract features (e.g., EN or NSP) from the images, we tried to apply the pattern-recognition method directly to the images. The original images had a size of 640 × 480 pixels; the input for the ANN should be a vector, which would have resulted in a size of 307,200 elements. In order to reduce the number of elements for the input vector, we resized the original image to a size of 126 × 94 pixels, resulting in an input vector of 11,844 elements. The samples were divided in training and testing datasets in the same manner as described in the previous paragraph. The target vector and output layer were the same as in the previous paragraph. An example of a successful training performance for 30 neurons in the hidden layer is presented in Figure 18.

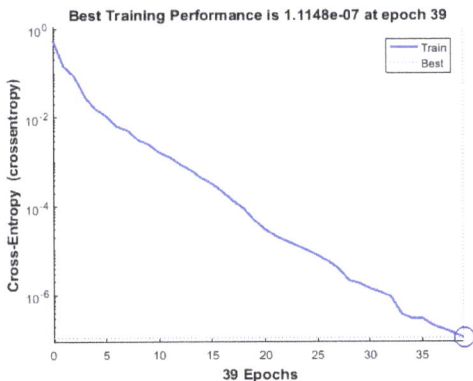

Figure 18. Performance diagram for the artificial neural network (ANN).

The confusion matrix in this case was similar to that in Figure 16. We also tested the TSR for this type of network. At first, we trained networks with a number of neurons from 10 to 1000 with 10 training sessions for each case, to see the overall trend of the TSR as a function of neuron numbers. The results are presented in the diagram in Figure 19. The trend has been established using a first-order (linear) approximation of the set of TSRs (the red line in Figure 19), which shows that the TSR decreases slightly with the increase in the number of neurons in the hidden layer.

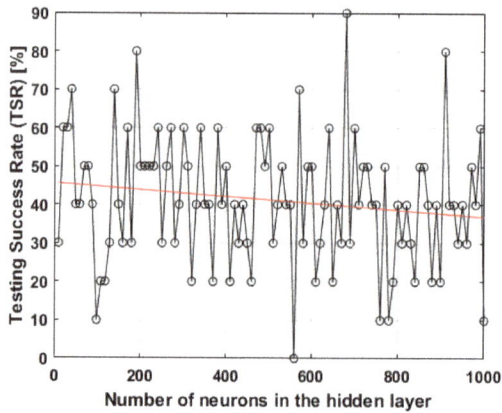

Figure 19. Training success rate (TSR) for networks with 10 to 1000 neurons in the hidden layer (10 training sessions for each network; each point in the diagram).

Secondly, we raised the number of training sessions to 100 for the range of 10 to 200 neurons. The TSR obtained is shown in the diagram in Figure 20.

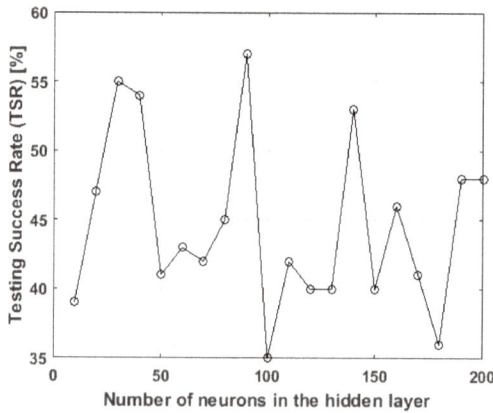

Figure 20. Training success rate (TSR) for networks with 10 to 200 neurons in the hidden layer (100 training sessions for each network; each point in the diagram).

It can be observed from Figure 20 that the TSR was much less stable in this case than the TSR referenced in the previous paragraph (Figure 17). This means that there was a smaller chance to successfully train a network directly on the image data (e.g., 55% for 30 neurons) than on NSP data (for 30 neurons is 100%). In the studied range, the best TSR was for 90 neurons, at 57%.

4.3. Image Classification with Autoencoders on Image Data

Deep learning is a new approach that has been introduced in the field of ANNs in the last decades by a number of researchers (e.g., [25,26]). There are a relatively large number of methods classified as deep learning, such as deep belief networks, restricted Boltzmann machines, deep autoencoders, and others. Deep learning algorithms, which use more than one hidden layer, have been successfully applied to image classification. In order to find out whether this type of network would be better than the above presented networks, we used the MATLAB Neural Network Toolbox's Deep Learning section, which contains a module for autoencoder networks [24]. Autoencoders use methods to separately train each layer; they then stack these together in a single network with multiple layers and train the final network as a whole. As input and target data, we used the same sets as in Section 4.2. The network we used was composed of two autoencoder layers and one softmax layer as output. An example of successful training performance for 30 neurons in the first hidden layer, 10 neurons in the second hidden layer and 2 neurons in the output layer is presented in Figure 21.

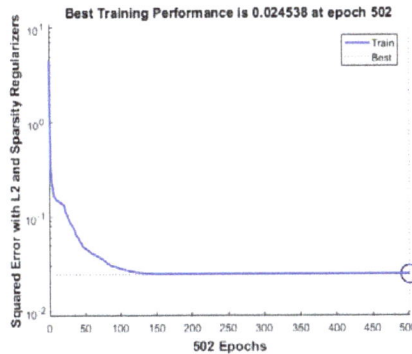

Figure 21. Performance diagram for the second layer of a two-hidden-layer autoencoder artificial neural network (ANN).

Regarding the TSR for this type of network, we trained the network for 10 to 520 neurons in the first hidden layer with a step of 10 (10, 20, 30, ..., 520) and a constant number of 10 neurons for the second layer. For each number of neurons, we trained the network 10 times, and we obtained the diagram in Figure 22.

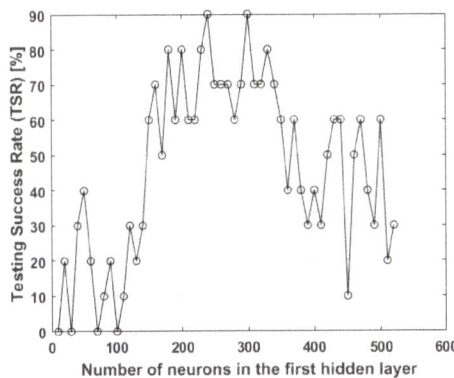

Figure 22. Training success rate (TSR) for networks with 10 to 520 neurons in the first hidden layer and 10 neurons in the second layer (10 training sessions for each network; each point in the diagram).

As we can see in the diagram in Figure 22, the TSR rises abruptly at 150 neurons in the first hidden layer (increasing by more than 60%), then falls again to a mean value of about 40%. This shows that the probability to successfully train an autoencoder network, for the types of images we studied, is higher for a range of 150 to 350 neurons in the first layer. For a second set of experiments, we chose to have a fixed number of 300 neurons in the first hidden layer, which was a maximal value of 90%

TSR (Figure 22), and we ran the trainings for a range from 10 to 200 neurons in the second layer, with a step of 10, each 10 times. The results are presented in Figure 23.

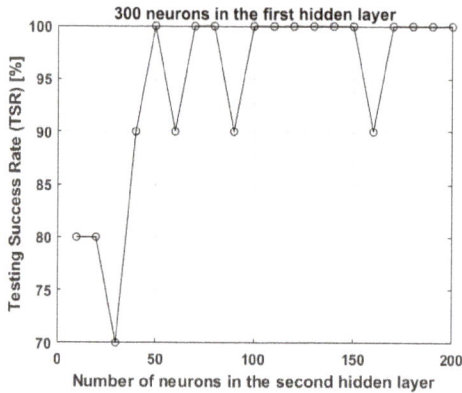

Figure 23. Training success rate (TSR) for networks with 300 neurons in the first hidden layer and 10 to 200 neurons in the second layer (10 training sessions for each network; each point in the diagram).

The results show that for 300 neurons in the first layer and a range of 100 to 150 neurons in the second layer, we could obtain a TSR of 100%.

5. Conclusions

Our goal in this study was to analyze a tool-flank-wear monitoring system according to the paradigm described in Section 2 (Figure 1). The hardware needed to develop such a system is relatively simple to implement; there are a large number of suppliers for machine vision systems and positioning devices. The tool database can be easily extended with the specific parameters for tool-wear monitoring, with the observation that for a large database, the number of parameters has to be kept as low as possible. The main questions we focused on were the following:

1. What kind of image processing and classification method would be successful?
2. What are the costs for such a system to be implemented?

Regarding the first question, we developed and tested three methods, which used ANNs to classify the tool-flank images: one based on image-processing feature extraction followed by ANN classification, and the other two methods applying ANNs directly on the image data. A comparison of the three methods' performances is presented in Table 1.

For method A, the image processing time to extract features was 4.2 s for all images of the image set (21 images). The computational times for network training are given in Table 1. For method A, the preprocessing time had to be added (4.2 s). As our focus was to compare only the presented algorithms, these are the only important computational times. Method A has the highest TSR, which suggests that this is the most reliable of the three methods tested. It also has the smallest training time and the smallest number of epochs needed for training. Method B is less attractive mainly because of the smaller TSR, which means that there may be a greater number of training sessions needed to develop a successful network. Even so, its advantage is that it is simpler to implement and requires a small number of neurons. Comparing the TSR for methods B and C, both working directly on the image data, we can conclude that autoencoders perform better than single-hidden-layer networks for large numbers of neurons. Method C requires the largest number of neurons, the largest number of epochs and the longest time to train.

Table 1. Comparative network training parameters.

Type of Network	No. of Neurons	Average No. of Training Epochs	Average Training Success Rate	Average Training Time (s)
A. Single hidden layer on image features	10	15–20	100	0.20
	200	25–30	96	0.25
B. Single hidden layer on image data	10	30–40	46	0.75
	200	30–40	42	12
	L1, L2	L1, L2	L1 + L2	L1 + L2
C. Two autoencoder hidden layers (L1, L2) on image data	10 to 140, 10	1000, 1000	15	280
	150 to 200, 10	1000, 1000	70	1900
	300, 100 to 150	1000, 1000	100	1900

Our experiments show that the TSR increases with the increase in the number of neurons.

During the training of the analyzed networks, we counted the percentage of misclassifications for each image in the testing set from the total number of misclassifications occurring for each of the three network types, with the same number of training sessions as in the case of the TSR. The result is presented in Figure 24. It can be observed that image number 3 from the training set (Figure 25) had the highest number of misclassifications regardless of the network type or number of neurons, which means that greater care has to be taken in the selection of the training set in order to produce a good result during operation.

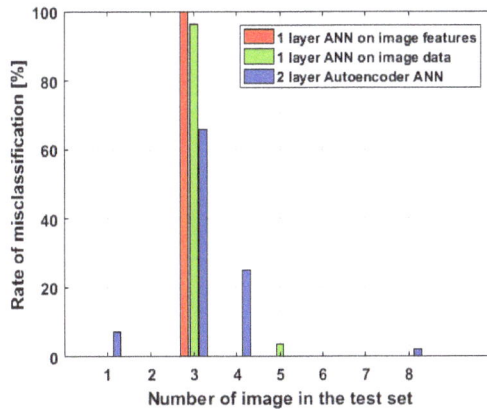

Figure 24. Number of misclassifications of tool-flank images from the total number of misclassifications.

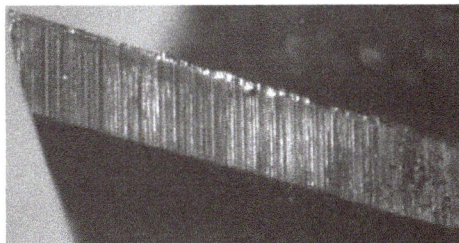

Figure 25. Image of sample number 3 from the test set labeled as unworn tool flank but misclassified, in some cases, as worn tool flank.

Regarding the second question, our assessment is that the major parts of the costs to implement this tool-wear monitoring solution are the cost of image acquisition, the cost of developing the software, and training the decision-making system to discriminate between the unworn and worn tools. The image-acquisition process costs increase with the number of acquisitioned images. To reduce these costs, the method that is successful when trained with a small number of images is recommended. Although it seems that method A gives the best results, an image-processing specialist may be needed to develop the feature extraction algorithm, as there is a large amount of work spent on the manual adjustment of parameters. Additionally, the feature extraction algorithm may be different for different types of tools, further increasing the development costs.

For methods B and C, no image processing is needed before applying the ANN classification. If ANN software is available (MATLAB Neural Network Toolbox or other similar software product), the development process is relatively straightforward with few parameters to adjust (e.g., number of neurons in the hidden layer); thus the training process does not require highly qualified software experts. A discussion has to be made also on how many networks should be employed: Is it possible to use one network for a larger set of tools or does each tool have to be provided with its own network? The implementation of the system is made gradually as the manufacturing system is running, without interrupting the production process. The only interruption of the production is for the time needed to install the camera and the positioning device. After installing the camera and the positioning device, images are acquired without interrupting the production process. In time, as the tools are used to process the parts, an image database is developed. As soon as the image database for a specific tool is complete, the training stage is accomplished during the runtime of the manufacturing system. When the training has produced a reliable neural network, this is implemented in the system. In this paper, we described the concept of a new system based on image acquisition and processing of the tool flank. The system is capable of automatically detecting tool wear in an early stage. Regarding the image processing methods, we present two new methods to obtain image features, which make the discrimination between worn and unworn tool-flank images possible (EN and NSP). We applied, to the best of our knowledge for the first time, a classification of the worn and unworn tool-flank images using a two-hidden-layer autoencoder ANN, which proved to be 100% successful for a large range of the number of nodes. We also present a detailed comparison of three ANNs in order to establish their capabilities to classify worn and unworn tool-flank images.

Acknowledgments: 1. This research was financed by: POSDRU 107/1.5/S/77265 (2010) of the Ministry of Labor, Family and Social Protection, Romania, co-financed by the European Social Fund—"Investing in people". 2. The publishing sponsor was R & D center "Cercetare Dezvoltare Agora Oradea".

Author Contributions: The contribution of the first three authors was in engineering issues and of the last two authors was ANN modeling.

Conflicts of Interest: The authors declare no conflict of interest.

References

1. Denkena, B.; Krüger, M.; Schmidt, J. Condition-based tool management for small batch production. *Int. J. Adv. Manuf. Technol.* **2014**, *74*, 471–480.
2. Çelik, Y.H.; Kilickap, E.; Güney, M.J. Investigation of cutting parameters affecting on tool wear and surface roughness in dry turning of Ti-6Al-4V using CVD and PVD coated tools. *J. Braz. Soc. Mech. Sci. Eng.* **2016**, 1–9, doi:10.1007/s40430-016-0607-6.
3. Jain, V.; Raj, T. Tool life management of unmanned production system based on surface roughness by ANFIS. *Int. J. Syst. Assur. Eng. Manag.* **2016**, 1–10, doi:10.1007/s13198-016-0450-2.
4. Bhat, N.N.; Dutta, S.; Vashisth, T. Tool condition monitoring by SVM classification of machined surface images in turning. *Int. J. Adv. Manuf. Technol.* **2016**, *83*, 1487–1502.
5. Huang, P.B. An intelligent neural-fuzzy model for an in-process surface roughness monitoring system in end milling operations. *J. Intell. Manuf.* **2016**, *27*, 689–700.

6. Li, N.; Chen, Y.; Kong, D. Force-based tool condition monitoring for turning process using v-support vector regression. *Int. J. Adv. Manuf. Technol.* **2016**, 1–11, doi:10.1007/s00170-016-9735-5.
7. Rimpault, X.; Chatelain, J.-F.; Klemberg-Sapieha, J.E.; Balazinski, M. Tool wear and surface quality assessment of CFRP trimming using fractal analyses of the cutting force signals. *CIRP J. Manuf. Sci. Technol.* **2017**, *16*, 72–80.
8. Wang, G.; Guo, Z.; Qian, L. Tool wear prediction considering uncovered data based on partial least square regression. *J. Mech. Sci. Technol.* **2014**, *28*, 317–322.
9. Khajavi, M.N.; Nasernia, E.; Rostaghi, M. Milling tool wear diagnosis by feed motor current signal using an artificial neural network. *J. Mech. Sci. Technol.* **2016**, *30*, 4869–4875.
10. Arruda, E.M.; Ribeiro Filho, S.L.M.; Assunção, J.T.; Brandão, L.C. Online prediction of tool wear in the milling of the AISI P20 steel through electric power of the main motor. *Arab. J. Sci. Eng.* **2015**, *40*, 3321–3328.
11. Postnov, V.V.; Idrisova, Y.V.; Fetsak, S.I. Influence of machine-tool dynamics on the tool wear. *Russ. Eng. Res.* **2015**, *35*, 936–940.
12. Stavropoulos, P.; Papacharalampopoulos, A.; Vasiliadis, E.; Chryssolouris, G. Tool wear predictability estimation in milling based on multi-sensorial data. *Int. J. Adv. Manuf. Technol.* **2016**, *82*, 509–521.
13. Zhang, G.; To, S.; Zhang, S.H. Evaluation for tool flank wear and its influences on surface roughness in ultra-precision raster fly cutting. *Int. J. Mech. Sci.* **2016**, *118*, 125–134.
14. Cerce, L.; Pusavec, F.; Kopac, J. 3D cutting tool-wear monitoring in the process. *J. Mech. Sci. Technol.* **2015**, *29*, 3885–3895.
15. Garcia-Ordás, M.T.; Alegre, E.; González-Castro, V. A computer vision approach to analyze and classify tool wear level in milling processes using shape descriptors and machine learning techniques. *Int. J. Adv. Manuf. Technol.* **2016**, 1–15, doi:10.1007/s00170-016-9541-0.
16. Javed, K.; Gouriveau, R.; Li, X. Tool wear monitoring and prognostics challenges: A comparison of connectionist methods toward an adaptive ensemble model. *J. Intell. Manuf.* **2016**, 1–18, doi:10.1007/s10845-016-1221-2.
17. Kong, D.; Chen, Y.; Li, N. Tool wear monitoring based on kernel principal component analysis and v-support vector regression. *Int. J. Adv. Manuf. Technol.* **2016**, 1–16, doi:10.1007/s00170-016-9070-x.
18. Chetan, E.; Narasimhulu, A.; Ghosh, S.; Rao, P.V. Study of tool wear mechanisms and mathematical modeling of flank wear during machining of Ti alloy (Ti6Al4V). *J. Inst. Eng. (India)* **2015**, *96*, 279–285.
19. Mia, M.; Al Bashir, M.; Dhar, N.R. Modeling of principal flank wear: An empirical approach combining the effect of tool. *Environ. Workpiece Hardness J. Inst. Eng. (India)* **2016**, *97*, 517–526.
20. Yang, W.A.; Zhou, W.; Liao, W. Prediction of drill flank wear using ensemble of co-evolutionary particle swarm optimization based-selective neural network ensembles. *J. Intell. Manuf.* **2016**, *27*, 343–361.
21. Muratov, K.R. Influence of rigid and frictional kinematic linkages in tool–workpiece contact on the uniformity of tool wear. *Russ. Eng. Res.* **2016**, *36*, 321–323.
22. Park, K.H.; Yang, G.D.; Lee, D.Y. Tool wear analysis on coated and uncoated carbide tools in inconel machining. *Int. J. Precis. Eng. Manuf.* **2015**, *16*, 1639–1645.
23. Yingfei, G.; Muñoz de Escalona, P.; Galloway, A. Influence of cutting parameters and tool wear on the surface integrity of cobalt-based stellite 6 alloy when machined under a dry cutting environment. *J. Mater. Eng. Perform.* **2017**, *26*, 312–326.
24. Mathworks®. MATLAB, Neural Network Toolbox, Image Processing Toolbox, R2016b, User's Guide. Available online: https://www.mathworks.com/help/ (accessed on 4 November 2017).
25. Hinton, G.E.; Osindero, S.; Teh, Y.W. A fast learning algorithm for deep belief nets. *Neural Comput.* **2006**, *18*, 1527–1554.
26. Bengio, Y.; Lamblin, P.; Popovici, D.; Larochelle, H. Greedy layer-wise training of deep networks. In Proceedings of the Twentieth Annual Conference on Neural Information Processing Systems, Vancouver, BC, Canada, 4–7 December 2006.

MDPI

Article

Novel Integrated Multi-Criteria Model for Supplier Selection: Case Study Construction Company

Željko Stević [1,*] [iD], Dragan Pamučar [2] [iD], Marko Vasiljević [1], Gordan Stojić [3] [iD] and Sanja Korica [4]

1   Faculty of Transport and Traffic Engineering, University of East Sarajevo, Vojvode Mišića 52, 74000 Doboj, Bosnia and Herzegovina; drmarkovasiljevic@gmail.com
2   Department of logistics, University of Defence in Belgrade, Pavla Jurisica Sturma 33, 11000 Belgrade, Serbia; dpamucar@gmail.com
3   Faculty of Technical Sciences, University of Novi Sad, Trg Dositeja Obradovica 6, 21000 Novi Sad, Serbia; gordan@uns.ac.rs
4   Faculty for ecology and environmental protection, University Union—Nikola Tesla, Cara Dusana 62-64, 11000 Belgrade, Serbia; koricasanja@gmail.com
*   Correspondence: zeljkostevic88@yahoo.com; Tel.: +387-66-795-413

Received: 12 October 2017; Accepted: 9 November 2017; Published: 17 November 2017

Abstract: Supply chain presents a very complex field involving a large number of participants. The aim of the complete supply chain is finding an optimum from the aspect of all participants, which is a rather complex task. In order to ensure optimum satisfaction for all participants, it is necessary that the beginning phase consists of correct evaluations and supplier selection. In this study, the supplier selection was performed in the construction company, on the basis of a new approach in the field of multi-criteria model. Weight coefficients were obtained by DEMATEL (Decision Making Trial and Evaluation Laboratory) method, based on the rough numbers. Evaluation and the supplier selection were made on the basis of a new Rough EDAS (Evaluation based on Distance from Average Solution) method, which presents one of the latest methods in this field. In order to determine the stability of the model and the applicability of the proposed Rough EDAS method, an extension of the COPRAS and MULTIMOORA method by rough numbers was also performed in this study, and the findings of the comparative analysis were presented. Besides the new approaches based on the extension by rough numbers, the results are also compared with the Rough MABAC (MultiAttributive Border Approximation area Comparison) and Rough MAIRCA (MultiAttributive Ideal-Real Comparative Analysis). In addition, in the sensitivity analysis, 18 different scenarios were formed, the ones in which criteria change their original values. At the end of the sensitivity analysis, SCC (Spearman Correlation Coefficient) of the obtained ranges was carried out, confirming the applicability of the proposed approaches.

Keywords: construction; rough number; EDAS; DEMATEL; MCDM; supply chain

1. Introduction

Supplier selection, according to Soheilirad et al. [1], presents a very important issue in decision making about management, and assumes several qualitative and quantitative criteria. The importance of this process in companies is reflected through establishing the final price of the product. The price of the raw material, as the main part of the product, is very important in the final product [2,3]. Supplier selection is one of the important tasks for the management of the supply [4]. At the same time, the management and development of strategic relationship with suppliers is crucial for the achievement of competitive market leadership [5]. Taking into account the fact that supplier selection in the supply chain is a group decision made based on several criteria, according to Zolfani et al. [6], the managers are supposed to know the most suitable method for choosing the right supplier. This is of vital importance because the modern supply chains need to meet strict requirements, so the managers find it very difficult to choose the right evaluation for the potential suppliers. This will ensure efficient production and final price formation which will be competitive on the market. In order to maximize the business values of products and services, efficient strategy for supply chain management has become a key component for a large number of final customers as well [7].

When one considers the efficiency of the complete supply chain, it is impossible not to notice that, to a large degree, it depends on the correct choice of the supplier, because exactly this process represents one of the most important factors which has a direct impact on the company performance. By performing the correct evaluation and the right choice of the supplier, this subsystem of logistics can efficiently carry out assignments which refer to the company supply. The right suppliers can meet requirements and needs which are set in the subsystem of supply, and refer to the quality, price, quantity of goods, delivery deadline and other deadlines, flexibility, reliability etc. The searching for suppliers who can meet these requirements is a permanent and primary goal. In order to obtain the above mentioned, it is necessary to continuously collect and process data about suppliers, as well as to establish and maintain connections with them.

Despite numerous new models developed in MCDM, the question arises which method or approach to apply. The aim is to enable the decision makers to express their preferences as clearly as possible, and reduce the subjectivity and uncertainty which is always present in the process of decision making. Accordingly, this paper has developed a new approach using the advantages of rough numbers and the Evaluation based on Distance from Average Solution (EDAS) method. Rough numbers present modification of the traditional rough numbers and take care of the uncertainty in decision making. The uncertainty is always present in the process of the evaluation of the importance of alternatives or criteria. Rough numbers take care of such uncertainty through the possibility of the preference expression for each alternative or criteria. Such preferences are further converted in rough intervals, more precisely determining the preference. Fuzzy set theory, applied in a large number of research, recently faces the criticism, according to Saaty and Tran, (2007) [8] of fuzzyfication of one of the most used methods in this field, i.e., AHP method, to not give good results, and they recommend elimination of the uncertainty by application of intermediate values. Fuzzyfication of numbers, besides increasing the complexity of manipulation, also makes the derivation process more difficult, and often leads to less desirable, than desirable results [9]. The application of rough numbers in the field of decision making is not appropriate until the precise conditions are defined as to when it works well and when it does not. Saaty and Tran, (2007) [8] consider that such conditions cannot be found in the field of decision making. The uncertainty present in the process of decision making is also considered in [10], according to it plays a dominant role, and it requires adequate handling in order to make adequate decisions. More precise and objective expression of uncertainty processes is possible if one applies intuitionistic multiplicative preference relation (IMPR) [11], intuitionistic fuzzy set theory (IFS) [12,13], interval-valued intuitionistic fuzzy soft set (IVIF) [14] or rough set theory, proposed in this manuscript. Integration of rough numbers in MCDM methods gives the possibility to explore subjective and unclear evaluation of the experts and to avoid the assumptions, which is not the case when applying fuzzy theory [15]. Therefore, we can conclude that the application of rough numbers and the development of R'DEMATEL–R'EDAS approach have significant importance. The advantages of a novel approach have been presented for each method in the following sections of the manuscript.

This paper has several objectives. The first is the development of the methodology for the treatment of uncertainty in the field of multi-criteria decision making through presenting new Rough EDAS, Rough COPRAS and Rough MULTIMOORA algorithms. The second goal is the popularization of the idea of rough numbers (RN) through their use in decision making in real business and economy system. The third goal is the improvement of the methodology for the evaluation and choice of the supplier through the new approach in treatment of uncertainty based on rough numbers.

The paper is organized into several sections. In the introduction, the importance and the influence of the supplier on the complete supply chain is presented. The second section contains the literature review, where the criteria for the supplier selection are emphasized. Besides, the methods used for solving these problems in the supply chain, in particular, in the construction, are presented. In third section, we present Rough DEMATEL algorithm and novel Rough EDAS methods, which are the proposed models in this study. The fourth section describes supplier selection in the construction company based on the previously defined methods. In the fifth section, the sensitivity analysis is performed, where different sets with different criteria values are defined and used for checking the stability of the proposed model. Besides the sensitivity analysis, in the same section, the discussion of the obtained results is carried out. In addition, for the stability check of the obtained ranges, expansion of the COPRAS and MULTIMOORA methods based on rough numbers is performed. The sixth section is the conclusion with the guidelines for the future research.

2. Literature Review

Multi-criteria decision making (MCDM) is one of the process for finding the optimal alternative from all the feasible alternatives according to some criteria or attributes [16–18]. Making a model of multi-criteria decision making for supplier selection requires a previous detailed analysis of the criteria on which the evaluation will be based. Since the evaluation of the supplier presents an everyday process in different supply chains, it is necessary to provide adequate input parameters in model. The right choice of the set of the criteria and determination of their relative weight, according to Lima and Carpinetti [19], is of great importance for the coordination of decision regarding buying and the strategic goals of the company. According to Vonderembse and Tracey [20], most companies consider the application of the criteria for supplier selection to be a very important part of the complete process of supplier selection. Dickson is regarded as the pioneer in this field. In his study [21], he introduced a model for supplier selection on the basis of set of 23 criteria. Following this, different authors have tried to continue in the same direction, with certain modifications and improvements, so different studies in this field propose the use of different criteria [22–27]. Table 1 gives an overview of the most commonly used criteria in the process of supplier selection.

Table 1. An overview of the most commonly used criteria in the process of supplier selection.

Criteria	References
Quality of material	[6,21,26–49]
Price of material	[6,21,26–31,33–40,42–56]
Certification of products	[26,31,40,42,44,50,52,57]
Delivery time	[21,26–31,33–40,42–49,51,53–56,58]
Reputation	[6,21,26,28,29,34,36,43,46,48,49,54,55,58–60]
Volume discounts	[37,40,42]
Warranty period	[21,26,31,35,37,40]
Reliability	[6,26,30,33,34,36,42,48,50,51,54,56,57,60]
Method of payment	[26,38–40,44,45,47,57]

An overview of the Table 1 shows that, regardless the constant changes and increasing use of the qualitative criteria, financial parameters, delivery, and quality present the criteria found in almost all research regarding supplier selection. This fact has been confirmed in the above-mentioned studies, where these parameters play an important role in decision making in the process of the evaluation of the supplier.

In construction business, planning of construction processes and effective management are extremely important for success [61]. Supplier selection, in any phase of the construction, presents very important part in this field [6]. Decision making in construction management is, according to Antuchevičiene et al. [62], always very complex and complicated, especially when more than one criterion is considered, which is often the case.

Decision making is one of the most important topics in various fields to make a right decision, so as to reach the final goal [63,64]. Therefore, the use of the MCDM method in the construction business is very popular, and helps in making extremely important decisions. According to Zavadskas et al. [65] and Tamošaitienė et al. [66], multi-criteria decision-making models are important tools in the construction management. In these kinds of studies, making decisions with as little subjectivity as possible and elimination of uncertainty is one of the main goals. This has been considered in [67], showing that uncertainty and imprecision are always present in the supplier selection process. Tamošaitienė et al. [48] have developed a hybrid model for supplier selection in the construction business which consists of three methods: AHP, ARAS, and Multiplicative Utility Function, while in the work of Izadikhah [68], TOPSIS method for a group decision making with Atanassov's interval-valued intuition fuzzy numbers was used for the same purposes. The evaluation and supplier selection in Iran [69] have been carried out by combination of AHP and ANP methods. The evaluation of the supplier for the construction material can be performed by using Fuzzy Principal Component Analysis [38].

In the work of Fouladgaret et al. [70], the authors use a new hybrid model for selecting the most appropriate working strategy in construction company. This model implies the application of fuzzy ANP and fuzzy COPRAS methods. The strategy of the management of the construction companies was also a subject of the study in [71], where the application of SWOT analysis and AHP method chooses the adequate solution. The application of the classical AHP method in [72] provides contractor selection for the defined project, while in [73], the combination of AHP and Fuzzy ARAS determines construction site selection. The combination of AHP and ARAS methods is also used for the assessment of project managers in construction [61], while for determination of the importance of the criteria, AHP method is used [74]. Besides, MCDM methods are constantly applied in other fields [75–77].

The AHP method, in its traditional or fuzzy form, is the most commonly used method for supplier selection [48,78–83]. Despite that, it has recently faced frequent criticism and disapproval from different authors. Therefore, in this paper, the DEMATEL method is used for determining the weights of criteria in their rough form. The DEMATEL method, in the field of the supply chains, has already been used in several different studies [40,84–90]. The determination of the weight coefficients by using DEMATEL method have, so far, shown good results. In their study, Chang et al. [91] suggest to the other researchers to use this method as well, with certain improvements, because it is useful for making decisions which include complex criteria used in group decision making. Implementation and evaluation of the most important criteria for supplier selection in the green supply chain has been performed in [92]. The evaluation of the criteria for supplier selection by using a combination of DEMATEL and ANP method has been performed in the work of Sarkar et al. [93].

It is noticeable that the rough set theory is, today, very often used for decision making in different fields. In the paper [94], the authors use a rough TOPSIS approach for failure mode and effects analysis in uncertain environments. The combination of rough AHP and MABAC method proposed in [95–97], and is used in [98] for selection of medical tourism sites. The combination of interval rough AHP and GIS is proposed for flood hazard mapping [99]. Rough AHP and rough TOPSIS approach are also used in the work of Song et al. [100]. From its beginnings until nowadays, the theory of rough sets has evolved through solving many problems by using rough sets [101–105], and through the use of rough numbers as in [106,107], while in the work [5], the authors use a grey based rough set. Supplier selection can also be evaluated by using a new rough set approach which has been developed by Chai and Liu [108] and, according to the authors, provides stable results.

3. Methods

3.1. R'DEMATEL Method

DEMATEL method (Decision Making Trial and Evaluation Laboratory) is very suitable for design and the analysis of the structural model. This is achieved by defining the cause-and-effect relation between complex factors [95]. Cause-and-effect relations are obtained on the basis of the total direct and indirect influences transferred from one factor to the other, but also received from the other factors. The implementation of DEMATEL method explores interdependent factors and determines the level of this dependency. The method is based on the graph theory, and it ensures visual planning and problem solving. In this way the relative factors can be divided into cause-and-effect in order to gain a better insight into their mutual relations. This also provides a better understanding of the complex structure of this problem, relations between the factors, and relations between the structure level and influence of the factors [99].

In order to provide a comprehensive investigation of any incorrectness present in the group decision making, the DEMATEL method modification is carried out in this study by using rough numbers. In this way, the additional information for the determination of the uncertain intervals is not necessary. In the following text, we will explain the steps of the R'DEMATEL method.

Step 1: Expert analysis of the factors. Suppose that there exists m experts and n factors (criteria) which are observed, and each expert should determine the degree of the influence of the factor i on the factor j. Comparative analysis of the set of i-factor and j-factor observed from the e-expert is denoted by $x_{ij}{}^e$, where $i = 1, 2, \ldots, n; j = 1, 2, \ldots, n$. The value of each $x_{ij}{}^e$ set takes one integer value of the following point scale: 0—no influence, 1—low influence, 2—moderate influence, 3—high influence, 4—very high influence. The response of the e-expert is expressed by the nonnegative matrix of the $n \times n$ range, whereas each element of the e-matrix in the expression $X^e = [x^e{}_{ij}]_{n \times n}$ presents integer nonnegative number $x_{ij}{}^e$, where $1 \le e \le m$.

$$X^e = \begin{bmatrix} 0 & x_{12}^e & \cdots & x_{1n}^e \\ x_{21}^e & 0 & \cdots & x_{2n}^e \\ \vdots & \vdots & \ddots & \vdots \\ x_{n1}^e & x_{n2}^e & \cdots & 0 \end{bmatrix}_{n \times n} \quad ; 1 \le i, j \le n; 1 \le e \le m \quad (1)$$

where $x_{ij}{}^e$ presents linguistic expressions from the previously defined linguistic scale used by the expert to present his comparison in the set of the criteria.

Thus, matrices $X^1, X^2, \ldots, X^m$ are response matrices of the each of m experts. The diagonal elements of the response matrices of all experts take value zero because the same factors have no influence.

Step 2: Determination of the averaged response matrices of the experts. On the basis of the response matrix $X^e = [x^e{}_{ij}]_{n \times n}$ ($1 \le e \le m$) of all m experts, we obtain aggregate sequence matrix of experts X^*.

$$X^* = \begin{bmatrix} x_{11}^1, x_{11}^2, \ldots, x_{11}^m & x_{12}^1; x_{12}^2; \ldots; x_{12}^m, & \cdots, & x_{1n}^1; x_{1n}^2; \ldots, x_{1n}^m \\ x_{21}^1, x_{21}^2, \ldots, x_{21}^m & x_{22}^1; x_{22}^2; \ldots; x_{22}^m, & \cdots, & x_{2n}^1; x_{2n}^2; \ldots, x_{2n}^m \\ \cdots & \cdots & \cdots & \cdots \\ x_{n1}^1, x_{n1}^2, \ldots, x_{n1}^m & x_{n2}^1; x_{n2}^2; \ldots; x_{n2}^m, & \cdots, & x_{nn}^1; x_{nn}^2; \ldots, x_{nn}^m \end{bmatrix} \quad (2)$$

where $x_{ij} = \left\{ x_{ij}^1, x_{ij}^2, \ldots, x_{ij}^m \right\}$ presents sequences defining relative importance of the criteria i in comparison with the criteria j. According to Zhu et al. (2015) [109], the sequence x_{ij}^e ($1 \le e \le m$) is transformed into rough sequence $RN(x_{ij}^e) = \left[\underline{Lim}(x_{ij}^e), \ \overline{Lim}(x_{ij}^e) \right]$, where i presents the lower limit $\underline{Lim}(x_{ij}^e)$ and j the upper limit $\overline{Lim}(x_{ij}^e)$ of the rough sequence $RN(x_{ij}^e)$, respectively.

These rough sequences are defined in the matrix (2). With that, we obtain rough matrices X^1, X^2, ..., X^m (where m is the number of the experts). In this way, for the group of rough matrices X^1, X^2, ..., X^m on the position (ij) we obtain rough sequence $RN(x_{ij}) = \left\{ \left[\underline{Lim}(x_{ij}^1), \overline{Lim}(x_{ij}^1) \right], \left[\underline{Lim}(x_{ij}^2), \overline{Lim}(x_{ij}^2) \right], \cdots, \left[\underline{Lim}(x_{ij}^m), \overline{Lim}(x_{ij}^m) \right] \right\}$.

Applying the Expression (3) leads to the averaged rough sequence:

$$RN(z_{ij}) = RN(x_{ij}^1, x_{ij}^2, \cdots, x_{ij}^m) = \begin{cases} \underline{Lim}(z_{ij}) = \frac{1}{m} \sum_{e=1}^{m} x_{ij}^e \\ \overline{Lim}(z_{ij}) = \frac{1}{m} \sum_{e=1}^{m} x_{ij}^e \end{cases} \tag{3}$$

where e presents e-expert ($e = 1, 2, \ldots, m$) and $RN(z_{ij})$ presents rough sequence. In this way, we obtained an averaged rough matrix of the response:

$$Z = \begin{bmatrix} 0 & RN(z_{12}) & \cdots & RN(z_{1n}) \\ RN(z_{21}) & 0 & \cdots & RN(z_{2n}) \\ \vdots & \vdots & \ddots & \vdots \\ RN(z_{n1}) & RN(z_{n2}) & \cdots & 0 \end{bmatrix} \tag{4}$$

Matrix Z shows initial effects caused by factor j, as well as initial effects which factor j receives from other factors. Sum of each i-row of the matrix Z represents the total direct effect which i-factor has delivered to the other factors. Sum of each j-column of the matrix Z represents the total direct effect which j-column has delivered to the other factors.

Step 3: Normalize direct-relation matrix. From the matrix Z, the initial direct-relation matrix is estimated from the expression $D = [RN(d_{ij})]_{n \times n}$ (17). So normalized, each element of the matrix D takes value between zero and one. Matrix D is obtained when each element $RN(z_{ij})$ of the matrix Z is divided by rough number $RN(s)$ (Expressions (5)–(8)).

$$D = \begin{bmatrix} 0 & RN(d_{12}) & \cdots & RN(d_{1n}) \\ RN(d_{21}) & 0 & \cdots & RN(d_{2n}) \\ \vdots & \vdots & \ddots & \vdots \\ RN(d_{n1}) & RN(d_{n2}) & \cdots & 0 \end{bmatrix} \tag{5}$$

where $RN(d_{ij})$ is obtained using the Expression (18)

$$RN(d_{ij}) = \frac{RN(z_{ij})}{s} = RN\left(\frac{\underline{Lim}(z_{ij})}{s}, \frac{\overline{Lim}(z_{ij})}{s} \right) \tag{6}$$

where

$$s = \max\left(\sum_{j=1}^{n} RN(z_{ij}) \right) = \max\left(\sum_{j=1}^{n} \underline{Lim}(z_{ij}), \sum_{j=1}^{n} \overline{Lim}(z_{ij}) \right) \tag{7}$$

i.e.,

$$s = \max\left[\max\left(\sum_{j=1}^{n} \underline{Lim}(z_{ij}) \right), \max\left(\sum_{j=1}^{n} \overline{Lim}(z_{ij}) \right) \right] \tag{8}$$

Step 4: Determination of the total relation matrix. By using the Expressions (9) and (10), the total relation matrix ($T = [RN(t_{ij})]_{n \times n}$) of the range $n \times n$ is estimated. Element $RN(t_{ij})$ represents the direct influence of the factor i on the factor j, whereas matrix T represents the total relations between each pair of factors.

Since each rough number consists of two sequences, i.e., lower and upper approximation, then the normalized matrix of the average perception $D = [RN(d_{ij})]_{n \times n}$ can be divided into two sub-matrices, i.e., $D = \left[D^L, D^U\right]$, where $D^L = \left[\underline{Lim}(d_{ij})\right]_{n \times n}$ and $D^U = \left[\overline{Lim}(d_{ij})\right]_{n \times n}$. Moreover, $\lim\limits_{m \to \infty} \left(D^L\right)^m = O$ and $\lim\limits_{m \to \infty} \left(D^U\right)^m = O$ where O presents zero matrix.

$$\left.\begin{aligned} & \lim_{m \to \infty} \left(I + D^L + D^{2L} + \ldots + D^{mL}\right) = \left(I - D^L\right)^{-1} \\ & and \\ & \lim_{m \to \infty} \left(I + D^U + D^{2U} + \ldots + D^{mU}\right) = \left(I - D^U\right)^{-1} \end{aligned}\right\} \tag{9}$$

So, the total relation matrix T is obtained from the estimation of the following elements:

$$\left.\begin{aligned} & T^L = \lim_{m \to \infty} \left(I + D^L + D^{2L} + \ldots + D^{mL}\right) = \left(I - D^L\right)^{-1} = \left[\underline{Lim}(t_{ij}^L)\right]_{n \times n} \\ & and \\ & T^U = \lim_{m \to \infty} \left(I + D^U + D^{2U} + \ldots + D^{mU}\right) = \left(I - D^U\right)^{-1} = \left[\underline{Lim}(t_{ij}^U)\right]_{n \times n} \end{aligned}\right\} \tag{10}$$

where $D^L = \left[\underline{Lim}(d_{ij})\right]_{n \times n}$ and $D^U = \left[\overline{Lim}(d_{ij})\right]_{n \times n}$

Sub-matrices T^L and T'^U together represent rough total relation matrix $T = (T^L, T^U)$. On the basis of the Expressions (9) and (10), one can obtain the rough total relation matrix:

$$T = \begin{bmatrix} RN(t_{11}) & RN(t_{12}) & \cdots & RN(t_{1n}) \\ RN(t_{21}) & RN(t_{22}) & \cdots & RN(t_{2n}) \\ \vdots & \vdots & \ddots & \vdots \\ RN(t_{n1}) & RN(t_{n2}) & \cdots & RN(t_{nn}) \end{bmatrix} \tag{11}$$

where $RN(t_{ij}) = \left[\underline{Lim}(t_{ij}), \overline{Lim}(t_{ij})\right]$ is a rough number describing the indirect effects of the factor i on factor j. Then the matrix T describes mutual dependence of each pair of factors.

Step 5: Estimation of the sum of rows and columns of the total relation matrix T. In the total relation matrix T, the sum of rows and columns is represented by vectors R and C of range $n \times 1$:

$$RN(R_i) = \left[\sum_{j=1}^{n} RN(t_{ij})\right]_{n \times 1} = \left[\sum_{j=1}^{n} t_{ij}^L, \sum_{j=1}^{n} t_{ij}^U\right]_{n \times 1} \tag{12}$$

$$RN(C_i) = \left[\sum_{i=1}^{n} RN(t_{ij})\right]_{1 \times n} = \left[\sum_{i=1}^{n} t_{ij}^L, \sum_{i=1}^{n} t_{ij}^U\right]_{1 \times n} \tag{13}$$

To effectively determine the "Prominence" and the "Relation", the sum of rows R_i to the sum of columns C_i in the total relation matrix T need to be converted into the crisp forms R_i^{crisp} and C_i^{crisp} by applying Equations (14)–(16).

$$RN(\hat{R}_i) = \left[\underline{Lim}(\hat{R}_i), \overline{Lim}(\hat{R}_i)\right] = \begin{cases} \underline{Lim}(\hat{R}_i) = \dfrac{\underline{Lim}(R_i) - \min\limits_{i}\left\{\underline{Lim}(R_i)\right\}}{\max\limits_{i}\left\{\overline{Lim}(R_i)\right\} - \min\limits_{i}\left\{\underline{Lim}(R_i)\right\}} \\[4mm] \overline{Lim}(\hat{R}_i) = \dfrac{\overline{Lim}(R_i) - \min\limits_{i}\left\{\underline{Lim}(R_i)\right\}}{\max\limits_{i}\left\{\overline{Lim}(R_i)\right\} - \min\limits_{i}\left\{\underline{Lim}(R_i)\right\}} \end{cases} \tag{14}$$

where $\underline{Lim}(R_i)$ and $\overline{Lim}(R_i)$ represent the lower limit and upper limit of the rough number $RN(R_i)$, respectively; $\underline{Lim}(\hat{R}_i)$ and $\overline{Lim}(\hat{R}_i)$ are the normalized forms of $\underline{Lim}(R_i)$ and $\overline{Lim}(R_i)$.

After normalization, we obtain a total normalized crisp value:

$$\beta_i = \frac{\underline{Lim}(\hat{R}_i) \cdot \left\{1 - \underline{Lim}(\hat{R}_i)\right\} + \overline{Lim}(\hat{R}_i) \cdot \overline{Lim}(\hat{R}_i)}{1 - \underline{Lim}(\hat{R}_i) + \overline{Lim}(\hat{R}_i)} \tag{15}$$

Finally, crisp form R_i^{crisp} for $RN(R_i)$ is obtained by applying Equation (16):

$$R_i^{crisp} = \min_i \left\{\underline{Lim}(R_i)\right\} + \beta_i \cdot \left[\max_i\{\overline{Lim}(R_i)\} - \min_i\left\{\underline{Lim}(R_i)\right\}\right] \tag{16}$$

The value R_i^{crisp} shows the total direct and indirect effects which criteria i has enabled to other criteria. The value C_i^{crisp} shows the total direct and indirect effects which criteria j has obtained from other criteria. In the case when $i = j$ then the expression $(R_i^{crisp} + C_i^{crisp})$ represents the importance of the criteria, and the expression $(R_i^{crisp} - C_i^{crisp})$ represents the intensity of the influence of the criteria in comparison to the others [85].

Step 6: Determination of the threshold value (α) and creation of the cause-and-effect relationship—CERD. The threshold value (α) is estimated as the average of the elements of the matrix T (17):

$$\alpha = \frac{\sum_{i=1}^{n} \sum_{j=1}^{n} \left[RN(t_{ij})\right]}{N} \tag{17}$$

where N presents the total number of the elements of matrix (11).

CERD is produced in order to visually present complex relations, and provide information about establishing the most important factors and the way they affect each other. Factors t_{ij} with values higher than a threshold value α, are chosen to present cause-and-effect relations.

Values of the elements of the matrix T, which have value higher than a threshold value are inserted and put in diagram. In CERD, x-axis is $(R_i^{crisp} + C_i^{crisp})$, y-axis is $(R_i^{crisp} - C_i^{crisp})$. These values are used to present relations between the two factors. When presenting the relations between factors, the arrow of the cause-and-effect relation is directed from the factor which has lower value than α, to the factor which has higher value than α. After determination of the criteria and presentation in CERD, in the next step, the weight coefficients of the criteria are estimated.

Step 7: Determination of the weight coefficients of the criteria. The weight coefficients of the criteria are obtained by using the Expression (18) [110]:

$$RN(W_j) = \begin{cases} \underline{Lim}(W_j) = \sqrt{\left(\underline{Lim}(R_i) + \underline{Lim}(C_i)\right)^2 + \left(\underline{Lim}(R_i) - \overline{Lim}(C_i)\right)^2} \\ \overline{Lim}(W_j) = \sqrt{\left(\overline{Lim}(R_i) + \overline{Lim}(C_i)\right)^2 + \left(\overline{Lim}(R_i) - \underline{Lim}(C_i)\right)^2} \end{cases} \tag{18}$$

Normalization of the weight coefficients is performed by using the Expression (19):

$$RN(w_j) = \frac{RN(W_j)}{\max\limits_{j}\left\{\underline{Lim}(W_j), \overline{Lim}(W_j)\right\}} \tag{19}$$

where w_j represents the final weights of criteria [95].

3.2. Rough EDAS Method

EDAS (Evaluation based on Distance from Average Solution) method belongs to the group of newer methods of multi-criteria decision making. In a very short time, it has found its way through the wide application in solving engineering problems, as well as problems in business decision making. This method [111] has a number of extensions, and the extension by fuzzy logics [112] is performed exactly in the field of supply chain for supplier selection. Several studies have already been published in different fields, where this method has been applied in its traditional form or some other forms [113–122]. It resembles a very important support in decision making in everyday conflict situations. The estimation of the alternatives in this method is based on the measurements of the positive and negative deviations from the average solution, estimated on the basis of all criteria. In this study, the extension of EDAS method by rough numbers has been performed. After defining the problem and forming multi-criteria model which consists of n alternatives and m criteria, it is necessary to define the set of k experts who will be evaluating alternatives for each of the criteria. After formulating multi-criteria model of n alternative and m criteria with k experts, Rough EDAS consists of the following steps.

Step 1. Converting individual matrices into the group rough matrix. If each expert matrix is noted by $k_1, k_2, \ldots, k_n$ then the group rough matrix is obtained according to Zhai et al. [109]:

$$RGM = \begin{bmatrix} \left[x_{11}^L, x_{11}^U\right] & \left[x_{12}^L, x_{12}^U\right] & \cdots & \left[x_{1m}^L, x_{1m}^U\right] \\ \left[x_{21}^L, x_{21}^U\right] & \left[x_{22}^L, x_{22}^U\right] & \cdots & \left[x_{2m}^L, x_{2m}^U\right] \\ \vdots & \vdots & \ddots & \vdots \\ \left[x_{m1}^L, x_{m1}^U\right] & \left[x_{m2}^L, x_{m2}^U\right] & \cdots & \left[x_{mm}^L, x_{mm}^U\right] \end{bmatrix} \tag{20}$$

Step 2. Finding the average solution for all criteria as shown in (21):

$$RN(AV) = \left[AV_j^L; \, AV_j^U\right]_{1 \times m} \tag{21}$$

based on the Equation (22):

$$\sum_{i=1}^{n} \frac{X_{ij}}{n} = \left[\frac{X_{il}^L}{n}; \, \frac{X_{il}^U}{n}\right] \tag{22}$$

Step 3. Determination of positive deviation RN(PDA) and negative deviation RN(NDA) from the average solution RN(AV) on the basis of all criteria by applying the Equations (23)–(24):

$$RN(PDA) = \left[PDA_{ij}^L; \, PDA_{ij}^U\right]_{n \times m} \tag{23}$$

$$RN(NDA) = \left[NDA_{ij}^L; \, NDA_{ij}^U\right]_{n \times m} \tag{24}$$

If the criterion belongs to the Benefit group, then RN(PDA) and RN(NDA) are estimated as follows:

$$RN(PDA) = \frac{\max\left(0, \, \left[X_{il}^L - AV_j^U; \, X_{il}^U - AV_j^L\right]\right)}{\left[AV_j^L; \, AV_j^U\right]} \tag{25}$$

$$RN(NDA) = \frac{\max\left(0, \, \left[AV_j^L - X_{il}^U; \, AV_j^U - X_{il}^L\right]\right)}{\left[AV_j^L; \, AV_j^U\right]} \tag{26}$$

If the criterion belongs to the Expenses group, then:

$$RN(PDA) = \frac{\max\left(0, \, \left[AV_j^L - X_{il}^U; \, AV_j^U - X_{il}^L\right]\right)}{\left[AV_j^L; \, AV_j^U\right]} \tag{27}$$

$$RN(NDA) = \frac{\max\left(0, \ \left[X_{il}^L - AV_j^U; \ X_{il}^U - AV_j^L\right]\right)}{\left[AV_j^L; \ AV_j^U\right]} \tag{28}$$

Since we are dealing with the rough numbers having upper and lower limit, often it can happen that lower limit of rough number has a negative value and upper positive, or, even that both values are less than zero. Since it is necessary to reduce these values to zero or positive values, the following equations then need to be applied:

$$if \ PDA_{ij}^L + PDA_{ij}^U \leq 0 \rightarrow RN(PDA) = 0 \tag{29}$$

$$if \ PDA_{ij}^L + PDA_{ij}^U > 0, \ and \ (PDA_{ij}^L > 0; \ PDA_{ij}^U > 0) \ then \ RN(PDA) = \left[PDA_{ij}^L; \ PDA_{ij}^U\right] \tag{30}$$

$$if \ PDA_{ij}^L + PDA_{ij}^U > 0, but \ (PDA_{ij}^L < 0; \ PDA_{ij}^U > 0) \ then \ RN(PDA) = \left|PDA_{ij}^L; PDA_{ij}^U\right| \tag{31}$$

The same applies for $RN(NDA)$.

Equations (29)–(31) include the following cases. If the sum of lower and upper limit (PDA) is less than zero, than rough number has zero value. If the sum of two limits is higher than zero, and lower and upper limit are higher than zero, than (PDA) stays constant (keeps its value). If the sum of two limits is higher than zero, but lower limit has negative value, than rough number (PDA) takes its absolute value, i.e., lower limit becomes positive value.

Step 4. Weighting of matrices RN(PDA) and RN(NDA) by using the equations:

$$VP_i = \left[vp_{ij}^L; vp_{ij}^U\right]_{mxn}$$
$$vp_{ij}^L = w_j^L \times pda_{ij}^L, \ i = 1, 2, \ldots m, j \tag{32}$$
$$vp_{ij}^U = w_j^U \times pda_{ij}^U, \ i = 1, 2, \ldots m, j$$

$$VN_i = \left[vn_{ij}^L; vn_{ij}^U\right]_{mxn}$$
$$vn_{ij}^L = w_j^L \times nda_{ij}^L, \ i = 1, 2, \ldots m, j \tag{33}$$
$$vn_{ij}^U = w_j^U \times nda_{ij}^U, \ i = 1, 2, \ldots m, j$$

where w_j^L and w_j^U are lower and upper limit of the criteria weight expressed as rough number.

Step 5. Determination of the sum of the previously weighted matrix:

$$RN(SP_i) = \left[sp_i^L; sp_i^U\right] = \sum_{j=1}^{m}\left[vp_{ij}^L; vp_{ij}^U\right] \tag{34}$$

$$RN(SN_i) = \left[sn_i^L; \ sn_i^U\right] = \sum_{j=1}^{m}\left[vn_{ij}^L; \ vn_{ij}^U\right] \tag{35}$$

Step 6. Normalization of RN(SP) and RN(SN) for all alternatives:

$$RN(NSP_i) = \frac{\left[sp_i^L; \ sp_i^U\right]}{max\left[sp_i^L; \ sp_i^U\right]} \tag{36}$$

$$RN(NSN_i) = 1 - \frac{\left[sn_i^L; sn_i^U\right]}{max\left[sn_i^L; \ sn_i^U\right]} \tag{37}$$

Step 7. Estimation of the values of all alternatives RN(ASi) and their ranging.

$$RN(AS_i) = \frac{1}{2}[NSP_i + NSN_i] \tag{38}$$

Alternative which has highest value presents the best solution.

4. Case Study

Supplier selection in the construction company was carried out on the basis of nine criteria presented in the Table 1: quality of the material, price of the material, certification of the products, delivery time, reputation, volume discounts, warranty period, reliability, and the method of payments. The second and the fourth criteria (the price of the material and delivery time) are the Expenses criteria, and the others are the Benefit criteria. Figure 1 shows the proposed model for the supplier selection in this study.

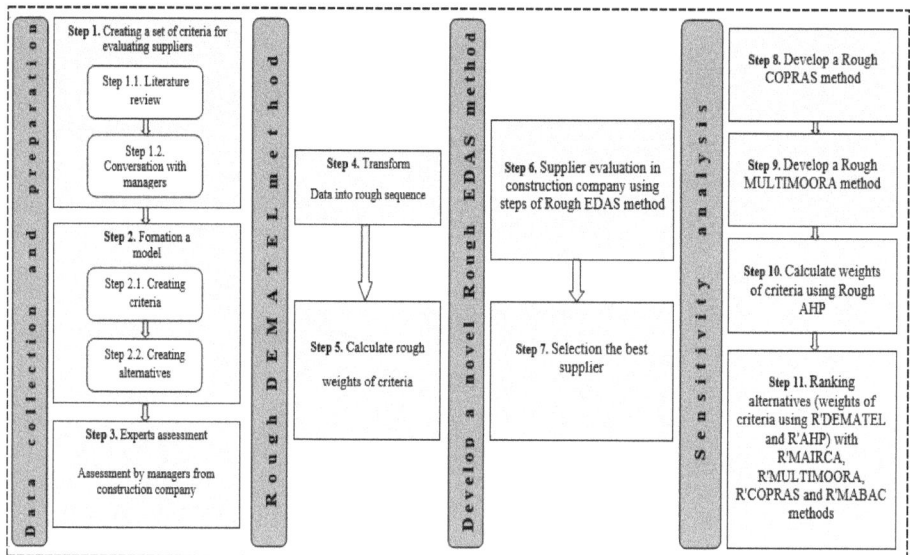

Figure 1. Proposed model for the supplier selection.

Figure 1 shows the proposed model for the supplier selection, which consists of 4 phases and 11 steps. The first phase assumes the collection and preparation of the date in three steps. The first step is defining the set of the criteria for evaluation of the supplier on the basis of the other studies in this field, and an interview with the managers with long-lasting experience on management functions in the supply business. Then, a multi-criteria model has been adopted, consisting of nine criteria and six alternatives evaluated by team of seven experts. The second phase of the model assumes the application of rough DEMATEL method for estimation of the relative weight of the criteria. The steps of this method are presented in detail in section Methods, and the procedure for the weight estimation is described in following. The third and the central phase of this study is development of a new novel Rough EDAS method used for evaluation and the best supplier selection. The last, fourth phase includes sensitivity analysis extended by two additional methods for multi-criteria decision making, COPRAS and MULTIMORA. Comparison of the obtained results and alternative ranges has been performed, applying the above-mentioned two methods, as well with the R'MAICA and R'MABAC, previously extended by rough numbers.

4.1. Estimation of the Criteria Weight by Applying R'DEMATEL Method

In this study, the team of seven experts took part in the process of determination of weight coefficients of criteria. The experts with the minimum of five-year experience in the supply chain management were chosen. After the interview with the experts, the collected data were processed, and the aggregation of the expert opinion was obtained. The collecting of data through the interview with the experts was carried out in the period from March 2017 until June 2017.

Step 1 Expert Analysis of the Factors

In the first step of application of DEMATEL method for the determination of the criteria, experts have used the point scale: 0—no influence, 1—very low influence, 2—low influence, 3—medium influence, 4—high influence, 5—very high influence. After the experts' evaluation, seven matrices of dimensions 9 × 9 have been constructed, in order to compare the set of the criteria. This is presented in Table 2.

Table 2. Experts' comparison of the evaluation criteria.

	E_1									E_2								
	C_1	C_2	C_3	C_4	C_5	C_6	C_7	C_8	C_9	C_1	C_2	C_3	C_4	C_5	C_6	C_7	C_8	C_9
C_1	0	5	5	5	2	4	4	4	3	0	5	5	4	3	5	4	4	3
C_2	4	0	4	2	2	3	5	4	3	4	0	4	4	1	5	5	3	3
C_3	4	3	0	3	3	4	2	4	3	4	1	0	5	3	5	3	4	3
C_4	4	3	5	0	2	3	3	3	5	4	2	5	0	2	5	3	3	5
C_5	5	4	5	4	0	4	5	5	3	5	3	5	5	0	5	5	5	3
C_6	4	4	3	3	1	0	2	4	4	4	2	4	4	1	0	2	4	4
C_7	4	4	3	4	2	4	0	5	5	4	2	4	5	2	5	0	5	5
C_8	3	3	4	2	1	2	2	0	3	4	2	5	4	1	4	2	0	3
C_9	3	3	4	2	2	2	3	3	0	3	1	4	3	2	4	3	3	0

$\cdots$

	E_6									E_7								
	C_1	C_2	C_3	C_4	C_5	C_6	C_7	C_8	C_9	C_1	C_2	C_3	C_4	C_5	C_6	C_7	C_8	C_9
C_1	0	5	4	4	3	5	5	5	2	0	5	5	4	4	4	4	5	3
C_2	4	0	3	2	1	3	4	3	3	4	0	4	1	5	5	4	2	2
C_3	5	3	0	3	3	3	2	3	3	4	3	0	5	4	5	2	4	2
C_4	3	3	4	0	2	4	3	3	4	3	3	5	0	3	5	3	4	5
C_5	3	4	4	4	0	3	4	3	4	5	4	5	5	0	5	5	4	3
C_6	4	3	3	3	1	0	2	3	5	3	3	3	4	2	0	2	4	4
C_7	5	3	3	4	2	3	0	3	5	4	3	3	5	2	4	0	5	4
C_8	4	3	4	2	1	3	1	0	4	3	3	4	4	1	5	2	0	3
C_9	3	3	4	2	1	3	3	5	0	2	3	5	3	1	5	3	4	0

Step 2 Determination of the matrix of the average response of the experts

On the basis of the response expert matrix (Table 2), matrix of the aggregated sequences of experts is constructed (2). By applying the matrix according to Zhai et al. [109], each of the shown sequences is transformed into rough sequence. So, for the sequence $x_{12} = \{5; 5; 4; 4; 4; 5; 5\}$ we get:

$$\underline{Lim}(4) = 4.00, \quad \overline{Lim}(4) = \frac{1}{7}(5+5+4+5+5+4+4) = 4.57$$

$$\underline{Lim}(5) = \frac{1}{7}(5+5+4+5+5+4+4) = 4.57, \quad \overline{Lim}(5) = 5.00$$

$$\underline{Lim}(4) = 4.00, \quad \overline{Lim}(4) = \frac{1}{7}(5+5+4+5+5+4+4) = 4.57$$

On the basis of the obtained values, each sequence is transformed into rough sequence:

$$RN(x_{12}^1) = [4.57, 5.00];$$

$$RN(x_{12}^2) = [4.57, 5.00];$$

$$RN(x_{12}^3) = [4.00, 4.57];$$

$$RN(x_{12}^4) = [4.00, 4.57];$$
$$RN(x_{12}^5) = [4.00, 4.57];$$
$$RN(x_{12}^6) = [4.57, 5.00];$$
$$RN(x_{12}^7) = [4.57, 5.00].$$

The obtained rough sequences present the uncertainty of the group of experts, which is the result of nonconformity in the criteria evaluation.

By applying the Expression (3), the averaging of the rough sequences is performed. Therefore, we obtain average rough sequence:

$$RN(z_{12}) = \left[\underline{Lim}(z_{12}), \overline{Lim}(z_{12}) \right] = \begin{cases} \underline{Lim}(z_{12}) = \frac{4.57+4.57+4+4+4+4.57+4.57}{7} = 4.33 \\ \overline{Lim}(z_{12}) = \frac{5+5+4.57+4.57+4.57+5+5}{7} = 4.82 \end{cases}$$

In this way, we get the final rough sequence $RN(z_{12}) = [4.33, 4.82]$. The application of the described procedure for the other elements of the matrix of the aggregated sequence of the experts (2) provides the average rough matrix of the average responses (4):

$$Z = \begin{bmatrix} [0.00,\ 0.00] & [4.33,\ 4.82] & [4.33,\ 4.82] & [4.25,\ 4.76] & ,\cdots, & [2.33,\ 2.82] \\ [3.74,\ 3.98] & [0.00,\ 0.00] & [3.18,\ 3.67] & [2.92,\ 4.08] & ,\cdots, & [2.51,\ 2.92] \\ [4.18,\ 4.67] & [2.04,\ 3.43] & [0.00,\ 0.00] & [3.59,\ 4.65] & ,\cdots, & [2.74,\ 2.98] \\ [3.25,\ 3.76] & [2.56,\ 3.19] & [4.67,\ 4.97] & [0.00,\ 0.00] & ,\cdots, & [4.67,\ 4.97] \\ \cdots & \cdots & \cdots & \cdots & \ddots & \cdots \\ [2.51,\ 2.92] & [2.01,\ 3.13] & [4.08,\ 4.49] & [2.55,\ 3.89] & ,\cdots, & [0.00,\ 0.00] \end{bmatrix}$$

Step 3   Normalization of the group direct relation matrix

On the basis of the matrix Z, the elements of the initial direct-relation matrix are determined (5):

$$Z = \begin{bmatrix} [0.00,\ 0.00] & [0.12,\ 0.16] & [0.12,\ 0.15] & [0.12,\ 0.15] & ,\cdots, & [0.06,\ 0.09] \\ [0.09,\ 0.12] & [0.00,\ 0.00] & [0.09,\ 0.12] & [0.08,\ 0.13] & ,\cdots, & [0.07,\ 0.09] \\ [0.11,\ 0.15] & [0.06,\ 0.11] & [0.00,\ 0.00] & [0.10,\ 0.15] & ,\cdots, & [0.07,\ 0.09] \\ [0.09,\ 0.12] & [0.07,\ 0.10] & [0.13,\ 0.16] & [0.00,\ 0.00] & ,\cdots, & [0.13,\ 0.16] \\ \cdots & \cdots & \cdots & \cdots & \ddots & \cdots \\ [0.07,\ 0.09] & [0.06,\ 0.1] & [0.11,\ 0.14] & [0.07,\ 0.12] & ,\cdots, & [0.00,\ 0.00] \end{bmatrix}$$

The elements of the matrix Z are obtained from the Expression (6):

$$RN(d_{12}) = \frac{RN(z_{12})}{s} = \left[\frac{4.56}{36.71}, \frac{4.94}{36.71} \right] = [0.12, 0.16]$$

where the value s is obtained from the Expression (8):

$$s = \max\left[\max\left(\sum_{j=1}^{n} \underline{Lim}(z_{ij}) \right), \max\left(\sum_{j=1}^{n} \overline{Lim}(z_{ij}) \right) \right]$$
$$= \max[\max(30.74; 24.13; 24.88; 27.06; 31.66; \cdots ; 21.31), \max(35.10; 28.76; 30.37; 30.55; 36.71; \cdots ; 27.72)]$$
$$= \max[31.66; 36.71] = 36.71$$

Step 4   Determination of the total relation matrix

The total relation matrix T (11) of the range 9×9 is determined from the Expressions (9) and (10):

$$Z = \begin{bmatrix}
[0.242, 2.903] & [0.306, 2.814] & [0.368, 3.210] & [0.344, 3.313] & ,\cdots, & [0.293, 2.851] \\
[0.277, 2.519] & [0.156, 2.225] & [0.290, 2.662] & [0.266, 2.757] & ,\cdots, & [0.255, 2.387] \\
[0.303, 2.689] & [0.219, 2.461] & [0.219, 2.717] & [0.289, 2.935] & ,\cdots, & [0.261, 2.527] \\
[0.294, 2.655] & [0.237, 2.444] & [0.346, 2.841] & [0.211, 2.795] & ,\cdots, & [0.320, 2.568] \\
\cdots & \cdots & \cdots & \cdots & \ddots & \cdots \\
[0.234, 2.4] & [0.191, 2.226] & [0.286, 2.582] & [0.235, 2.648] & ,\cdots, & [0.165, 2.210]
\end{bmatrix}$$

Step 5   Determination of the sum of the rows and columns of the total relation matrix T

In the total relation matrix T, the sum of the rows and columns is presented by vectors R and C (Expressions (12) and (13)):

$$RN(R_i) = \begin{bmatrix}
[2.740, 25.844] \\
[2.190, 21.469] \\
[2.278, 22.797] \\
[2.424, 22.705] \\
[2.841, 27.027] \\
[2.122, 20.699] \\
[2.522, 23.897] \\
[1.935, 19.686] \\
[1.939, 20.628]
\end{bmatrix} ; RN(C_i) = \begin{bmatrix}
[2.551, 23.891] \\
[2.067, 22.051] \\
[2.779, 25.344] \\
[2.541, 26.193] \\
[1.387, 13.850] \\
[2.477, 26.275] \\
[2.171, 19.896] \\
[2.554, 24.467] \\
[2.464, 22.784]
\end{bmatrix}$$

In order to obtain the most reliable cause-and-effect relation between criteria and efficient production in CERD, rough values of vectors R and C are transformed into crisp values by using (14)–(16). By applying the Expression (26), the normalization of the vector R is performed:

$$RN(\widehat{R}_1) = \left[\underline{Lim}(\widehat{R}_1), \overline{Lim}(\widehat{R}_1) \right] = \begin{cases} \underline{Lim}(\widehat{R}_1) = \frac{2.740-1.935}{27.027-1.935} = 0.032 \\ \overline{Lim}(\widehat{R}_1) = \frac{25.844-2.740}{27.027-1.935} = 0.953 \end{cases}$$

After normalization, by applying (15), a total normalized crisp value is obtained:

$$\beta_1 = \frac{\underline{Lim}(\widehat{R}_1) \cdot \left\{ 1 - \underline{Lim}(\widehat{R}_1) \right\} + \overline{Lim}(\widehat{R}_1) \cdot \overline{Lim}(\widehat{R}_1)}{1 - \underline{Lim}(\widehat{R}_1) + \overline{Lim}(\widehat{R}_1)} = \frac{0.032 \cdot (1 - 0.032) + 0.953 \cdot 0.953}{1 - 0.032 + 0.953} = 0.489$$

Finally, applying the Expression (16), crisp values of the vector R_1^{crisp} are obtained:

$$R_1^{crisp} = \min_i \left\{ \underline{Lim}(R_i) \right\} + \beta_1 \cdot \left[\max_i \{ \overline{Lim}(R_i) \} - \min_i \left\{ \underline{Lim}(R_i) \right\} \right] = 1.935 + 0.489 \cdot (27.027 - 1.935) = 14.201$$

In the similar way, crisp values of other vectors (R_i^{crisp} and C_i^{crisp}) are obtained and shown in Table 3.

Table 3. Crisp values of vectors R_i^{crisp} and C_i^{crisp}.

Criterion	R_i^{crisp}	C_i^{crisp}	$R_i^{crisp} + C_i^{crisp}$	$R_i^{crisp} - C_i^{crisp}$
C_1	14.201	12.939	27.141	1.262
C_2	10.678	11.270	21.948	−0.592
C_3	11.663	14.171	25.835	−2.508
C_4	11.708	14.628	26.336	−2.920
C_5	15.157	5.545	20.702	9.612
C_6	10.105	14.642	24.747	−4.538
C_7	12.624	9.870	22.495	2.754
C_8	9.290	13.360	22.650	−4.070
C_9	9.919	12.082	22.000	−2.163

Step 6 Determination of the threshold value (α) and creation of the cause-and-effect diagram

Before determination of the α value, conversion of the elements of the rough matrix T into crisp values was performed using Expressions (14)–(16). After obtaining the crisp values, the value of the α is obtained, $\alpha = 1.3935$ (17). Then, α value is used for determination of the cause-and-effect relation between the evaluation criteria. Cause-and-effect relations are presented in CERD (Figure 2). In CERD, the complex relations are visually presented, and provide information about decision making, i.e., which criteria are the most important and how they affect each other. Criteria from the T matrix having values higher than the threshold value, α, are chosen to present cause-and-effect relations.

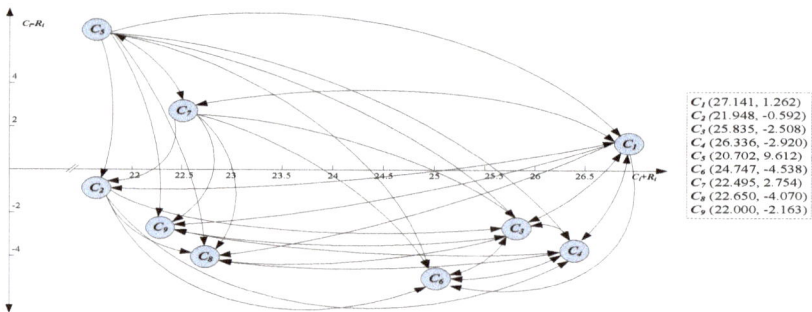

Figure 2. Cause-and-effect relations between criteria—CERD.

Step 7 Determination of the weight coefficients of the criteria

The weight coefficients of the criteria are estimated on the basis of the rough values of the vectors $RN(R_i) + RN(C_i)$ and $RN(R_i) - RN(C_i)$ which are defined in Table 4.

Table 4. Rough values of vectors $RN(R_i)$ and $RN(C_i)$.

Criterion	$RN(R_i)$	$RN(C_i)$	$RN(R_i) + RN(C_i)$	$RN(R_i) - RN(C_i)$
C1	[2.740, 25.844]	[2.551, 23.891]	[5.292, 49.734]	[−21.150, 23.292]
C2	[2.190, 21.469]	[2.067, 22.051]	[4.257, 43.521]	[−19.861, 19.402]
C3	[2.278, 22.797]	[2.779, 25.344]	[5.057, 48.141]	[−23.067, 20.018]
C4	[2.424, 22.705]	[2.541, 26.193]	[4.964, 48.898]	[−23.770, 20.164]
C5	[2.841, 27.027]	[1.387, 13.850]	[4.228, 40.876]	[−11.008, 25.640]
C6	[2.122, 20.699]	[2.477, 26.275]	[4.599, 46.975]	[−24.153, 18.223]
C7	[2.522, 23.897]	[2.171, 19.896]	[4.693, 43.793]	[−17.374, 21.725]
C8	[1.935, 19.686]	[2.554, 24.467]	[4.488, 44.153]	[−22.532, 17.133]
C9	[1.939, 20.628]	[2.464, 22.784]	[4.403, 43.412]	[−20.845, 18.164]

By applying the Expressions (30) and (31), the values of the weight coefficients have been obtained:

$$RN(W_1) = \begin{cases} Lim(W_1) = \sqrt{\left(\underline{Lim}(R_1) + \underline{Lim}(C_1)\right)^2 + \left(\underline{Lim}(R_1) - \overline{Lim}(C_1)\right)^2} = \sqrt{(2.740 + 2.551)^2 + (2.740 - 23.891)^2} = 21.80 \\ \overline{Lim}(W_1) = \sqrt{\left(\overline{Lim}(R_1) + \overline{Lim}(C_1)\right)^2 + \left(\overline{Lim}(R_1) - \underline{Lim}(C_1)\right)^2} = \sqrt{(25.844 + 23.891)^2 + (25.844 - 2.551)^2} = 54.92 \end{cases}$$

Thus, we obtain rough weight vectors:

$$RN(W_j) = \begin{bmatrix} [21.802, \ 54.918] \\ [20.313, \ 47.650] \\ [23.615, \ 52.138] \\ [24.282, \ 52.893] \\ [11.792, \ 48.253] \\ [24.587, \ 50.385] \\ [17.997, \ 48.885] \\ [22.975, \ 47.361] \\ [21.305, \ 47.059] \end{bmatrix}$$

By applying the Expression (31), additive normalization of the obtained rough weight vectors is performed. In other words, rough weight coefficients are down to the interval [0, 1]:

$$RN(w_1) = \left[\frac{\underline{Lim}(W_1)}{\max\limits_{j}\left\{\underline{Lim}(W_j), \overline{Lim}(W_j)\right\}}, \ \frac{\overline{Lim}(W_1)}{\max\limits_{j}\left\{\underline{Lim}(W_j), \overline{Lim}(W_j)\right\}} \right] = \left[\frac{21.80}{54.92}, \frac{54.92}{54.92} \right] = [0.397, \ 1.00]$$

Thus, we obtain rough normalized values of the weight coefficients of the criteria:

$$RN(w_j) = \begin{bmatrix} [0.397, \ 1.000] \\ [0.370, \ 0.868] \\ [0.430, \ 0.949] \\ [0.442, \ 0.963] \\ [0.215, \ 0.879] \\ [0.448, \ 0.917] \\ [0.328, \ 0.890] \\ [0.418, \ 0.862] \\ [0.388, \ 0.857] \end{bmatrix}$$

4.2. Supplier Selection Using Rough EDAS Method

After obtaining the weight values of the criteria, the expert team performed evaluation of the alternatives which is shown in Table 5.

Table 5. Evaluation of the alternatives based on the criteria of seven experts.

	E_1									E_2								
	C_1	C_2	C_3	C_4	C_5	C_6	C_7	C_8	C_9	C_1	C_2	C_3	C_4	C_5	C_6	C_7	C_8	C_9
A_1	7	1	3	9	1	3	5	3	3	9	1	3	7	9	7	5	9	7
A_2	7	3	7	9	3	5	5	5	3	7	3	9	5	7	7	5	7	5
A_3	5	7	7	5	7	7	7	5	7	3	9	7	1	5	5	7	5	5
A_4	5	3	3	5	7	3	9	5	5	3	7	3	1	5	3	7	5	5
A_5	5	9	9	3	7	5	9	5	7	3	9	9	1	5	5	9	5	5
A_6	3	7	7	3	5	3	3	3	3	5	7	7	3	5	5	3	5	3

Table 5. *Cont.*

	C₁	C₂	C₃	C₄	C₅	C₆	C₇	C₈	C₉	C₁	C₂	C₃	C₄	C₅	C₆	C₇	C₈	C₉
				E₃										**E₄**				
A₁	3	1	1	9	1	3	5	1	3	5	3	3	9	3	1	3	1	1
A₂	3	3	5	9	3	5	5	3	1	5	1	5	7	5	3	1	1	3
A₃	5	3	3	7	7	3	5	3	3	7	5	5	7	5	5	3	3	5
A₄	5	5	1	7	5	5	3	3	5	7	3	3	5	5	3	3	3	3
A₅	5	5	5	5	9	7	5	5	5	3	5	7	5	7	7	3	5	5
A₆	3	7	3	5	3	3	3	3	7	5	5	5	5	3	5	3	5	7
				E₅										**E₆**				
A₁	7	1	1	9	3	7	5	7	7	5	3	3	9	1	5	3	5	5
A₂	7	3	7	9	5	7	5	9	5	5	3	7	9	1	3	5	5	5
A₃	5	7	5	5	9	9	7	9	7	5	5	7	7	9	7	5	7	3
A₄	5	5	1	5	9	9	9	9	9	3	3	3	7	7	5	3	5	5
A₅	5	9	9	1	9	5	9	9	9	7	5	9	7	9	7	5	7	3
A₆	3	7	7	3	7	7	3	7	3	5	5	7	7	1	5	3	5	5
				E₇														
A₁	5	3	3	9	1	5	3	5	7									
A₂	5	3	7	9	1	7	5	5	5									
A₃	5	5	7	7	9	7	5	7	5									
A₄	5	3	3	7	7	5	3	5	7									
A₅	5	5	9	7	9	3	5	7	7									
A₆	5	5	7	7	1	5	3	5	7									

Step 1 Conversion of individual matrices into group rough matrix

After computing alternatives by the expert team and converting linguistic values intonumerical, it is necessary to translate individual matrices of all experts into group matrix by applying the matrix according to Zhai et al. [109]. The example of the group matrix elements evaluation is presented in Table 6.

$$\tilde{x}_{11} = \{7,9,3,5,7,5,5\}$$

$$\underline{Lim}(3) = 3.00, \overline{Lim}(3) = \frac{1}{7}(7+9+3+5+7+5+5) = 5.86$$

$$\underline{Lim}(5) = \frac{1}{4}(3+5+5+5) = 4.50, \overline{Lim}(5) = \frac{1}{6}(7+9+5+7+5+5) = 6.33$$

$$\underline{Lim}(7) = \frac{1}{6}(7+3+5+7+5+5) = 5.33, \overline{Lim}(7) = \frac{1}{3}(7+9+7) = 7.67$$

$$\underline{Lim}(9) = \frac{1}{7}(7+9+3+5+7+5+5) = 5.86, \overline{Lim}(9) = 9.00$$

$$RN(x_{11}^1) = RN(x_{11}^5) = [5.33; 7.67]; RN(x_{11}^2) = [5.86; 9.00]; RN(x_{11}^3) = [3.00; 5.86];$$
$$RN(x_{11}^4) = RN(x_{11}^6) = RN(x_{11}^7) = [4.50; 6.33];$$

$$x_{11}^L = \frac{x_{11}^1 + x_{11}^2 + x_{11}^5 + x_{11}^4 + x_{11}^5 + x_{11}^6 + x_{11}^7}{S} = \frac{5.33+5.86+3.00+4.50+5.33+4.50+4.50}{7} = 4.72$$

$$x_{11}^U = \frac{x_{11}^1 + x_{11}^2 + x_{11}^5 + x_{11}^4 + x_{11}^5 + x_{11}^6 + x_{11}^7}{S} = \frac{7.67+9.00+5.86+6.33+7.67+6.33+6.33}{7} = 7.03$$

Table 6. Group rough matrix.

	A_1	A_2	A_3	A_4	A_5	A_6
C_1	[4.72, 7.03]	[4.74, 6.37]	[4.48, 5.52]	[4.00, 5.43]	[4.00, 5.43]	[3.65, 4.63]
C_2	[1.37, 2.35]	[2.47, 2.96]	[4.72, 7.03]	[3.35, 4.99]	[5.73, 7.69]	[5.65, 6.63]
C_3	[2.02, 2.84]	[6.00, 7.43]	[5.01, 6.65]	[2.02, 2.84]	[7.39, 8.83]	[5.39, 6.83]
C_4	[8.47, 8.96]	[7.39, 8.83]	[4.37, 6.60]	[4.12, 6.33]	[2.56, 5.67]	[3.72, 5.73]
C_5	[1.44, 4.26]	[2.22, 4.95]	[6.27, 8.28]	[5.63, 7.26]	[7.01, 8.65]	[3.05, 5.78]
C_6	[3.05, 5.78]	[4.09, 5.91]	[4.97, 7.28]	[3.67, 5.88]	[4.74, 6.37]	[4.00, 5.43]
C_7	[3.37, 4.35]	[3.94, 4.92]	[4.74, 6.37]	[3.77, 6.78]	[5.11, 7.78]	[3.00, 3.00]
C_8	[2.54, 6.38]	[3.39, 6.61]	[4.22, 6.95]	[4.05, 6.03]	[5.35, 6.99]	[4.00, 5.43]
C_9	[3.26, 6.12]	[3.01, 4.65]	[4.09, 5.91]	[4.57, 6.65]	[4.72, 7.03]	[3.93, 6.07]

Step 2   Evaluation of the average solution compared to all criteria

The average solution compared to all criteria is obtained after applying Equation (22):

$$RN(AV) \begin{bmatrix} [3.66,\ 5.74] \\ [3.33,\ 5.28] \\ [3.98,\ 5.90] \\ [4.38,\ 7.02] \\ [3.66,\ 6.53] \\ [3.50,\ 6.11] \\ [3.42,\ 5.53] \\ [3.36,\ 6.40] \\ [3.37,\ 6.07] \end{bmatrix}$$

Step 3   Evaluation of the positive deviation RN(PDA) and negative deviation RN(NDA) from the average solution compared to all criteria

In order to obtain the values of the positive deviation *RN(PDA)* (23), shown in Table 7, and the negative deviation *RN(NDA)* (24) from the average solution *RN(AV)*, shown in Table 8, compared to all criteria, it is necessary to apply the Equations (25)–(31), and take care if criteria belong to the Expenses or Benefit type. The procedure for the Benefit type evaluation is as follows:

First, it is necessary to apply the Equation (25). Since it is the first criterion which belongs to Benefit type and first alternative, it is necessary to calculate the difference between alternative one on criterion one and the average solution for the first criterion:

$$\left[X_{il}^L - AV_j^U; \ X_{il}^U - AV_j^L \right] = [4.72 - 5.74; \ 7.03 - 3.66] = [-1.02; \ 3.37]$$

In this case, the lower limit of the obtained rough number has negative value and it is necessary to apply the Equation (31) and then rough number $[-1.02; \ 3.37]$ goes to its positive value $[1.02; \ 3.37]$. After that, it is necessary that the obtained absolute value is divided by the average solution on first criterion (25):

$$\frac{[1.02;\ 3.37]}{[3.66;\ 5.74]} = \left[\frac{1.02}{5.74}; \ \frac{3.37}{3.66} \right] = [0.18; \ 0.92]$$

For example, for the value of alternative one compared to third criterion, evaluating the difference between that value and average solution, both negative values of rough number are obtained (for upper and lower limit).By applying the Equation (29), values of rough number will be equal to zero.

On the other hand, for the value of alternative five compared to third criterion, both positive values of rough number are obtained (for upper and lower limit). And according to the Equation (30), they keep the same value.

Evaluation of the *RN(PDA)* value for the Expenses type of the criteria is performed in the same way, except that at the beginning, it is necessary to define the deviation of the average solution from the alternative value of the criteria discussed in Equation (27). Evaluation of the *RN(NDA)* value is performed in the same way, only Equations (26) and (28)–(31) are applied.

Table 7. Values of the positive deviation from average solution.

PDA	A_1		A_2		A_3		A_4		A_5		A_6	
C_1	0.18	0.92	0.17	0.74	0.22	0.51	0.30	0.49	0.30	0.49	0.00	0.00
C_2	0.19	1.17	0.07	0.84	0.00	0.00	0.32	0.58	0.00	0.00	0.00	0.00
C_3	0.00	0.00	0.02	0.87	0.15	0.67	0.00	0.00	0.25	1.22	0.09	0.72
C_4	0.00	0.00	0.00	0.00	0.32	0.61	0.28	0.66	0.18	1.02	0.19	0.75
C_5	0.00	0.00	0.00	0.00	0.04	1.26	0.14	0.98	0.07	1.36	0.00	0.00
C_6	0.00	0.00	0.33	0.69	0.19	1.08	0.00	0.00	0.22	0.82	0.00	0.00
C_7	0.00	0.00	0.00	0.00	0.14	0.86	0.32	0.98	0.08	1.28	0.00	0.00
C_8	0.00	0.00	0.47	0.96	0.34	1.07	0.37	0.79	0.16	1.08	0.00	0.00
C_9	0.00	0.00	0.00	0.00	0.33	0.75	0.25	0.97	0.22	1.09	0.35	0.80

Table 8. Values of the negative deviation from average solution.

NDA	A_1		A_2		A_3		A_4		A_5		A_6	
C_1	0.00	0.00	0.00	0.00	0.00	0.00	0.00	0.00	0.00	0.00	0.17	0.57
C_2	0.00	0.00	0.00	0.00	0.11	1.11	0.00	0.00	0.09	1.31	0.07	0.99
C_3	0.19	0.98	0.00	0.00	0.00	0.00	0.19	0.98	0.00	0.00	0.00	0.00
C_4	0.21	1.05	0.05	1.02	0.00	0.00	0.00	0.00	0.00	0.00	0.00	0.00
C_5	0.09	1.39	0.20	1.18	0.00	0.00	0.00	0.00	0.00	0.00	0.32	0.95
C_6	0.37	0.87	0.00	0.00	0.00	0.00	0.39	0.70	0.00	0.00	0.32	0.60
C_7	0.17	0.63	0.27	0.47	0.00	0.00	0.00	0.00	0.00	0.00	0.08	0.74
C_8	0.47	1.15	0.00	0.00	0.00	0.00	0.00	0.00	0.00	0.00	0.32	0.71
C_9	0.45	0.83	0.21	0.91	0.00	0.00	0.00	0.00	0.00	0.00	0.00	0.00

Step 4 Weighting of the matrices RN(PDA) and RN(NDA)

By adopting the fourth step of the Rough EDAS method, i.e., Equation (32), the weighted matrix for the positive deviation from the average value *VPi* is obtained, and is presented in Table 9.

Table 9. Weighted matrix *VPi* for the positive deviation.

VPI	A_1		A_2		A_3		A_4		A_5		A_6	
C_1	0.07	0.92	0.07	0.74	0.09	0.51	0.12	0.49	0.12	0.49	0.00	0.00
C_2	0.07	1.02	0.03	0.73	0.00	0.00	0.12	0.50	0.00	0.00	0.00	0.00
C_3	0.00	0.00	0.01	0.82	0.07	0.64	0.00	0.00	0.11	1.16	0.04	0.68
C_4	0.00	0.00	0.00	0.00	0.14	0.58	0.12	0.64	0.08	0.98	0.09	0.73
C_5	0.00	0.00	0.00	0.00	0.01	1.11	0.03	0.86	0.02	1.20	0.00	0.00
C_6	0.00	0.00	0.15	0.63	0.08	0.99	0.00	0.00	0.10	0.75	0.00	0.00
C_7	0.00	0.00	0.00	0.00	0.05	0.77	0.10	0.88	0.03	1.14	0.00	0.00
C_8	0.00	0.00	0.20	0.83	0.14	0.92	0.15	0.68	0.07	0.93	0.00	0.00
C_9	0.00	0.00	0.00	0.00	0.13	0.65	0.10	0.83	0.09	0.93	0.14	0.69

By applying the Equation (33), weighted matrix for the negative deviation from the average value is obtained and is presented in Table 10.

Table 10. Weighted matrix *VNi* for the negative deviation.

VNI	A$_1$		A$_2$		A$_3$		A$_4$		A$_5$		A$_6$	
C$_1$	0.00	0.00	0.00	0.00	0.00	0.00	0.00	0.00	0.00	0.00	0.07	0.57
C$_2$	0.00	0.00	0.00	0.00	0.04	0.97	0.00	0.00	0.03	1.14	0.03	0.86
C$_3$	0.08	0.93	0.00	0.00	0.00	0.00	0.08	0.93	0.00	0.00	0.00	0.00
C$_4$	0.09	1.01	0.02	0.98	0.00	0.00	0.00	0.00	0.00	0.00	0.00	0.00
C$_5$	0.02	1.22	0.04	1.03	0.00	0.00	0.00	0.00	0.00	0.00	0.07	0.84
C$_6$	0.17	0.80	0.00	0.00	0.00	0.00	0.17	0.64	0.00	0.00	0.14	0.55
C$_7$	0.06	0.56	0.09	0.41	0.00	0.00	0.00	0.00	0.00	0.00	0.02	0.66
C$_8$	0.20	0.99	0.00	0.00	0.00	0.00	0.00	0.00	0.00	0.00	0.14	0.61
C$_9$	0.18	0.72	0.08	0.78	0.00	0.00	0.00	0.00	0.00	0.00	0.00	0.00

Step 5 Determination of the Sum of the Previously Weighted Matrix

By adopting the step five, six, and seven, i.e., the Equations (34)–(38), the final results are obtained and are presented in Table 11. Step 5 summarizes values of the previously weighted matrices *RN(SPi)* (34), *RN(NPi)* (35).

$$RN(SPi) = \begin{bmatrix} [0.14, 1.94] \\ [0.45, 3.76] \\ [0.70, 6.16] \\ [0.74, 4.88] \\ [0.61, 7.57] \\ [0.26, 2.10] \end{bmatrix}; RN(SNi) = \begin{bmatrix} [0.79, 6.23] \\ [0.24, 3.21] \\ [0.04, 0.97] \\ [0.26, 1.57] \\ [0.03, 1.14] \\ [0.46, 4.09] \end{bmatrix}$$

Step 6 Normalization of the RN(SP) and RN(NP) values for all the alternatives

By applying the Equation (36)

$$RN(NSP_1) = \frac{[sp_1^L; sp_1^U]}{max[sp_5^L; sp_5^U]} = \left[\frac{0.14}{7.67}; \frac{1.94}{0.61}\right] = [0.02; 3.20]$$

and (37)

$$RN(NSN_1) = 1 - \frac{[sn_1^L; sn_1^U]}{max[sn_1^L; sn_1^U]} = 1 - \left[\frac{0.79}{6.23}; \frac{6.23}{0.79}\right] = [0.87; -6.89]$$

normalized values are obtained, and are presented in Table 11.

Step 7 Estimation of all RN(ASi) alternatives values and their ranging

By applying the Equation (38), the values *RN(ASi)* are obtained and are presented in Table 11. In the final, seventh step, the ranging towards the falling series of numbers was performed as well. The highest value presents the best solution, whereas the lowest value is the worst solution.

Table 11. Results and ranging of the alternatives.

	SPi	SNi	NSPi	NSNi	ASi	Rank
A$_1$	[0.14, 1.94]	[0.79, 6.23]	[0.02, 3.20]	[0.87, −6.89]	−1.40	5
A$_2$	[0.45, 3.76]	[0.24, 3.21]	[0.06, 6.21]	[0.96, −3.07]	2.08	4
A$_3$	[0.70, 6.16]	[0.04, 0.97]	[0.09, 10.17]	[0.99, −0.22]	5.52	2
A$_4$	[0.74, 4.88]	[0.26, 1.57]	[0.10, 8.06]	[0.96, −0.99]	4.07	3
A$_5$	[0.61, 7.57]	[0.03, 1.14]	[0.08, 12.49]	[0.99, −0.44]	6.56	1
A$_8$	[0.26, 2.10]	[0.46, 4.09]	[0.03, 0.28]	[0.93, −4.19]	−1.48	6

Alternative five, according to the results, presents the best choice.

5. Sensitivity Analysis and Discussion

In order to determine the stability of the obtained results, sensitivity analysis has been performed. In the first part, it assumes the change of the criteria weight through 18 different sets. In the first nine sets, the value of each criterion is reduced by 16%, while the value of the rest is increased by 2%, respectively. In the tenth set, the first, the third, and the fourth criteria are reduced by 12%, and the rest are increased by 6%, while in the eleventh set of values, the same criteria are reduced by 24%, and the rest are increased by 12%. Since the ranges of the alternatives have not considerably changed through previously formed sets, the next sets present more significant change of the criteria weigh value expressed in percentage. Thus, in the 12th set, the first, the third, the fourth and the sixth criteria are decreased by 20%, while the remaining five are increased by 16%, while in the 13th set, the first and the fourth criteria are decreased by even 35%, and the rest are decreased by 10%. The second and the seventh criteria in the 14th set are reduced by 28%, while the rest are increased by 8%. The 15th set is based on the five criteria in total, because the first, third, fourth and sixth have been assigned zero values, and the values of the rest of the criteria stay constant. In the 16th set, all criteria have the same importance, while the lower and the upper limit of rough number is also equal. Then the second, the fifth, and the seventh criteria in the 17th set have been assigned zero values, and the rest stay constant. The last, 18th set has the same values of all criteria (assigned maximum value of the first criterion in the basic calculation). In Figure 3, the ranges of the alternatives of all sets have been presented.

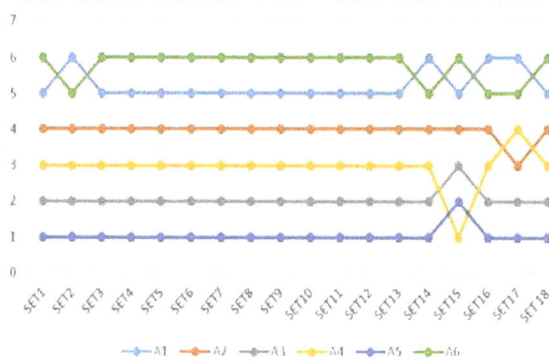

Figure 3. Ranking of alternatives through the scenario.

In Figure 3, the range of alternatives through formed sets has been presented and, as one can see, alternative five presents the best choice in 17 of the 18 formed scenarios. Only in the 15th set, when the individual criteria are eliminated, does it take the second position. The same is true for alternative two, which takes the third position only in the 17th set, while in other scenarios is in the fourth position. Besides the stability shown in the first part of the sensitivity analysis, the comparison of the proposed model with the other hybrid multi-criteria models has also been performed. The hybrid models used for the comparison of the results are shown in Table 14. In order to validate the proposed model for determination of the criteria weight, besides R'DEMATEL method, Rough AHP method [123] has been applied as well. AHP model is chosen for the comparison, since this is the method in which literature has already been used for the ranging of alternatives and determination of the criteria weight [109].

For experts' comparison of criteria sets in Rough AHP method, the same experts as in Rough DEMATEL method participated. After the experts' evaluation of criteria, by applying Saaty's scale, seven matrices of comparison of dimensions 9×9 were constructed in criteria sets. This is shown in Table 12.

Table 12. Experts' comparison of the criteria.

E₁

	C1	C2	C3	C4	C5	C6	C7	C8	C9
C1	1	4	9	7	8	3	6	5	2
C2	0.25	1	5	3	4	0.50	3	2	0.33
C3	0.11	0.20	1	0.33	0.50	0.16	0.25	0.33	0.12
C4	0.14	0.33	3	1	2	0.25	0.50	0.33	0.16
C5	0.12	0.25	2	0.50	1	0.20	0.33	0.25	0.14
C6	0.33	2	6	4	5	1	3	3	0.50
C7	0.16	0.33	4	2	3	0.33	1	1	0.25
C8	0.20	0.50	3	3	4	0.33	2	1	0.33
C9	0.50	3	8	6	7	2	4	3	1

E₂

	C1	C2	C3	C4	C5	C6	C7	C8	C9
C1	1	3	8	6	7	2	5	4	2
C2	0.33	1	5	3	4	0.500	3	2	0.33
C3	0.12	0.20	1	0.33	0.500	0.16	0.25	0.33	0.12
C4	0.16	0.33	3	1	2	0.25	0.500	0.33	0.16
C5	0.14	0.25	2	0.500	1	0.20	0.33	0.25	0.14
C6	0.500	2	6	4	5	1	3	3	0.500
C7	0.20	0.33	4	2	3	0.33	1	0.500	0.25
C8	0.25	0.500	3	3	4	0.33	2	1	0.33
C9	0.50	3	8	6	7	2	4	3	1

...

E₆

	C1	C2	C3	C4	C5	C6	C7	C8	C9
C1	1	4	9	5	8	3	6	5	2
C2	0.25	1	5	2	4	0.50	3	2	0.33
C3	0.11	0.20	1	0.25	0.50	0.16	0.25	0.33	0.12
C4	0.20	0.50	4	1	4	0.33	2	1	0.25
C5	0.12	0.25	2	0.25	1	0.20	0.33	0.25	0.14
C6	0.33	2	6	3	5	1	3	3	0.50
C7	0.16	0.33	4	0.50	4	0.33	1	1	0.25
C8	0.20	0.50	3	1	4	0.33	2	1	0.33
C9	0.50	3	8	4	7	2	4	3	1

E₇

	C1	C2	C3	C4	C5	C6	C7	C8	C9
C1	1	3	8	5	7	2	5	4	3
C2	0.33	1	6	3	4	0.50	4	2	1
C3	0.12	0.16	1	0.25	0.50	0.16	0.33	0.25	0.20
C4	0.20	0.33	4	1	3	0.33	1	0.50	0.33
C5	0.14	0.25	2	0.33	1	0.20	0.33	0.25	0.25
C6	0.50	2	6	3	5	1	3	2	2
C7	0.20	0.25	3	1	3	0.33	1	0.50	0.33
C8	0.25	0.50	4	2	4	0.50	2	1	0.50
C9	0.33	1	5	3	4	0.50	3	2	1

By applying the expressions according to Zhai et al. [109], each of presented sequences is transformed in rough sequence. The example of the group matrix elements evaluation is presented in Table 13.

$$\underline{Lim}(2) = 2.00, \ \overline{Lim}(2) = \frac{1}{7}(4+3+2+3+3+4+3) = 3.14$$

$$\underline{Lim}(3) = \frac{1}{5}(3+2+3+3+3) = 2.80, \ \overline{Lim}(3) = \frac{1}{6}(4+3+3+3+4+3) = 3.33$$

$$\underline{Lim}(4) = \frac{1}{7}(4+3+2+3+3+4+3) = 3.14, \ \overline{Lim}(4) = 4.00$$

$$RN(x_{12}^1) = RN(x_{12}^6) = [3.14; 4.00]; RN(x_{12}^2) = RN(x_{12}^4) = RN(x_{12}^5) = RN(x_{12}^7) = [2.80; 3.33];$$
$$RN(x_{12}^3) = [2.00; 3.14]$$

$$x_{12}^L = \frac{x_{12}^1 + x_{12}^2 + x_{12}^s + x_{12}^4 + x_{12}^5 + x_{12}^6 + x_{12}^7}{S} = \frac{3.14 + 2.80 + 2.00 + 2.80 + 2.80 + 3.14 + 2.80}{7} = 2.78$$

$$x_{12}^U = \frac{x_{12}^1 + x_{12}^2 + x_{12}^s + x_{12}^4 + x_{12}^5 + x_{12}^6 + x_{12}^7}{S} = \frac{4.00 + 3.33 + 3.14 + 3.33 + 3.33 + 4.00 + 3.33}{7} = 3.49$$

Table 13. Group rough matrix.

	A₁	A₂	A₃	A₄	A₅	A₆
C_1	[1, 1]	[2.78, 3.49]	[7.78, 8.49]	[5.51, 6.22]	[7.08, 7.49]	[2.08, 2.49]
C_2	[0.3, 0.38]	[1, 1]	[5.18, 5.99]	[2.78, 3.83]	[4.18, 4.99]	[0.42, 0.63]
C_3	[0.12, 0.13]	[0.17, 0.19]	[1, 1]	[0.29, 0.33]	[0.51, 0.63]	[0.16, 0.17]
C_4	[0.16, 0.18]	[0.28, 0.38]	[3.08, 3.49]	[1, 1]	[2.18, 2.99]	[0.23, 0.29]
C_5	[0.14, 0.14]	[0.2, 0.24]	[1.74, 1.98]	[0.36, 0.47]	[1, 1]	[0.18, 0.20]
C_6	[0.42, 0.49]	[1.78, 2.49]	[6.04, 6.53]	[3.53, 4.51]	[5.06, 5.8]	[1, 1]
C_7	[0.19, 0.22]	[0.26, 0.32]	[3.74, 3.98]	[1.35, 1.92]	[3.02, 3.26]	[0.31, 0.33]
C_8	[0.23, 0.29]	[0.39, 0.49]	[2.27, 2.94]	[2.19, 2.91]	[3.74, 3.98]	[0.34, 0.38]
C_9	[0.50, 1.08]	[2.19, 3.52]	[7.17, 8.22]	[4.66, 6.15]	[6.19, 7.45]	[1.34, 2.23]

After defining group rough matrix (Table 13), it is necessary to define geometric middle of upper and lower limit of group matrix of criteria, i.e., geometric middle of rows is calculated. From the obtained matrix maximum value, the upper limit is chosen, and all other values are divided by that one. In that way, we obtain the final values of the criteria weight:

$$w_j \begin{bmatrix} [0.837, 1.000] \\ [0.354, 0.462] \\ [0.066, 0.073] \\ [0.132, 0.176] \\ [0.083, 0.093] \\ [0.495, 0.610] \\ [0.173, 0.196] \\ [0.229, 0.269] \\ [0.636, 0.880] \end{bmatrix}$$

For ranging of alternatives, the following methods were used: R'MAIRCA [104], R'MULTIMOORA (extended in this research), R'COPRAS (extended in this research), R'MABAC [98]. Combining rough AHP and rough DEMATEL, which were used for the evaluation of the criteria weight with R'MAIRCA, R'MULTIMOORA, R'Ctable

OPRAS, R'MABAC and R'EDAS models, ten hybrid models were developed (in total), which are shown in Table 14. The same input data were used for all models which assumes criteria weight obtained from R′DEMATEL and R′AHP, and the same values from the group matrix by R′EDAS method (Table 5).

As already mentioned in Section 3.1 of the paper, DEMATEL method is very suitable for the design and analysis of the structural model. This is achieved by defining the cause-and-effect relation between complex factors [95]. Cause-and-effect relations are obtained on the basis of the total direct and indirect influences transferred from one factor to the other, but also received from the other factors. The implementation of DEMATEL method explores interdependent factors and determines the level of this dependency. The method is based on the graph theory, and it ensures visual planning and problem solving. In this way, the relative factors can be divided into cause-and-effect, in order to gain a better insight into their mutual relations. This also provides a better understanding of the complex structure of this problem, relations between the factors, and relations between the structure level and influence of the factors [99]. Based on all of the above, DEMATEL gives more objective insight into weight coefficient values. Besides, the proposed modification of the DEMATEL using RN makes it possible to take into account doubts that occur during the expert evaluation of criteria, thus bridging the existing gap in the methodology in the treatment of uncertainty based on RN. In Section 3.2, one of the reasons for using EDAS method is given: mathematical apparatus which assumes evaluation of alternatives on the basis of positive and negative deviations from the average solution. Such model presents very important support in making decisions in everyday conflict situations. Therefore, in a very short time, EDAS method has found its way in wide application in engineering and business problems. This method [111] has a number of extensions, and the extension by fuzzy logics [114] is performed exactly in the field of supply chain for supplier selection. Several studies have already been published in different fields, where this method has been applied in its traditional form or some other forms [113–122]. Besides the mentioned advantages of EDAS method, we can point to additional advantages important for this paper: (1) stability of the solution on change of nature and character of the criteria; (2) provides well-structured analytical frame for ranging of alternatives; (3) the number of steps stays the same no matter of the number of criteria; (4) very useful in the case of large number of alternatives and criteria; (5) applicable for the qualitative and quantitative type of criteria; (6) provides the possibility of the stability analysis of the model regarding weight factor interval change.

Some of the advantages mentioned for the EDAS model are also present in the three other multi-criteria models (MABAC, COPRAS and MULTIMOORA) used for the validation of the research results. According to the research of Pamučar and Ćirović (2015), MABAC, COPRAS and MULTIMOORA belong to the group of multi-criteria models providing stable solutions no matter of change of nature and character of criteria. Besides, they offer: (1) very well structured analytical frame for ranging of alternatives; (2) the number of steps stays the same no matter of the number of criteria; (3) applicable even when the information of certain attributes is missing; (4) provides ranging of alternatives in coordinate scales without normalization process; (5) the possibility of using the limit value for obtaining the preferences; (6) can be used for both qualitative and quantitative type of the criteria and offers the possibility of the stability analysis regarding the weight coefficients change. Taking all this into account, we can conclude that these models are very credible and have wide application in solving a number of multi-criteria problems.

Table 14. The comparative analysis of the ranges by applying different approaches.

Alternative	R'D-R-MAIRCA		R'D-R-MULTIMOORA		R'D-R-COPRAS		R'D-R-MABAC		R'D-R-EDAS	
	Value	Rank	Value	Rank	Value	Rank	Value	Rank	Value	Rank
A1	−0.243	6	4.055	6	63.32	6	8.469	6	−3.690	5
A2	−0.324	4	4.392	3	79.70	3	8.875	4	3.139	4
A3	−0.543	2	5.080	2	88.60	2	9.968	2	9.949	2
A4	−0.402	3	4.406	4	78.93	4	9.263	3	7.073	3
A5	−0.696	1	5.402	1	100.00	1	10.73	1	12.053	1
A6	−0.170	5	4.137	5	77.23	5	8.106	5	−3.913	6

Alternative	R'A-R-MAIRCA		R'A-R-MULTIMOORA		R'A-R-COPRAS		R'A-R-MABAC		R'A-R-EDAS	
	Value	Rank	Value	Rank	Value	Rank	Value	Rank	Value	Rank
A1	0.024	3	0.711	4	90.72	3	−2.638	3	0.255	5
A2	0.115	5	0.796	5	93.58	5	−2.470	5	2.312	4
A3	0.064	2	0.928	2	98.60	2	−2.334	2	4.461	2
A4	0.075	4	0.810	3	94.25	4	−2.489	4	3.928	3
A5	0.034	1	0.966	1	100.00	1	−2.295	1	4.844	1
A6	0.225	6	0.820	6	83.13	6	−2.746	6	−1.461	6

* R'D-R-MAIRCA (Rough DEMATEL rough MAIRCA); R'D-R-MULTIMOORA (Rough DEMATEL rough MULTIMOORA); R'D-R-COPRAS (Rough DEMATEL rough COPRAS); R'D-R-EDAS (Rough DEMATEL rough EDAS); R'A-R-MAIRCA (Rough AHP rough MAIRCA); R'A-R-MULTIMOORA (Rough AHP rough MULTIMOORA); R'A-R-COPRAS (Rough AHP rough COPRAS); R'A-R-MABAC (Rough AHP rough MABAC); R'A-R-EDAS (Rough AHP rough EDAS).

Ranges of alternatives and the values of multi-criteria functions of hybrid models are shown in Table 14. In Figure 4, the changes of the ranges of hybrid models are graphically presented.

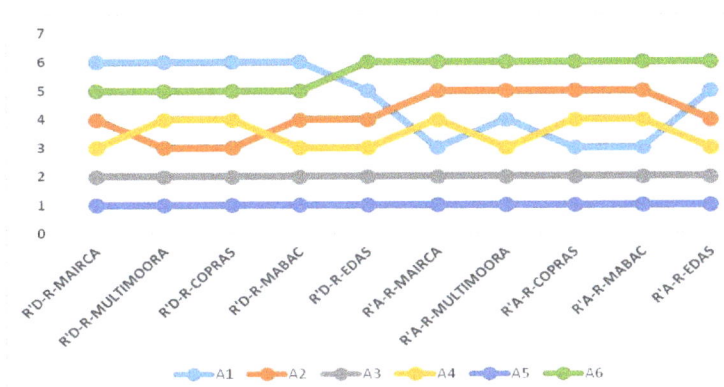

Figure 4. Alternative ranges in combination of R'DEMATEL and R'AHP.

Figure 4 shows that alternative five turned out to be the best choice in all 10 models, adequately verifying the proposed model. Alternative A3 is also in all formulated models in the same position—second place. In other alternatives, certain differences are present, as well as dependence from the model applied. Alternative one has the largest variation of all, because it appeared in the last position four times, two times in the fifth position, three times in the third position, and once at the fourth position. Alternative six is, in all models, placed in the fifth or sixth place. Alternative two varies from third to fifth place, and alternative four is in third and fourth place in half of the cases.

For the statistical comparison of the ranges, Spirman coefficient of correlation was used (r_k). The comparison of the ranges was performed through mutual comparison of all 10 hybrid models (Table 15).

Regarding Table 15, one can notice a good correlation of ranges between the considered approaches, since the total average value of r_k is 0.893. The least values of the correlation are obtained by comparison of the ranges R'A-R-MAIRCA model with R'D-R-MAIRCA, R'D-R-MULTIMOORA, R'D-R-COPRAS, and R'D-R-MABAC models, where the obtained values of r_k are 0.657, 0.600, 0.600, and 0.657, respectively. The similar values are also obtained by comparison of R'A-R-COPRAS and R'A-R-MABAC with R'A-R-MAIRCA, R'D-R-MULTIMOORA, R'D-R-COPRAS, and R'D-R-MABAC models. These variations in r_k values arise from application of different approaches for estimation of the weight coefficients. So, the obtained values of the weight coefficients have further influenced the changes in the ranges of the observed models. Therefore, in further analysis, we grouped the models using the same approaches for evaluation of the weight coefficients (R'AHP and R'DEMATEL) and analyzed their mutual correlation. So, for R'AHP model we obtain average r_k = 0.951, while for R'DEMATEL model, we obtain r_k = 0.962. Taking into account that all values of r_k in the frame of approaches (R'AHP and R'DEMATEL) are considerably higher than 0.8, as well as the average value of r_k is 0.893, we can conclude that there is a very high correlation of the ranges, and that the proposed model has been confirmed.

Table 15. Correlation of the ranges of tested models.

Methods	R'D-R-MAI RCA	R'D-R-MULTI MOORA	R'D-R-COP RAS	R'D-R-MA BAC	R'D-R- EDAS	R'A-R-MA IRCA	R'A-R-MULTI MOORA	R'A-R-COP RAS	R'A-R-MA BAC	R'A-R- EDAS	Average
R'D-R-MAIRCA	1.000	0.943	0.943	1.000	0.943	0.657	0.829	0.657	0.657	0.943	0.857
R'D-R-MULTIMOORA	-	1.000	1.000	0.943	0.886	0.600	0.714	0.600	0.600	0.886	0.803
R'D-R-COPRAS	-	-	1.000	0.943	0.886	0.600	0.714	0.600	0.600	0.886	0.779
R'D-R-MABAC	-	-	-	1.000	0.943	0.657	0.829	0.657	0.657	0.943	0.812
R'D-R-EDAS	-	-	-	-	1.000	0.829	0.943	0.829	0.829	1.000	0.905
R'A-R-MAIRCA	-	-	-	-	-	1.000	0.943	1.000	1.000	0.829	0.954
R'A-R-MULTIMOORA	-	-	-	-	-	-	1.000	0.943	0.943	0.943	0.957
R'A-R-COPRAS	-	-	-	-	-	-	-	1.000	1.000	0.829	0.943
R'A-R-MABAC	-	-	-	-	-	-	-	-	1.000	0.829	0.915
R'A-R-EDAS	-	-	-	-	-	-	-	-	-	1.000	1.000
Overall average											0.893

6. Conclusions

A very important aspect for objective decision making in multi-criteria models is taking into account the uncertainty and imprecision. What often happens are difficulties in presenting information, regarding the attributes of some decisions, through very precise (correct) numerical values. These difficulties are the consequence of certain doubts in decision making, as well as the complexity and uncertainty of many real factors. In this paper, R'DEMATEL–R'EDAS model has been presented. It provides the quantification of the imprecision in group decision making by applying rough numbers. The main idea is based on the interval approach, which assumes application of interval numbers for the presentation of the attribute value. The advantages for the application of rough numbers are numerous. Rough numbers use exclusively intern findings for the presentation of the attribute value. In this way, subjectivity and assumptions are eliminated, since they can considerably influence the attribute value and the final choice of the alternative. In the application of rough numbers, instead of additional/external parameters, only the given data are used. In this way, the uncertainties already present in the date are used, which has considerable influence on the objectivity of the decision-making process. One of the additional advantages of using rough numbers is their application on the sets with small amounts of data, for which traditional statistical models are not suitable.

The application of rough numbers in multi-criteria decision making is presented through the R'DEMATEL and R'EDAS hybrid model. The case study for the evaluation of the supplier in the construction company has been performed. This study shows that rough numbers can be efficiently applied in multi-criteria decision-making models, and can very well handle the doubts appearing in decision making. Another important part of this paper is presenting new R'DEMATEL and R'EDAS models, developed by the authors, which is a contribution to the present MCDM literature. The hybrid R'DEMATEL–R'EDAS model provides objective aggregation of experts' decisions and takes into account subjectivity and uncertainty present in group decision making. Besides, two new approaches have been introduced based on the combination of MCDM and rough numbers: COPRAS and Rough MULTIMOORA. Development of these models additionally contributes to the literature considering theoretical and practical aspects of multi-criteria techniques.

Besides the general contribution in the field of MCDM, the proposed models help in the field of the supplier selection in the supply chains. It is shown that R'DEMATEL–R'EDAS model enables evaluation of the supplier, despite the imprecision and lack of quantitative information present in decision making process. In this way, the evaluation methodology and the supplier selection in the construction company is improved. To our knowledge, there is no application of such or similar approaches in the literature.

Since our approach is new and not much explored yet, future research will be in the direction of the application of rough numbers in the traditional models for estimation of the weight coefficients (for example, BestWorst Method). Also, one of the future aspects will be the integration of rough numbers in fuzzy numbers, and application of fuzzy-rough numbers in MCDM. This would considerably improve the exploitation of uncertainty and subjectivity always present in the decision-making process.

Author Contributions: Each author has participated and contributed sufficiently to take public responsibility for appropriate portions of the content.

Conflicts of Interest: The authors declare no conflict of interest.

References

1. Soheilirad, S.; Govindan, K.; Mardani, A.; Zavadskas, E.K.; Nilashi, M.; Zakuan, N. Application of data envelopment analysis models in supply chain management: A systematic review and meta-analysis. *Ann. Oper. Res.* **2017**, 1–55. [CrossRef]
2. Bai, C.; Sarkis, J. Integrating sustainability into supplier selection with grey system and rough set methodologies. *Int. J. Prod. Econ.* **2010**, *124*, 252–264. [CrossRef]

3. Ramanathan, R. Supplier selection problem: Integrating DEA with the approaches of total cost of ownership and AHP. *Supply Chain Manag. Int. J.* **2007**, *12*, 258–261. [CrossRef]
4. Zhong, L.; Yao, L. An ELECTRE I-based multi-criteria group decision making method with interval type-2 fuzzy numbers and its application to supplier selection. *Appl. Soft Comput.* **2017**, *57*, 556–576. [CrossRef]
5. Bai, C.; Sarkis, J. Evaluating supplier development programs with a grey based rough set methodology. *Expert Syst. Appl.* **2011**, *38*, 13505–13517. [CrossRef]
6. Zolfani, S.H.; Chen, I.S.; Rezaeiniya, N.; Tamošaitienė, J. A hybrid MCDM model encompassing AHP and COPRAS-G methods for selecting company supplier in Iran. *Technol. Econ. Dev. Econ.* **2012**, *18*, 529–543. [CrossRef]
7. Cox, A.; Ireland, P. Managing construction supply chains: The common sense approach. *Eng. Constr. Archit. Manag.* **2002**, *9*, 409–418. [CrossRef]
8. Saaty, T.L.; Tran, L.T. On the invalidity of fuzzifying numerical judgments in the Analytic Hierarchy Process. *Math. Comput. Model.* **2007**, *46*, 962–975. [CrossRef]
9. Wang, Y.M.; Luo, Y.; Hua, Z. On the extent analysis method for fuzzy AHP and its applications. *Eur. J. Oper. Res.* **2008**, *186*, 735–747. [CrossRef]
10. Garg, H.; Arora, R. Generalized and group-based generalized intuitionistic fuzzy soft sets with applications in decision-making. *Appl. Intell.* **2017**, 1–14. [CrossRef]
11. Garg, H. Generalized interaction aggregation operators in intuitionistic fuzzy multiplicative preference environment and their application to multicriteria decision-making. *Appl. Intell.* **2017**, 1–17. [CrossRef]
12. Garg, H. Some Picture Fuzzy Aggregation Operators and Their Applications to Multicriteria Decision-Making. *Arab. J. Sci. Eng.* **2017**, *42*, 5275–5290. [CrossRef]
13. Garg, H. Confidence levels based Pythagorean fuzzy aggregation operators and its application to decision-making process. *Comput. Math. Organ. Theory* **2017**, *23*, 546–571. [CrossRef]
14. Garg, H.; Arora, R. A nonlinear-programming methodology for multi-attribute decision-making problem with interval-valued intuitionistic fuzzy soft sets information. *Appl. Intell.* **2017**, 1–16. [CrossRef]
15. Lee, C.; Lee, H.; Seol, H.; Park, Y. Evaluation of new service concepts using rough set theory and group analytic hierarchy process. *Expert Syst. Appl.* **2012**, *39*, 3404–3412. [CrossRef]
16. Garg, H. A new generalized improved score function of interval-valued intuitionistic fuzzy sets and applications in expert systems. *Appl. Soft Comput.* **2016**, *38*, 988–999. [CrossRef]
17. Garg, H. A novel accuracy function under interval-valued pythagorean fuzzy environment for solving multicriteria decision making problem. *J. Intell. Fuzzy Syst.* **2016**, *31*, 529–540. [CrossRef]
18. Garg, H. Generalized Pythagorean Fuzzy Geometric Aggregation Operators Using Einstein t-Norm and t-Conorm for Multicriteria Decision-Making Process. *Int. J. Intell. Syst.* **2017**, *32*, 597–630. [CrossRef]
19. Lima-Junior, F.R.; Carpinetti, L.C.R. A multicriteria approach based on fuzzy QFD for choosing criteria for supplier selection. *Comput. Ind. Eng.* **2016**, *101*, 269–285. [CrossRef]
20. Vonderembse, M.A.; Tracey, M. The impact of supplier selection criteria and supplier involvement on manufacturing performance. *J. Supply Chain Manag.* **1999**, *35*, 33–39. [CrossRef]
21. Dickson, G.W. An analysis of vendor selection and the buying process. *J. Purch.* **1966**, *2*, 5–17. [CrossRef]
22. Teeravaraprug, J. Outsourcing and vendor selection model based on Taguchi loss function. *Songklanakarin J. Sci. Technol.* **2008**, *30*, 523–530.
23. Liao, C.N. Supplier selection project using an integrated Delphi, AHP and Taguchi loss function. *Probstat Forum* **2010**, *3*, 118–134.
24. Parthiban, P.; Zubar, H.A.; Garge, C.P. A multi criteria decision making approach for suppliers selection. *Procedia Eng.* **2012**, *38*, 2312–2328. [CrossRef]
25. Mehralian, G.; Rajabzadeh Gatari, A.; Morakabati, M.; Vatanpour, H. Developing a suitable model for supplier selection based on supply chain risks: An empirical study from Iranian pharmaceutical companies. *Iran. J. Pharm. Res.* **2012**, *11*, 209–219. [PubMed]
26. Cristea, C.; Cristea, M. A multi-criteria decision making approach for supplier selection in the flexible packaging industry. *MATEC Web Conf.* **2017**, *94*, 06002. [CrossRef]
27. Fallahpour, A.; Olugu, E.U.; Musa, S.N. A hybrid model for supplier selection: Integration of AHP and multi expression programming (MEP). *Neural Comput. Appl.* **2017**, *28*, 499–504. [CrossRef]

28. Weber, C.A.; Current, J.R.; Benton, W.C. Vendor selection criteria and methods. *Eur. J. Oper. Res.* **1991**, *50*, 2–18. [CrossRef]
29. Tam, M.C.; Tummala, V.R. An application of the AHP in vendor selection of a telecommunications system. *Omega* **2001**, *29*, 171–182. [CrossRef]
30. Muralidharan, C.; Anantharaman, N.; Deshmukh, S.G. A multi-criteria group decision making model for supplier rating. *J. Supply Chain Manag.* **2002**, *38*, 22–33. [CrossRef]
31. Simpson, P.M.; Siguaw, J.A.; White, S.C. Measuring the performance of suppliers: An analysis of evaluation processes. *J. Supply Chain Manag.* **2002**, *38*, 29–41. [CrossRef]
32. Kannan, V.R.; Choon Tan, K. Buyer-supplier relationships: The impact of supplier selection and buyer-supplier engagement on relationship and firm performance. *Int. J. Phys. Distrib. Logist. Manag.* **2006**, *36*, 755–775. [CrossRef]
33. Gencer, C.; Gürpinar, D. Analytic network process in supplier selection: A case study in an electronic firm. *Appl. Math. Model.* **2007**, *31*, 2475–2486. [CrossRef]
34. Chan, F.T.; Kumar, N. Global supplier development considering risk factors using fuzzy extended AHP-based approach. *Omega* **2007**, *35*, 417–431. [CrossRef]
35. Guo, X.; Yuan, Z.; Tian, B. Supplier selection based on hierarchical potential support vector machine. *Expert Syst. Appl.* **2009**, *36*, 6978–6985. [CrossRef]
36. Lee, A.H. A fuzzy supplier selection model with the consideration of benefits, opportunities, costs and risks. *Expert Syst. Appl.* **2009**, *36*, 2879–2893. [CrossRef]
37. Wang, W.P. A Fuzzy linguistic computing approach to supplier selection. *Appl. Math. Model.* **2010**, *34*, 3130–3141. [CrossRef]
38. Lam, K.C.; Tao, R.; Lam, M.C.K. A material supplier selection model for property developers using fuzzy principal component analysis. *Autom. Constr.* **2010**, *19*, 608–618. [CrossRef]
39. Balezentis, A.; Balezentis, T. An innovative multi-criteria supplier selection based on two-tuple MULTIMOORA and hybrid data. *Econ. Comput. Econ. Cybern. Stud. Res.* **2011**, *45*, 37–56.
40. Raut, R.D.; Bhasin, H.V.; Kamble, S.S. Evaluation of supplier selection criteria by combination of AHP and fuzzy DEMATEL method. *Int. J. Bus. Innov. Res.* **2011**, *5*, 359–392. [CrossRef]
41. Zeydan, M.; Çolpan, C.; Çobanoğlu, C. A combined methodology for supplier selection and performance evaluation. *Expert Syst. Appl.* **2011**, *38*, 2741–2751. [CrossRef]
42. Jamil, N.; Besar, R.; Sim, H.K. A Study of Multicriteria Decision Making for Supplier Selection in Automotive Industry. *J. Ind. Eng.* **2013**, *2013*, 841584. [CrossRef]
43. Kilic, H.S. An integrated approach for supplier selection in multi-item/multi-supplier environment. *Appl. Math. Model.* **2013**, *37*, 7752–7763. [CrossRef]
44. Uygun, Ö.; Kaçamak, H.; Ayşim, G.; Şimşir, F. Supplier selection for automotive industry using multi-criteria decision making techniques. *Tojsat Online J. Sci. Technol.* **2013**, *3*, 126–137.
45. Hruška, R.; Průša, P.; Babić, D. The use of AHP method for selection of supplier. *Transport* **2014**, *29*, 195–203. [CrossRef]
46. Özbek, A. Supplier Selection with Fuzzy. *Tojsat J. Econ. Sustain. Dev.* **2015**, *6*, 114–125.
47. Stević, Ž.; Tanackov, I.; Vasiljević, M.; Novarlić, B.; Stojić, G. An integrated fuzzy AHP and TOPSIS model for supplier evaluation. *Serbian J. Manag.* **2016**, *11*, 15–27. [CrossRef]
48. Tamošaitienė, J.; Zavadskas, E.K.; Šileikaitė, I.; Turskis, Z. A novel hybrid MCDM approach for complicated supply chain management problems in construction. *Procedia Eng.* **2017**, *172*, 1137–1145. [CrossRef]
49. Wang, T.K.; Zhang, Q.; Chong, H.Y.; Wang, X. Integrated Supplier Selection Framework in a Resilient Construction Supply Chain: An Approach via Analytic Hierarchy Process (AHP) and Grey Relational Analysis (GRA). *Sustainability* **2017**, *9*, 289. [CrossRef]
50. Birgün Barla, S. A case study of supplier selection for lean supply by using a mathematical model. *Logist. Inf. Manag.* **2003**, *16*, 451–459. [CrossRef]
51. Wang, G.; Huang, S.H.; Dismukes, J.P. Product-driven supply chain selection using integrated multi-criteria decision-making methodology. *Int. J. Prod. Econ.* **2004**, *91*, 1–15. [CrossRef]
52. Ting, S.C.; Cho, D.I. An integrated approach for supplier selection and purchasing decisions. *Supply Chain Manag.* **2008**, *13*, 116–127. [CrossRef]
53. Sawik, T.; Single, V.S. Multiple objective supplier selection in make to order environment. *Omega* **2010**, *38*, 203–212. [CrossRef]

54. Yücenur, G.N.; Vayvay, Ö.; Demirel, N.Ç. Supplier selection problem in global supply chains by AHP and ANP approaches under fuzzy environment. *Int. J. Adv. Manuf. Technol.* **2011**, *56*, 823–833. [CrossRef]

55. Rezaei, J.; Fahim, P.B.; Tavasszy, L. Supplier selection in the airline retail industry using a funnel methodology: Conjunctive screening method and fuzzy AHP. *Expert Syst. Appl.* **2014**, *41*, 8165–8179. [CrossRef]

56. Büyüközkan, G.; Göçer, F. Application of a new combined intuitionistic fuzzy MCDM approach based on axiomatic design methodology for the supplier selection problem. *Appl. Soft Comput.* **2017**, *52*, 1222–1238. [CrossRef]

57. Hudymáčová, M.; Benková, M.; Pócsová, J.; Škovránek, T. Supplier selection based on multi-criterial AHP method. *Acta Montan. Slovaca* **2010**, *15*, 249–255.

58. Lin, H.T.; Chang, W.L. Order selection and pricing methods using flexible quantity and fuzzy approach for buyer evaluation. *Eur. J. Oper. Res.* **2008**, *187*, 415–428. [CrossRef]

59. Ellram, L.M. The supplier selection decision in strategic partnerships. *J. Purch. Mater. Manag.* **1990**, *26*, 8–14. [CrossRef]

60. Çebi, F.; Bayraktar, D. An integrated approach for supplier selection. *Logist. Inf. Manag.* **2003**, *16*, 395–400. [CrossRef]

61. Zavadskas, E.K.; Vainiūnas, P.; Turskis, Z.; Tamošaitienė, J. Multiple criteria decision support system for assessment of projects managers in construction. *Int. J. Inf. Technol. Decis. Mak.* **2012**, *11*, 501–520. [CrossRef]

62. Antuchevičiene, J.; Zavadskas, E.K.; Zakarevičius, A. Multiple criteria construction management decisions considering relations between criteria. *Technol. Econ. Dev. Econ.* **2010**, *16*, 109–125. [CrossRef]

63. Garg, H. Generalized Intuitionistic Fuzzy Entropy-Based Approach for Solving Multi-attribute Decision-Making Problems with Unknown Attribute Weights. *Proc. Natl. Acad. Sci. India Sect. A Phys. Sci.* **2017**, 1–11. [CrossRef]

64. Garg, H. Generalized intuitionistic fuzzy interactive geometric interaction operators using Einstein t-norm and t-conorm and their application to decision making. *Comput. Ind. Eng.* **2017**, *101*, 53–69. [CrossRef]

65. Zavadskas, E.K.; Turskis, Z.; Tamošaitiene, J. Risk assessment of construction projects. *J. Civ. Eng. Manag.* **2010**, *16*, 33–46. [CrossRef]

66. Tamošaitienė, J.; Zavadskas, E.K.; Turskis, Z. Multi-criteria risk assessment of a construction project. *Procedia Comput. Sci.* **2013**, *17*, 129–133. [CrossRef]

67. Yao, M.; Minner, S. Review of multi-supplier inventory models in supply chain management: An update. *SSRN Electron. J.* **2017**. [CrossRef]

68. Izadikhah, M. Group decision making process for supplier selection with TOPSIS method under interval-valued intuitionistic fuzzy numbers. *Adv. Fuzzy Syst.* **2012**, *2012*, 407942. [CrossRef]

69. Eshtehardian, E.; Ghodousi, P.; Bejanpour, A. Using ANP and AHP for the supplier selection in the construction and civil engineering companies; case study of Iranian company. *KSCE J. Civ. Eng.* **2013**, *17*, 262–270. [CrossRef]

70. Fouladgar, M.M.; Yazdani-Chamzini, A.; Zavadskas, E.K.; Haji Moini, S.H. A new hybrid model for evaluating the working strategies: Case study of construction company. *Technol. Econ. Dev. Econ.* **2012**, *18*, 164–188. [CrossRef]

71. Zavadskas, E.K.; Turskis, Z.; Tamosaitiene, J. Selection of construction enterprises management strategy based on the SWOT and multi-criteria analysis. *Arch. Civ. Mech. Eng.* **2011**, *11*, 1063–1082. [CrossRef]

72. Erdogan, S.A.; Šaparauskas, J.; Turskis, Z. Decision Making in Construction Management: AHP and Expert Choice Approach. *Procedia Eng.* **2017**, *172*, 270–276. [CrossRef]

73. Turskis, Z.; Lazauskas, M.; Zavadskas, E.K. Fuzzy multiple criteria assessment of construction site alternatives for non-hazardous waste incineration plant in Vilnius city, applying ARAS-F and AHP methods. *J. Environ. Eng. Landsc. Manag.* **2012**, *20*, 110–120. [CrossRef]

74. Zavadskas, E.K.; Vilutienė, T.; Turskis, Z.; Šaparauskas, J. Multi-criteria analysis of Projects' performance in construction. *Arch. Civ. Mech. Eng.* **2014**, *14*, 114–121. [CrossRef]

75. Petković, D.; Madić, M.; Radovanović, M.; Gečevska, V. Application of the performance selection index method for solving machining MCDM problems. *FU Mech. Eng.* **2017**, *15*, 97–106.

76. Ristić, M.; Manić, M.; Mišić, D.; Kosanović, M.; Mitković, M. Implant material selection using expert system. *FU Mech. Eng.* **2017**, *15*, 133–144.

77. Stefanović-Marinović, J.; Troha, S.; Milovančević, M. An application of multicriteria optimization to the two-carrier two-speed planetary cear trains. *FU Mech. Eng.* **2017**, *15*, 85–95.

78. Eraslan, E.; Atalay, K.D. A Comparative holistic fuzzy approach for evaluation of the chain performance of suppliers. *J. Appl. Math.* **2014**, *2014*, 109821. [CrossRef]
79. Liao, C.N.; Fu, Y.K.; Wu, L.C. Integrated FAHP, ARAS-F and MSGP methods for green supplier evaluation and selection. *Technol. Econ. Dev. Econ.* **2016**, *22*, 651–669. [CrossRef]
80. Saad, S.M.; Kunhu, N.; Mohamed, A.M. A fuzzy-AHP multi-criteria decision-making model for procurement process. *Int. J. Logist. Syst. Manag.* **2016**, *23*, 1–24. [CrossRef]
81. Bali, S.; Amin, S.S. An analytical framework for supplier evaluation and selection: A multi-criteria decision making approach. *Int. J. Adv. Oper. Manag.* **2017**, *9*, 57–72.
82. Kabi, A.A.; Hussain, M.; Khan, M. Assessment of supplier selection for critical items in public organisations of Abu Dhabi. *World Rev. Sci. Technol. Sustain. Dev.* **2017**, *13*, 56–73. [CrossRef]
83. Secundo, G.; Magarielli, D.; Esposito, E.; Passiante, G. Supporting decision-making in service supplier selection using a hybrid fuzzy extended AHP approach: A case study. *Bus. Process Manag. J.* **2017**, *23*, 196–222. [CrossRef]
84. Yang, J.L.; Tzeng, G.H. An integrated MCDM technique combined with DEMATEL for a novel cluster-weighted with ANP method. *Expert Syst. Appl.* **2011**, *38*, 1417–1424. [CrossRef]
85. Gharakhani, D. The evaluation of supplier selection criteria by fuzzy DEMATEL method. *J. Basic Appl. Sci. Res.* **2012**, *2*, 3215–3224.
86. Ho, L.H.; Feng, S.Y.; Lee, Y.C.; Yen, T.M. Using modified IPA to evaluate supplier's performance: Multiple regression analysis and DEMATEL approach. *Expert Syst. Appl.* **2012**, *39*, 7102–7109. [CrossRef]
87. Hsu, C.W.; Kuo, T.C.; Chen, S.H.; Hu, A.H. Using DEMATEL to develop a carbon management model of supplier selection in green supply chain management. *J. Clean. Prod.* **2013**, *56*, 164–172. [CrossRef]
88. Lin, R.J. Using fuzzy DEMATEL to evaluate the green supply chain management practices. *J. Clean. Prod.* **2013**, *40*, 32–39. [CrossRef]
89. Mangla, S.; Kumar, P.; Barua, M.K. An evaluation of attribute for improving the green supply chain performance via DEMATEL method. *Int. J. Mech. Eng. Robot. Res.* **2014**, *1*, 30–35.
90. Wu, K.J.; Tseng, M.L.; Chiu, A.S.; Lim, M.K. Achieving competitive advantage through supply chain agility under uncertainty: A novel multi-criteria decision-making structure. *Int. J. Prod. Econ.* **2017**, *190*, 96–107. [CrossRef]
91. Chang, B.; Chang, C.W.; Wu, C.H. Fuzzy DEMATEL method for developing supplier selection criteria. *Expert Syst. Appl.* **2011**, *38*, 1850–1858. [CrossRef]
92. Iirajpour, A.; Hajimirza, M.; Alavi, M.G.; Kazemi, S. Identification and evaluation of the most effective factors in green supplier selection using DEMATEL method. *J. Basic Appl. Sci. Res.* **2012**, *2*, 4485–4493.
93. Sarkar, S.; Lakha, V.; Ansari, I.; Maiti, J. Supplier Selection in Uncertain Environment: A Fuzzy MCDM Approach. In *Proceedings of the First International Conference on Intelligent Computing and Communication*; Springer: Singapore, 2017; pp. 257–266.
94. Song, W.; Ming, X.; Wu, Z.; Zhu, B. A rough TOPSIS approach for failure mode and effects analysis in uncertain environments. *Qual. Reliab. Eng. Int.* **2014**, *30*, 473–486. [CrossRef]
95. Pamučar, D.; Ćirović, G. The selection of transport and handling resources in logistics centers using Multi-Attributive Border Approximation area Comparison (MABAC). *Expert Syst. Appl.* **2015**, *42*, 3016–3028. [CrossRef]
96. Gigović, L.; Pamučar, D.; Božanić, D.; Ljubojević, S. Application of the GIS-DANP-MABAC multi-criteria model for selecting the location of wind farms: A case study of Vojvodina, Serbia. *Renew. Energy* **2017**, *103*, 501–521. [CrossRef]
97. Pamučar, D.; Petrović, I.; Ćirović, G. Modification of the Best-Worst and MABAC methods: A novel approach based on interval-valued fuzzy-rough numbers. *Expert Syst. Appl.* **2017**, *91*, 89–106. [CrossRef]
98. Roy, J.; Chatterjee, K.; Bandhopadhyay, A.; Kar, S. Evaluation and selection of Medical Tourism sites: A rough AHP based MABAC approach. *arXiv* **2016**, arXiv:1606.08962.
99. Gigović, L.; Pamučar, D.; Bajić, Z.; Drobnjak, S. Application of GIS-Interval Rough AHP Methodology for Flood Hazard Mapping in Urban Areas. *Water* **2017**, *9*, 360. [CrossRef]
100. Khoo, L.-P.; Zhai, L.-Y. A prototype genetic algorithm enhanced rough set-based rule induction system. *Comput. Ind.* **2001**, *46*, 95–106. [CrossRef]

101. Zou, Z.; Tseng, T.L.B.; Sohn, H.; Song, G.; Gutierrez, R. A rough set based approach to distributor selection in supply chain management. *Expert Syst. Appl.* **2011**, *38*, 106–115. [CrossRef]

102. Nauman, M.; Nouman, A.; Yao, J.T. A three-way decision making approach to malware analysis using probabilistic rough sets. *Inf. Sci.* **2016**, *374*, 193–209. [CrossRef]

103. Liang, D.; Xu, Y.; Liu, D. Three-way decisions with intuitionistic fuzzy decision-theoretic rough sets based on point operators. *Inf. Sci.* **2017**, *375*, 183–201. [CrossRef]

104. Pamučar, D.; Mihajlović, M.; Obradović, R.; Atanasković, P. Novel approach to group multi-criteria decision making based on interval rough numbers: Hybrid DEMATEL-ANP-MAIRCA model. *Expert Syst. Appl.* **2017**, *88*, 58–80. [CrossRef]

105. Tiwari, V.; Jain, P.K.; Tandon, P. Product design concept evaluation using rough sets and VIKOR method. *Adv. Eng. Inf.* **2016**, *30*, 16–25. [CrossRef]

106. Shidpour, H.; Cunha, C.D.; Bernard, A. Group multi-criteria design concept evaluation using combined rough set theory and fuzzy set theory. *Expert Syst. Appl.* **2016**, *64*, 633–644. [CrossRef]

107. Chai, J.; Liu, J.N. A novel believable rough set approach for supplier selection. *Expert Syst. Appl.* **2014**, *41*, 92–104. [CrossRef]

108. Zhu, G.N.; Hu, J.; Qi, J.; Gu, C.C.; Peng, J.H. An integrated AHP and VIKOR for design concept evaluation based on rough number. *Adv. Eng. Inf.* **2015**, *29*, 408–418. [CrossRef]

109. Zhai, L.Y.; Khoo, L.P.; Zhong, Z.W. A rough set based QFD approach to the management of imprecise design information in product development. *Adv. Eng. Inf.* **2009**, *23*, 222–228. [CrossRef]

110. Gigović, L.; Pamučar, D.; Lukić, D.; Marković, S. Application of the GIS-Fuzzy DEMATEL MCDA model for ecotourism development site evaluation: A case study of "Dunavski ključ", Serbia. *Land Use Policy* **2016**, *58*, 348–365. [CrossRef]

111. Keshavarz Ghorabaee, M.; Zavadskas, E.K.; Olfat, L.; Turskis, Z. Multi-Criteria Inventory Classification Using a New Method of Evaluation Based on Distance from Average Solution (EDAS). *Informatica* **2015**, *26*, 435–451. [CrossRef]

112. Keshavarz Ghorabaee, M.; Zavadskas, E.K.; Amiri, M.; Turskis, Z. Extended EDAS Method for Fuzzy Multi-criteria Decision-making: An Application to Supplier Selection. *Int. J. Comput. Commun. Control* **2016**, *11*, 358–371. [CrossRef]

113. Turskis, Z.; Juodagalvienė, B. A novel hybrid multi-criteria decision-making model to assess a stairs shape for dwelling houses. *J. Civ. Eng. Manag.* **2016**, *22*, 1078–1087. [CrossRef]

114. Stević, Ž.; Tanackov, I.; Vasiljević, M.; Vesković, S. Evaluation in logistics using combined AHP and EDAS method. In Proceedings of the XLIII International Symposium on Operational Research, Belgrade, Serbia, 20–23 September 2016; pp. 309–313.

115. Keshavarz Ghorabaee, M.; Amiri, M.; Olfat, L.; Khatami Firouzabadi, S.A. Designing a multi-product multi-period supply chain network with reverse logistics and multiple objectives under uncertainty. *Technol. Econ. Dev. Econ.* **2017**, *23*, 520–548. [CrossRef]

116. Kahraman, C.; Keshavarz Ghorabaee, M.; Zavadskas, E.K.; Cevik Onar, S.; Yazdani, M.; Oztaysi, B. Intuitionistic fuzzy EDAS method: An application to solid waste disposal site selection. *J. Environ. Eng. Landsc. Manag.* **2017**, *25*, 1–12. [CrossRef]

117. Keshavarz Ghorabaee, M.; Amiri, M.; Zavadskas, E.K.; Turskis, Z. Multi-criteria group decision-making using an extended EDAS method with interval type-2 fuzzy sets. *E+M Ekon. Manag.* **2017**, *20*, 48–68.

118. Ecer, F. Third-party logistics (3PLs) provider selection via Fuzzy AHP and EDAS integrated model. *Technol. Econ. Dev. Econ.* **2017**, 1–20. [CrossRef]

119. Peng, X.; Liu, C. Algorithms for neutrosophic soft decision making based on EDAS, new similarity measure and level soft set. *J. Intell. Fuzzy Syst.* **2017**, *32*, 955–968. [CrossRef]

120. Keshavarz Ghorabaee, M.; Amiri, M.; Zavadskas, E.K.; Turskis, Z.; Antucheviciene, J. A new hybrid simulation-based assignment approach for evaluating airlines with multiple service quality criteria. *J. Air Transp. Manag.* **2017**, *63*, 45–60. [CrossRef]

121. Zavadskas, E.K.; Cavallaro, F.; Podvezko, V.; Ubarte, I.; Kaklauskas, A. MCDM Assessment of a Healthy and Safe Built Environment According to Sustainable Development Principles: A Practical Neighborhood Approach in Vilnius. *Sustainability* **2017**, *9*, 702. [CrossRef]

122. Trinkūnienė, E.; Podvezko, V.; Zavadskas, E.K.; Jokšienė, I.; Vinogradova, I.; Trinkūnas, V. Evaluation of quality assurance in contractor contracts by multi-attribute decision-making methods. *Econ. Res.-Ekon. Istraž.* **2017**, *30*, 1152–1180. [CrossRef]

123. Song, W.; Ming, X.; Wu, Z. An integrated rough number-based approach to design concept evaluation under subjective environments. *J. Eng. Des.* **2013**, *24*, 320–341. [CrossRef]

symmetry

MDPI

Article

The Selection of Wagons for the Internal Transport of a Logistics Company: A Novel Approach Based on Rough BWM and Rough SAW Methods

Željko Stević [1], Dragan Pamučar [2] , Edmundas Kazimieras Zavadskas [3,*] , Goran Ćirović [4] and Olegas Prentkovskis [5]

[1]   Faculty of Transport and Traffic Engineering, University of East Sarajevo, Vojvode Mišića 52, 74000 Doboj, Bosnia and Herzegovina; zeljkostevic88@yahoo.com
[2]   Department of logistics, University of Defence in Belgrade, Pavla Jurisica Sturma 33, 11000 Belgrade, Serbia; dpamucar@gmail.com
[3]   Institute of Sustainable Construction, Faculty of Civil Engineering, Vilnius Gediminas Technical University, Sauletekio al. 11, LT-10223 Vilnius, Lithuania
[4]   College of Civil Engineering and Geodesy, Belgrade University, Hajduk Stankova 2, 11000 Belgrade, Serbia; cirovic@sezampro.rs
[5]   Department of Mobile Machinery and Railway Transport, Faculty of Transport Engineering, Vilnius Gediminas Technical University, Plytinės g. 27, LT-10105 Vilnius, Lithuania; olegas.prentkovskis@vgtu.lt
*   Correspondence: edmundas.zavadskas@vgtu.lt; Tel.: +370-5-274-4910

Received: 30 September 2017; Accepted: 1 November 2017; Published: 4 November 2017

Abstract: The rationalization of logistics activities and processes is very important in the business and efficiency of every company. In this respect, transportation as a subsystem of logistics, whether internal or external, is potentially a huge area for achieving significant savings. In this paper, the emphasis is placed upon the internal transport logistics of a paper manufacturing company. It is necessary to rationalize the movement of vehicles in the company's internal transport, that is, for the majority of the transport to be transferred to rail transport, because the company already has an industrial track installed in its premises. To do this, it is necessary to purchase at least two used wagons. The problem is formulated as a multi-criteria decision model with eight criteria and eight alternatives. The paper presents a new approach based on a combination of the Simple Additive Weighting (SAW) method and rough numbers, which is used for ranking the potential solutions and selecting the most suitable one. The rough Best–Worst Method (BWM) was used to determine the weight values of the criteria. The results obtained using a combination of these two methods in their rough form were verified by means of a sensitivity analysis consisting of a change in the weight criteria and comparison with the following methods in their conventional and rough forms: the Analytic Hierarchy Process (AHP), Technique for Ordering Preference by Similarity to Ideal Solution (TOPSIS) and MultiAttributive Border Approximation area Comparison (MABAC). The results show very high stability of the model and ranks that are the same or similar in different scenarios.

Keywords: internal transport; rough Best–Worst Method (BWM); rough Simple Additive Weighting (SAW); logistics; railway wagon

1. Introduction

In the past decade, companies have recognized the significance of logistics for their complete system, as well as its outstanding importance in the global environment, as confirmed by Koskinen and Hilmola [1], in which the rationalization of basic logistics subsystems plays a key role. Transport is the most expensive logistics subsystem; that is, it causes the highest percentage of logistics costs. These costs are kept to a daily minimum, especially in large companies that have a large amount of transport movements on a daily basis. The research carried out in this paper relates to a paper manufacturing company, which is the largest company in its region and in its field, both from the aspect of the level of production, and in its efficiency of operations. However, by monitoring all of its logistics subsystems for a number of months, certain shortcomings were observed, as well as possibilities for making savings. In addition to the warehouse system of the company that currently represents one of the problems due to insufficient space for storing finished products, the internal transport is dominated by a certain amount of irrational movements of vehicles. To bring the logistics systems into a state of high rational functioning, a project was carried out to centralize the warehouse system that opens up new opportunities. As part of the project, since the company has railway infrastructure, it is necessary to redirect almost all of the internal transport to railway traffic, because it is well known that rail transport is cheaper than road transport. To achieve this, it is necessary to purchase at least two wagons that would meet the needs of internal transport. The paper defines a model of multi-criteria decision making consisting of eight wagons that represent the alternatives and eight criteria for their selection. In addition, a heterogeneous team of experts was formed to evaluate the elements of this model.

This paper has several objectives. The first objective is to improve the methodology for dealing with imprecision in the field of multi-criteria decision making by presenting the new Rough SAW algorithm. The second goal of this paper is to affirm the idea of rough numbers (RN) through a detailed presentation of the arithmetic operations with RN that are characteristic for multi-criteria decision making. Finally, the third goal of this paper is to bridge the gap in the methodology for evaluating the elements of internal transport, i.e., railway wagons, through a new approach to dealing with imprecision based on RN.

In addition to the introduction and conclusion, the paper has four sections (Sections 2–5). Section 2 gives a literature review with an emphasis on the SAW method, while Section 3 is a description of the method used. The basic assumptions are given concerning rough numbers and the detailed algorithm is presented for the rough BWM and novel rough SAW methods. This section also presents a new linguistic scale for evaluating alternatives, depending on the type of criteria. Section 4 presents the selection of a wagon in the paper manufacturing company using the new rough SAW method. Section 5 is a sensitivity analysis that checks the stability of the model and the results obtained.

2. Literature Review

Multi-criteria decision making has wide application in all areas, and, when it comes to logistics, its transport subsystem is often used to select the type of transport [2,3]. A study on the evaluation and selection of sustainable transport means has been carried out [4] as well as the evaluation of transport systems in Brazil [5], the prioritization of the investments in transport infrastructure [6] and the selection of logistics providers [7,8], while, in another study [9], the evaluation of city logistics scenarios was carried out. The systematization of methods belonging to the field of multi-criteria decision making that are applied in the field of transport systems was carried out by Mardani et al. [10] in which the authors conclude that these methods are adequate and offer significant help when making decisions in the area of transportation. Turskis and Zavadskas [11] presented a new multi-criteria model in order to select a location for a logistics center, which is a commonly considered problem using multi-criteria methods [12,13]. Logistics systems are extremely important for the functioning of the complete supply chain, so almost every day the evaluation and selection of suppliers is carried out [8,14,15] and this is one of the most important steps in optimizing logistics systems.

The literature related to the application of different models of multi-criteria decision making considers two basic approaches: (1) multi-attribute decision making; and (2) multi-objective decision making. In each of these two approaches, there are a number of methods whose differences are primarily seen in different mathematical algorithms [16]. In addition to the difference in mathematical procedures, Pohekar and Ramachandran [16] point out that multi-criteria models can also be classified as deterministic, stochastic or fuzzy models. Clarsson and Fuller [17] presented the categorization of multi-criteria models by means of four basic units. One of the categorizations is value and utility theory approaches, which includes the application of models to determine the relative significance of the optimization criteria and the alternatives. Within this category, most of the methods rely on the weight values of the optimization criteria, and the basic representatives are: Simple Additive Weighting (SAW), Analytic Hierarchy Process (AHP), fuzzy conjuctive/disjunctive methods, fuzzy outranking methods and max-min methods. Of these methods, alongside the AHP method, the SAW method has had the most common modifications and the widest application in solving multi-criteria models [16,18]. The SAW method is still also well-known as the weighted linear combination or scoring method. Many authors decide to apply this simple mathematical apparatus to solve various multi-criteria problems [19–22]. In the literature, in addition to the traditional SAW method, its modifications based on fuzzy theory are also well-known [23–27].

Shameli et al. [20] presented the application of the traditional (crisp) SAW method for solving a practical model for information security risk assessment. In addition to the traditional SAW method, Shameli et al. [20] used the TOPSIS model and the fuzzy modification of the SAW method to compare the results obtained. Afshari et al. [28] used the traditional SAW model to solve the personnel selection problem in Iran. Deni et al. [24] used the fuzzy SAW (FSAW) model to select high achieving students at the faculty level. Gupta and Gupta [25] used the SAW model to analyze the existing system of vendor rating. Azzizollah et al. [29] applied the fuzzy Delphi method to the collection of expert opinions when selecting suitable maintenance strategies. In that study, the optimal maintenance strategy was proposed by applying the crisp SAW model. Chen [30] modified the SAW algorithm by applying interval-valued fuzzy (IVF) sets. The SAW method was also used to evaluate the efficiency of diesel locomotives [31], while Jakimavičius and Burinskiene [32] used it to evaluate development scenarios based on modeling transport systems. Because of its simple and reliable algorithm, the SAW method has found application in solving various problems, such as: selecting the optimal maintenance strategy [26], selecting locations [33,34], the machine tool selection problem [35], personnel selection [28,36], and comparative analysis without application [23,30].

This overview shows that the literature is familiar with the crisp SAW, FSAW and IVF SAW algorithms. Bearing in mind that the SAW method falls into the category of methods that have found the widest application in solving multi-criteria models [16,18], further development of the SAW method through the application of other uncertainty approaches is justified. To achieve the greatest objectivity in deciding and appreciating uncertainty in decision making, numerous uncertainty approaches have been developed in the field of multi-criteria decision making: fuzzy theory [37], rough theory [38,39], grey theory [40,41], Z numbers [42,43], etc. Today, in addition to fuzzy theory, the most commonly used theory for dealing with imprecision in the decision-making process is rough theory, that is, rough numbers [44–47]. Having in mind all the advantages of using rough theory [38,39] in the decision-making process, the authors have decided in this paper to show the modification of the SAW algorithm and BWM using rough numbers (RBWM-SAW), which is an original contribution.

3. Methods

3.1. Operations with Rough Numbers

In rough set theory, any vague idea can be represented as a couple of exact concepts based on the lower and upper approximations. This is shown in Figure 1.

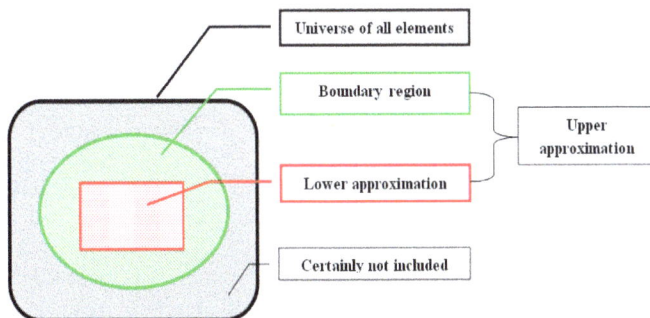

Figure 1. Elementary concept of rough set theory.

Suppose U is the universe which contains all the objects, Y is an arbitrary object of U, R is a set of t classes $\{G_1, G_2, \ldots, G_t\}$ that cover all the objects in U, $R = \{G_1, G_2, \ldots, G_t\}$. If these classes are ordered as $\{G_1 < G_2 < \ldots < G_t\}$, then $\forall Y \in U, G_q \in R, 1 \leq q \leq t$, by $R(Y)$ we mean the class to which the object belongs, the lower approximation $(\underline{Apr}(G_q))$, upper approximation $(\overline{Apr}(G_q))$ and boundary region $(\overline{Bnd}(G_q))$ of class G_q are, according to [48], defined as:

$$\underline{Apr}(G_q) = \{Y \in U/R(Y) \leq G_q\} \tag{1}$$

$$\overline{Apr}(G_q) = \{Y \in U/R(Y) \geq G_q\} \tag{2}$$

$$Bnd(G_q) = \{Y \in U/R(Y) \neq G_q\} = \{Y \in U/R(Y) > G_q\} \cup \{Y \in U/R(Y) < G_q\} \tag{3}$$

Then, G_q can be shown as rough number $(RN(G_q))$, which is determined by its corresponding lower limit $(\underline{Lim}(G_q))$ and upper limit $(\overline{Lim}(G_q))$ where:

$$\underline{Lim}(G_q) = \frac{1}{M_L} \sum \left\{ Y \in \underline{Apr}(G_q) \right\} R(Y) \tag{4}$$

$$\overline{Lim}(G_q) = \frac{1}{M_U} \sum \left\{ Y \in \overline{Apr}(G_q) \right\} R(Y) \tag{5}$$

$$RN(G_q) = \left[\underline{Lim}(G_q), \overline{Lim}(G_q) \right] \tag{6}$$

where M_L, M_U are the numbers of objects that contained in $\underline{Apr}(G_q)$ and $\overline{Apr}(G_q)$, respectively. The difference between them is expressed as rough boundary interval $(IRBnd(G_q))$:

$$IRBnd(G_q) = \overline{Lim}(G_q) - \underline{Lim}(G_q) \tag{7}$$

The operations for two rough numbers $RN(\alpha) = \left[\underline{Lim}(\alpha), \overline{Lim}(\alpha) \right]$ and $RN(\beta) = \left[\underline{Lim}(\beta), \overline{Lim}(\beta) \right]$ according to Zhai et al. [49] are:

Addition (+) of two rough numbers $RN(\alpha)$ and $RN(\beta)$

$$RN(\alpha) + RN(\beta) = \left[\underline{Lim}(\alpha) + \underline{Lim}(\beta), \overline{Lim}(\alpha) + \overline{Lim}(\beta) \right] \tag{8}$$

Subtraction (−) of two rough numbers $RN(\alpha)$ and $RN(\beta)$

$$RN(\alpha) - RN(\beta) = \left[\underline{Lim}(\alpha) - \overline{Lim}(\beta), \overline{Lim}(\alpha) - \underline{Lim}(\beta) \right] \tag{9}$$

Multiplication (×) of two rough numbers $RN(\alpha)$ and $RN(\beta)$

$$RN(\alpha) \times RN(\beta) = \left[\underline{Lim}(\alpha) \times \underline{Lim}(\beta), \overline{Lim}(\alpha) \times \overline{Lim}(\beta) \right] \tag{10}$$

Division (/) of two rough numbers $RN(\alpha)$ and $RN(\beta)$

$$RN(\alpha)/RN(\beta) = \left[\underline{Lim}(\alpha)/\overline{Lim}(\beta), \overline{Lim}(\alpha)/\underline{Lim}(\beta)\right] \qquad (11)$$

Scalar multiplication of rough number $RN(\alpha)$, where μ is a nonzero constant

$$\mu \times RN(\alpha) = \left[\mu \times \underline{Lim}(\alpha), \mu \times \overline{Lim}(\alpha)\right] \qquad (12)$$

3.2. Best–Worst Method

The BWM [50] is one of the more recent methods. Some of the advantages that cause authors decide to use BWM are as follows: (1) in comparison with the AHP method, which until the establishment of this method was in comparable and most commonly used to determine weight coefficients [48], it requires a smaller number of pairwise comparisons (in the AHP method, the number of comparisons is n(n − 1)/2, while, for the BWM, the number of comparisons is 2n − 3; (2) weight coefficients determined using the BWM are more reliable, since comparisons in this method are made with a higher degree of consistency compared with the AHP method; (3) with most MCDM models (e.g., AHP), the degree of consistency checks whether the comparison of criteria is consistent or not, while, in BWM, the degree of consistency is used to determine the level of consistency because the outputs from BWM are always consistent; and (4) the BWM for pairwise comparison of the criteria requires only integer values, which is not the case with other MCDM methods (e.g., AHP) which also require fractional numbers.

To more comprehensively take into account the imprecision that appears in the group decision making process, a modification of the Best–Worst (BWM) method was carried out using rough numbers (RN). By using rough numbers, the need for additional information to determine the uncertainty of the intervals of the numbers is eliminated. This maintains the quality of the existing data in group decision making and the perceptions of the experts are expressed objectively in aggregated Best-to-Others (BO) and Others-to-Worst (OW) vectors. Since this is a recent method [50], there are not many BWM modifications in the literature. In the literature until now, the majority of authors have applied the traditional (crisp) BWM algorithm [50–54] and a modification of the BW method carried out with fuzzy numbers [55,56]. The approach presented in this chapter introduces RN, which secure a more objective evaluation of the criteria in cases where there is imprecision in the expert decisions. The proposed modification of BWM using RN (RBWM (the Rough Best–Worst Method)) makes it possible to consider doubts that arise during the expert evaluation of the criteria. The RBWM makes it possible to bridge the existing gap that exists in the BWM methodology by applying a new approach in treating imprecision that is based on RN. The next section presents the algorithm for the RBWM that includes the following steps:

Step 1. Determining the set of evaluation criteria. This starts from the assumption that the process of decision making involves *m* experts. In this step, experts consider the set of evaluation criteria and select the final set of criteria $C = \{c_1, c_2, \dots c_n\}$, where *n* represents the total number of criteria.

Step 2. Determining the most significant (most influential) and worst (least significant) criteria. The experts decide on the best and the worst criteria from the set of criteria $C = \{c_1, c_2, \dots c_n\}$. If the experts decide on two or more criteria as the best, or worst, the best and worst criteria are selected arbitrarily.

Step 3. Determining the preferences of the most significant (most influential) criteria (B) from set C over the remaining criteria from the defined set. Under the assumption that there are *m* experts and *n* criteria under consideration, each expert should determine the degree of influence of the best criterion B on the criteria *j* ($j = 1, 2, \dots, n$). This is how we obtain a comparison between the best criterion and the other criteria. The preference of criterion B compared to the *j*-th criterion defined by the e-th expert is denoted with a^e_{Bj} ($j = 1, 2, \ldots, n; 1 \le e \le m$). The value of each pair a^e_{Bj} takes a value from the predefined scale in interval $a^e_{Bj} \in \{1,9\}$. As a result, a Best-to-Others (BO) vector is obtained:

$$A^e_B = (a^e_{B1}, a^e_{B2}, \ldots, a^e_{Bn}); \quad 1 \le e \le m \tag{13}$$

where a^e_{Bj} represents the influence (preference) of the best criterion B over criterion j, whereby $a^e_{BB} = 1$. This is how we obtain BO matrices $A^1_B, A^2_B, \ldots, A^m_B$ for each expert.

Step 4. Determining the preferences of the criteria from set C over the worst criterion (W) from the defined set. Each expert should determine the degree of influence of criterion j ($j = 1, 2, \ldots, n$) in relation to criterion W. The preference of criterion j in relation to criterion W defined by the e-th expert is denoted as a^e_{jW} ($j = 1, 2, \ldots, n; 1 \le e \le m$). The value of each pair a^e_{jW} takes a value from the predefined scale in interval. As a result, an Others-to-Worst (OW) vector is obtained:

$$A^e_W = (a^e_{1W}, a^e_{2W}, \ldots, a^e_{nW}); \quad 1 \le e \le m \tag{14}$$

where a^e_{jW} represents the influence (preference) of criterion j in relation to criterion W, whereby $a^e_{WW} = 1$. This is how we obtain OW matrices $A^1_W, A^2_W, \ldots, A^m_W$ for each expert.

Step 5. Determining the rough BO matrix for the average answers of the experts. Based on the BO matrices of the experts' answers $A^e_B = \left[a^e_{Bj}\right]_{1 \times mn}$, we form matrices of the aggregated sequences of experts A^{*e}_B

$$A^{*e}_B = \left[a^1_{B1}, a^2_{B1}, \ldots, a^m_{B1}; a^1_{B2}, a^2_{B2}; \ldots; a^m_{B2}, \ldots, a^1_{Bn}, a^2_{Bn}, \ldots, a^m_{Bn}\right]_{1 \times mn} \tag{15}$$

where $a^e_{Bj} = \left\{a^1_{Bj}, a^2_{Bj}, \ldots, a^m_{Bn}\right\}$ represents sequences by means of which the relative significance of criterion B is described in relation to criterion j. Using Equations (1)–(6), each sequence a^e_{Bj} is transformed into rough sequence $RN\left(a^e_{Bj}\right) = \left[\underline{Lim}(a^e_{Bj}), \overline{Lim}(a^e_{Bj})\right]$, where $\underline{Lim}(a^e_{Bj})$ represents the lower limit and $\overline{Lim}(a^e_{Bj})$ represents upper limit of the rough sequence $RN\left(a^e_{Bj}\right)$.

Thus, for sequence $RN\left(a^e_{Bj}\right)$, we obtain a BO matrix $A^{*1}_B, A^{*2}_B, \ldots, A^{*m}_B$. By applying Equation (16), we obtain the average rough sequence of the BO matrix:

$$RN(\bar{a}_{Bj}) = RN(a^1_{Bj}, \ldots, a^e_{Bj}) = \left[\bar{a}^L_{Bj}, \bar{a}^U_{Bj}\right] \tag{16}$$

where $\bar{a}^L_{Bj} = \frac{1}{m}\sum\limits_{e=1}^{m} a^{eL}_{Bj}$ and $\bar{a}^U_{Bj} = \frac{1}{m}\sum\limits_{e=1}^{m} a^{eU}_{Bj}$, e represents the e-th expert ($e = 1, 2, \ldots, m$), $RN\left(a^e_{Bj}\right)$ represents the rough sequences. We thus obtain the averaged rough BO matrix of average responses $\overline{A}_B$,

$$\overline{A}_B = [\bar{a}_{B1}, \bar{a}_{B2}, \ldots, \bar{a}_{Bn}]_{1 \times n} \tag{17}$$

Step 6. Determining the rough OW matrix of average expert responses. Based on the WO matrices of the expert responses $A^e_W = \left[a^e_{jW}\right]_{1 \times n}$, as with the rough BO matrices, for each element a^e_{jW}, we form matrices of the aggregated sequences of the experts A^{*e}_W

$$A^{*e}_W = \left[a^1_{1W}, a^2_{1W}, \ldots, a^m_{1W}; a^1_{2W}, a^2_{2W}; \ldots; a^m_{2W}, \ldots, a^1_{nW}, a^2_{nW}, \ldots, a^m_{nW}\right]_{1 \times mn} \tag{18}$$

where $a^e_{jW} = \left\{a^1_{jW}, a^2_{jW}, \ldots, a^m_{nW}\right\}$ represents the sequence with which the relative significance of criterion j is described in relation to criterion W.

As in Step 5, using Equations (1)–(6), the sequences a^e_{jW} are transformed into rough sequences $RN\left(a^e_{jW}\right) = \left[\underline{Lim}(a^e_{jW}), \overline{Lim}(a^e_{jW})\right]$. Thus, for each rough sequence of expert e ($1 \le e \le m$), a rough

BO matrix is formed. Equation (19) is used to average the rough sequences of the OW matrix of the experts to obtain an averaged rough OW matrix.

$$RN(\bar{a}_{jW}) = RN(a_{jW}^1, a_{jW}^2, \dots, a_{jW}^e) = \begin{cases} \bar{a}_{jW}^L = \frac{1}{m} \sum\limits_{e=1}^{m} a_{jW}^{eL} \\ \bar{a}_{jW}^U = \frac{1}{m} \sum\limits_{e=1}^{m} a_{jW}^{eU} \end{cases} \tag{19}$$

where e represents the e-th expert ($e = 1, 2, \dots, m$), $RN(a_{jW})$ represents the rough sequences. Thus, we obtain the averaged rough OW matrix of average responses $\bar{A}_W$

$$\bar{A}_W = [\bar{a}_{1W}, \bar{a}_{2W}, \dots, \bar{a}_{nW}]_{1 \times n} \tag{20}$$

Step 7. Calculation of the optimal rough values of the weight coefficients of the criteria $[RN(w_1), RN(w_2), \dots, RN(w_n)]$ from set C. The goal is to determine the optimal value of the evaluation criteria, which should satisfy the condition that the difference in the maximum absolute values (21)

$$\left| \frac{RN(w_B)}{RN(w_j)} - RN(a_{Bj}) \right| \text{ and } \left| \frac{RN(w_j)}{RN(w_W)} - RN(w_{jW}) \right| \tag{21}$$

for each value of j is minimized. To meet these conditions, the solution that satisfies the maximum differences according to the absolute value $\left| \frac{RN(w_B)}{RN(w_j)} - RN(a_{Bj}) \right|$ and $\left| \frac{RN(w_j)}{RN(w_W)} - RN(w_{jW}) \right|$ should be minimized for all values of j. For all values of the interval rough weight coefficients of the criteria $RN(w_j) = \left[Lim(w_j), \overline{Lim}(w_j) \right] = [w_j^L, w_j^U]$ the condition is met that $0 \le w_j^L \le w_j^U \le 1$ for each evaluation criterion $c_j \in C$. The weight coefficient w_j belongs to interval $[w_j^L, w_j^U]$, that is, $w_j^L \le w_j^U$ for each value $j = 1, 2, \dots, n$. On this basis, we can conclude that in the case of the rough values of the weight coefficients of the criteria, the condition is met that $\sum_{j=1}^n w_j^L \le 1$ and $\sum_{j=1}^n w_j^U \ge 1$. In this way, the condition is met that the weight coefficients are found at interval $w_j \in [0, 1]$, ($j = 1, 2, \dots, n$) and that $\sum_{j=1}^n w_j = 1$.

The previously defined limits will be presented in the following min-max model:

$$\min\max_j \left\{ \left| \frac{RN(w_B)}{RN(w_j)} - RN(a_{Bj}) \right|, \left| \frac{RN(w_j)}{RN(w_W)} - RN(w_{jW}) \right| \right\}$$

$$s.t.$$

$$\begin{cases} \sum_{j=1}^n w_j^L \le 1 \\ \sum_{j=1}^n w_j^U \ge 1; \\ w_j^L \le w_j^U, \; \forall j = 1, 2, \dots, n \\ w_j^L, w_j^U \ge 0, \; \forall j = 1, 2, \dots, n \end{cases} \tag{22}$$

where $RN(w_j) = \left[\underline{Lim}(w_j), \overline{Lim}(w_j) \right] = [w_j^L, w_j^U]$ is the rough weight coefficient of a criterion. Model (22) is equivalent to the following model:

$$\min \zeta$$

$$s.t.$$

$$\begin{cases} \left| \frac{w_B^L}{w_j^U} - \bar{a}_{Bj}^U \right| \le \zeta; \quad \left| \frac{w_B^U}{w_j^L} - \bar{a}_{Bj}^L \right| \le \zeta; \\ \left| \frac{w_j^L}{w_W^U} - \bar{a}_{jW}^U \right| \le \zeta; \quad \left| \frac{w_j^U}{w_W^L} - \bar{a}_{jW}^L \right| \le \zeta; \\ \sum_{j=1}^n w_j^L \le 1; \\ \sum_{j=1}^n w_j^U \ge 1; \\ w_j^L \le w_j^U, \; \forall j = 1, 2, \dots, n \\ w_j^L, w_j^U \ge 0, \; \forall j = 1, 2, \dots, n \end{cases} \tag{23}$$

where $RN(w_j) = [w_j^L, w_j^U]$ represents the optimum values of the weight coefficients, $RN(w_B) = [w_B^L, w_B^U]$ and $RN(w_W) = [w_W^L, w_W^U]$ represents the weight coefficients of the best and worst criterion respectively, while $RN(\bar{a}_{jW}) = [\bar{a}_j^L, \bar{a}_j^U]$ and $RN(\bar{a}_{Bj}) = [\bar{a}_{Bj}^L, \bar{a}_{Bj}^U]$, respectively, represent the values from the average rough OW and rough BO matrices (see Equations (17) and (20)).

By solving model (23) we obtain the optimal values of the weight coefficients for the evaluation criteria $[RN(w_1), RN(w_2), \ldots, RN(w_n)]$ and ξ^*.

The consistency ratio is a very important indicator by means of which we check the consistency of the pairwise comparison of the criteria in the rough BO and rough OW matrices.

Definition 1. *Comparison of the criteria is consistent when condition $RN(a_{Bj}) \times RN(a_{jW}) = RN(a_{BW})$ is fulfilled for all criteria j, where $RN(a_{Bj})$, $RN(a_{jW})$ and $RN(a_{BW})$ respectively represent the preference of the best criterion over criterion j, the preference of criterion j over the worst criterion, and the preference of the best criterion over the worst criterion.*

However, when comparing the criteria it can happen that some pairs of criteria j are not completely consistent. Therefore, the next section defines the consistency ratio (CR), which gives us information on the consistency of the comparison between the rough BO and rough OW matrices. To show how the CR is determined, we start from a calculation of the minimum consistency when comparing the criteria, which is explained in the following section.

As previously indicated, pairwise comparison of the criteria is carried out based on a predefined scale in which the highest value is 9 or any other maximum from a scale defined by the decision maker. The consistency of the comparison decreases when $RN(a_{Bj}) \times RN(a_{jW})$ is less or greater than $RN(a_{BW})$; that is, when $RN(a_{Bj}) \times RN(a_{jW}) \neq RN(a_{BW})$. It is clear that the greatest inequality occurs when $RN(a_{Bj})$ and $RN(a_{jW})$ have the maximum values that are equal $RN(a_{BW})$, which continues to affect the value of ξ. Based on these relationships we can conclude that

$$[RN(w_B)/RN(w_j)] \times [RN(w_j)/RN(w_W)] = RN(w_B)/RN(w_W) \tag{24}$$

As the largest inequality occurs when $RN(a_{Bj})$ and $RN(a_{jW})$ have their maximum values, then we need to subtract the value of ξ from $RN(a_{Bj})$ and $RN(a_{jW})$ and add $RN(a_{BW})$. Thus, we obtain Equation (25):

$$[RN(a_{Bj}) - \xi] \times [RN(a_{jW}) - \xi] = [RN(a_{BW}) + \xi] \tag{25}$$

Since for the minimum consistency $RN(a_{Bj}) = RN(a_{jW}) = RN(a_{BW})$ applies, we present Equation (25) as

$$[RN(a_{BW}) - \xi] \times [RN(a_{BW}) - \xi] = [RN(a_{BW}) + \xi] \Rightarrow \xi^2 - [1 + 2RN(a_{BW})]\xi + [RN(a_{BW})^2 - RN(a_{BW})] = 0 \tag{26}$$

Since we are using rough numbers, and if there is no consensus between the DM on their preferences of the best criterion over the worst criterion, then $RN(a_{BW})$ will not have a crisp value but we will use $RN(\bar{a}_{BW}) = [\bar{a}_{BW}^L, \bar{a}_{BW}^U]$. Since for the RN the condition $\bar{a}_{BW}^L \leq \bar{a}_{BW}^U$ applies, we can conclude that the preference of the best criterion over the worst cannot be greater than $\bar{a}_{BW}^U$. In this case, when we use upper limit $\bar{a}_{BW}^U$ for determining the value of CI, then all values connected with $RN(\bar{a}_{BW})$ can use the CI obtained for calculating the value of CR. We can conclude this from the fact that the consistency index, which corresponds to $\bar{a}_{BW}^U$, has the highest value in interval $[\bar{a}_{BW}^L, \bar{a}_{BW}^U]$. Based on this conclusion, we can transform Equation (26) in the following way

$$\xi^2 - \left(1 + 2\bar{a}_{BW}^U\right)\xi + \left(\bar{a}_{BW}^{U\,2} - \bar{a}_{BW}^U\right) = 0 \tag{27}$$

By solving Equation (27) for the different values of $\bar{a}_{BW}^U$, we can determine the maximum possible values of ξ, which is the CI for the R-BW method. Since we obtain the values of $RN(\bar{a}_{BW})$, i.e., $\bar{a}_{BW}^U$, based on the aggregated decisions of the DM, and these change the *RN* interval, it is not possible to predefine the values of ζ. The values of ζ depend on uncertainties in the decisions, since uncertainties change the *RN* interval. As explained in the algorithm for the R-BW method, interval $\left[a_{BW}^L, a_{BW}^U\right]$ changes depending on uncertainties in evaluating the criteria.

If the DM agree on their preference for the best criterion over the worst then a_{BW} represents the crisp value of a_{BW} from the defined scale and then the maximum values of ζ apply for different values of $a_{BW} \in \{1, 2, \ldots, 9\}$, Table 1.

Table 1. Values of the consistency index (*CI*).

a_{BW}	1	2	3	4	5	6	7	8	9
CI (maxζ)	0.00	0.44	1.00	1.63	2.30	3.00	3.73	4.47	5.23

In Table 1, the values a_{BW} are taken from the scale $\{1, 2, \ldots, 9\}$ which is defined in [50]. Based on *CI* (Table 1), we obtain the consistency ratio (*CR*)

$$CR = \frac{\zeta^*}{CI} \tag{28}$$

The *CR* takes values from interval $[0, 1]$, where values closer to zero show high consistency, while the values of *CR* closer to one show low consistency.

3.3. Rough SAW Method

As already mentioned in the previous section, the SAW method is a simple and easily applicable method of multi-criteria decision making. However, using only crisp numbers, it is impossible to obtain results that treat uncertainty and objectivity in an adequate way. Therefore, this paper continues by presenting a new approach that combines the SAW method and rough numbers. The Rough SAW method consists of the following steps:

Step 1: Define the problem that needs to be solved, which is made up of *m* alternatives and *n* criteria.

Step 2: Form a group of *k* experts, who evaluate the alternatives according to all the criteria using the following linguistic scale shown in Table 2.

Table 2. Linguistic scale for evaluating the alternatives depending on the type of criteria.

Linguistic Scale	For Criteria Max Type (Benefit Criteria)	For Criteria Min Type (Cost Criteria)
Very Poor (VP)	1	9
Poor (P)	3	7
Medium (M)	5	5
Good (G)	7	3
Very Good (VG)	9	1

Table 2 shows a new linguistic scale, based on which a group of experts evaluates the alternatives, taking into account the type of criteria (benefit or cost). In this method of solving engineering problems, it is very important that the evaluation of potential solutions is carried out in an adequate way, therefore implying the application of the mentioned scale. When we have criteria such as for example cost or income that can be shown quantitatively, there will be no complications in solving the problem as long as all criteria can be expressed quantitatively. However, this is not the case when evaluation is carried out using a linguistic scale (qualitative criteria) and when, right at the beginning of quantifying the criteria, there is an incorrect evaluation by experts, since the type of criteria has not been taken into account. If this is the case, it cannot be replaced later by normalization, and the application of a new scale is recommended.

Step 3: Convert individual matrices into a group rough matrix. It is necessary to transform each individual matrix of experts $k_1, k_2, \ldots, k_n$ into a rough group matrix using Equations (1)–(6):

$$RGN = \begin{bmatrix} [x_{11}^L, x_{11}^U] & [x_{12}^L, x_{12}^U] & \cdots & [x_{1n}^L, x_{1n}^U] \\ [x_{21}^L, x_{21}^U] & [x_{22}^L, x_{22}^U] & \cdots & [x_{2n}^L, x_{2n}^U] \\ \vdots & \vdots & \ddots & \vdots \\ [x_{m1}^L, x_{m1}^U] & [x_{m2}^L, x_{m2}^U] & \cdots & [x_{mn}^L, x_{mn}^U] \end{bmatrix} \tag{29}$$

Step 4: Normalize the group matrix using Equations (30) and (31):

$$r_{ij} = \frac{\left[x_{ij}^L; x_{ij}^U \right]}{\max \left[x_{ij}^{+L}; x_{ij}^{+U} \right]} \quad for \ C_1, C_2, \ldots, C_n \epsilon B \tag{30}$$

$\left[x_{ij}^L; x_{ij}^U \right]$ denotes the values of the alternatives according to criteria from the initial rough group matrix, while $\max \left[x_{ij}^{+L}; x_{ij}^{+U} \right]$ denotes maximum value of criterion if criterion belongs a set of benefit criteria.

$$r_{ij} = \frac{\min \left[x_{ij}^{-L}; x_{ij}^{-U} \right]}{\left[x_{ij}^L; x_{ij}^U \right]} \quad for \ C_1, C_2, \ldots, C_n \epsilon C \tag{31}$$

$\left[x_{ij}^L; x_{ij}^U \right]$ denotes the values of the alternatives according to criteria from the initial rough group matrix, while $\min \left[x_{ij}^{-L}; x_{ij}^{-U} \right]$ denotes minimal value of criterion if criterion belongs a set of cost criteria.

The values are marked with "+" and "−" to make it easier to recognize those which belong to different types of criteria.

The previously written equations can be more simply expressed as:

$$r_{ij} = \left[\frac{x_{ij}^L}{x_{ij}^{+U}}; \frac{x_{ij}^U}{x_{ij}^{+L}} \right] \quad for \ C_1, C_2, \ldots, C_n \epsilon B \tag{32}$$

$$r_{ij} = \left[\frac{x_{ij}^{-L}}{x_{ij}^U}; \frac{x_{ij}^{-U}}{x_{ij}^L} \right] \quad for \ C_1, C_2, \ldots, C_n \epsilon C \tag{33}$$

Then, a normalized matrix is obtained:

$$Rn = \begin{bmatrix} [r_{11}^L, r_{11}^U] & [r_{12}^L, r_{12}^U] & \cdots & [r_{1n}^L, r_{1n}^U] \\ [r_{21}^L, r_{21}^U] & [r_{22}^L, r_{22}^U] & \cdots & [r_{2n}^L, r_{2n}^U] \\ \vdots & \vdots & \ddots & \vdots \\ [r_{m1}^L, r_{m1}^U] & [r_{m2}^L, r_{m2}^U] & \cdots & [r_{mn}^L, r_{mn}^U] \end{bmatrix} \tag{34}$$

r_{ij} from matrix Rn denotes normalized values obtained using Equations (30) and (31)
Step 5: Weight the normalized matrix:

$$Vn = \left[v_{ij}^L; v_{ij}^U \right]_{mxn}$$
$$v_{ij}^L = w_j^L \times r_{ij}^L, i = 1, 2, \ldots m, j \tag{35}$$
$$v_{ij}^U = w_j^U \times r_{ij}^U, i = 1, 2, \ldots m, j$$

where w_j^L is the lower limit, and w_j^U is the upper limit of the weights of the criteria, expressed as rough numbers obtained using rough AHP or rough BWM as is the case in this paper.

Step 6: Sum all of the values of the alternatives obtained (summing by rows):

$$S = \left[s_{ij}^{L}; s_{ij}^{U} \right] \tag{36}$$

Step 7: Rank the alternatives in descending order; that is, the highest value is the best alternative. To rank the potential solutions more easily, the rough number can be converted into a crisp number using the average value.

4. Case Study

The company in which the research was carried out to select wagon for its internal transport is classified as a large company, since it has over 1000 workers and is an enormous elaborate complex that spreads over approximately 1,000,000 m². As already stated at the beginning of the paper, it is a manufacturing company in which the logistics subsystem and processes are its dominant activities, starting from the procurement of raw materials for the manufacture of paper, through the production process, transport and storage to the dispatch of finished products to end users. The company continuously works in three shifts, thus it achieves a large volume of production on a daily basis and sells its finished products worldwide to more than forty countries. This is also evidenced by the fact that it is currently fifth in its total amount of exports for the entire territory of Bosnia and Herzegovina. At the beginning of 2017, a centralization project for the warehouse system of the company was carried out, which could greatly contribute to achieving significant savings on an annual basis and which recommends changes in its internal transport. Since there are five production machines spatially located close to the proposed central warehouse, it is necessary to deliver the finished products to it in an optimal manner with, of course, the lowest possible costs. To achieve this, it is necessary to perform a complete rationalization of the movement of vehicles within the logistics subsystem of internal transport. Currently, most of the internal transport is carried out by means of road transport which burdens the company's logistics system with unnecessary costs. Most of this transport should be carried out by rail, since the company already has railway infrastructure installed as shown in Figure 2.

Figure 2. Industrial track within the company.

Currently, one small part of the internal transport is carried out in this way, for which the company uses one series S wagon (special open type flat wagon) which has been adapted to obtain a closed rail wagon. As a towing vehicle for this wagon, the company uses the loco tractor shown in Figure 3 which is the property of the company.

Figure 3. Loco tractor for carrying out internal transport.

In order for the company to achieve the anticipated savings and to rationalize the movement of transport vehicles, it is necessary to purchase at least two second-hand wagons for the undisturbed running of the majority of the internal transport by rail network. To this end, a heterogeneous expert team was formed consisting of managers in the logistics subsystem of the company, two experts from the Ministry of Transport and Communication and long-time professors in the field of railway transport and logistics from Serbia and Bosnia and Herzegovina. The expert team, whose members were familiar with the situation in the company and its current needs and requirements, performed the first assessment of the criteria shown in Table 3 based on nine point scale.

Table 3. The criteria defined for the selection of the wagon.

	Criteria	Characteristics and Meaning of the Criteria
C_1	Price of the wagon	The price of the second-hand wagon is the value expressed in monetary units
C_2	Maintenance conditions	The maintenance conditions include the ease of maintaining the wagon, the possibility of personal maintenance, and the cost of maintaining the wagon, etc.
C_3	Exploitation time	Since the wagons are second-hand, their age and the time they spent in use can play a role in their selection.
C_4	Load capacity	Load capacity (bearing capacity) is a value expressed in tons i.e., the total mass that it is possible to place inside the wagon.
C_5	Manipulative convenience	Manipulative convenience covers the ease of maneuverability by means of loading vehicles, in this case forklifts, and the possibility or impossibility of a forklift completely entering a wagon.
C_6	Time of last revision	This is the time passed since the last regular inspection of the wagon.
C_7	State of the bandages and flanges of the wheels	This is the quality and amount of wear and tear on the bandages and flanges of the wheels.
C_8	Ecological factor	The ecological factor includes the influence of the wagon on the environment, for example, noise produced by a wagon that can affect the psycho-physical condition of employees.

To implement the RBWM algorithm, the experts determined the best (B) and the worst (W) criterion by consensus. On this basis, the experts determined the BO vectors (15) in which the advantage of B criterion over other criteria from the defined set was considered.

$$
A_B^* = \begin{bmatrix}
2 & 2 & 2 & 2 & 2 & 2 & 3 \\
2 & 3 & 3 & 2 & 2 & 3 & 3 \\
6 & 5 & 6 & 5 & 5 & 6 & 5 \\
1 & 1 & 1 & 1 & 1 & 1 & 1 \\
5 & 4 & 5 & 5 & 4 & 4 & 5 \\
6 & 7 & 6 & 7 & 7 & 6 & 6 \\
7 & 8 & 8 & 8 & 7 & 8 & 7 \\
8 & 9 & 9 & 9 & 8 & 9 & 8
\end{bmatrix}
$$

After defining the BO vector, the experts determined the OW vectors (18) in which the advantage of the remaining criteria over W criterion was defined from the defined set.

$$
A_B^* = \begin{bmatrix}
8 & 8 & 8 & 8 & 8 & 7 & 8 \\
7 & 7 & 6 & 7 & 7 & 7 & 6 \\
5 & 4 & 5 & 5 & 4 & 4 & 5 \\
8 & 9 & 9 & 9 & 9 & 9 & 8 \\
6 & 6 & 6 & 5 & 5 & 6 & 6 \\
3 & 3 & 3 & 2 & 2 & 3 & 3 \\
3 & 2 & 3 & 2 & 2 & 2 & 3 \\
1 & 1 & 1 & 1 & 1 & 1 & 1
\end{bmatrix}
$$

Evaluation of the criteria was carried out using the scale $a_{Bj}^e, a_{jW}^e \in \{1,9\}$, where 1 indicates in significant domination, while 9 signifies exceptional domination. The expert comparisons through the BO and OW vectors are shown in Table 4.

Table 4. BO and OW vectors of the expert assessments.

Criteria	BO							OW						
	E_1	E_2	E_3	E_4	E_5	E_6	E_7	E_1	E_2	E_3	E_4	E_5	E_6	E_7
C_1	2	2	2	2	2	2	3	8	8	8	8	7	8	8
C_2	2	3	3	2	2	3	3	7	7	6	7	7	7	6
C_3	6	5	6	5	5	6	5	5	4	5	5	4	4	5
C_4	1	1	1	1	1	1	1	8	9	9	9	9	9	8
C_5	5	4	5	5	4	4	5	6	6	6	5	5	6	6
C_6	6	7	6	7	7	6	6	3	3	3	2	2	3	3
C_7	7	8	8	8	7	8	7	3	2	3	2	2	2	3
C_8	8	9	9	9	8	9	8	1	1	1	1	1	1	1

Using Equations (1)–(6), the crisp expert evaluation shown in the BO and OW vectors were transformed into rough numbers (Table 5).

Table 5. Rough BO and OW vectors of the expert assessments.

				BO			
Criteria	E_1	E_2	E_3	E_4	E_5	E_6	E_7
C_1	[2, 2.1]	[2, 2.2]	[2, 2.2]	[2, 2.2]	[2, 2.2]	[2, 2.2]	[2.2, 3]
C_2	[2, 2.6]	[2.6, 3]	[2.5, 3]	[2, 2.5]	[2, 2.5]	[2.5, 3]	[2.4, 3]
C_3	[5.4, 6]	[5, 5.3]	[5.4, 6]	[5, 5.3]	[5, 5.3]	[5.3, 6]	[5, 5.2]
C_4	[1, 1]	[1, 1]	[1, 1]	[1, 1]	[1, 1]	[1, 1]	[1, 1]
C_5	[4.6, 5]	[4, 4.5]	[4.6, 5]	[4.4, 5]	[4, 4.4]	[4, 4.4]	[4.4, 5]
C_6	[6, 6.4]	[6.5, 7]	[6, 6.4]	[6.4, 7]	[6.3, 7]	[6, 6.2]	[6, 6.2]
C_7	[7, 7.6]	[7.6, 8]	[7.5, 8]	[7.5, 8]	[7, 7.4]	[7.5, 8]	[7, 7.3]
C_8	[8, 8.6]	[8.6, 9]	[8.5, 9]	[8.5, 9]	[8, 8.4]	[8.5, 9]	[8, 8.3]

				OW			
Criteria	E_1	E_2	E_3	E_4	E_5	E_6	E_7
C_1	[7.9, 8]	[7.8, 8]	[7.8, 8]	[7.8, 8]	[7, 7.8]	[7.9, 8]	[7.7, 8]
C_2	[6.7, 7]	[6.7, 7]	[6, 6.6]	[6.7, 7]	[6.6, 7]	[6.5, 7]	[6, 6.5]
C_3	[4.6, 5]	[4, 4.5]	[4.6, 5]	[4.4, 5]	[4, 4.4]	[4, 4.4]	[4.4, 5]
C_4	[8, 8.7]	[8.8, 9]	[8.7, 9]	[8.6, 9]	[8.6, 9]	[8.5, 9]	[8, 8.5]
C_5	[5.7, 6]	[5.7, 6]	[5.6, 6]	[5, 5.6]	[5, 5.7]	[5.7, 6]	[5.5, 6]
C_6	[2.7, 3]	[2.7, 3]	[2.6, 3]	[2, 2.6]	[2, 2.7]	[2.7, 3]	[2.5, 3]
C_7	[2.4, 3]	[2, 2.3]	[2.4, 3]	[2, 2.3]	[2, 2.3]	[2, 2.3]	[2.3, 3]
C_8	[1, 1]	[1, 1]	[1, 1]	[1, 1]	[1, 1]	[1, 1]	[1, 1]

After transformation of the crisp values into RN, using Equations (16) and (19), the rough BO and OW expert matrices were transformed into aggregated RBO vectors (12)

$$\overline{A}_B = [[2.02,\ 2.28];[2.3,\ 2.79];[5.17,\ 5.59];[1,\ 1];[4.29,\ 4.76];[6.17,\ 6.6];[7.3,\ 7.75];[8.3,\ 8.75]]$$

and ROW vectors (20), Table 6.

$$\overline{A}_W = [[7.7,\ 7.97];[6.46,\ 6.87];[4.29,\ 4.76];[8.47,\ 8.89];[5.46,\ 5.89];[2.46,\ 2.89];[2.17,\ 2.61];[1,\ 1]]$$

Table 6. Aggregated RBO and ROW vectors.

Best: C4	RN	Worst: C8	RN
C1	[2.02, 2.28]	C1	[7.70, 7.97]
C2	[2.30, 2.79]	C2	[6.46, 6.87]
C3	[5.17, 5.59]	C3	[4.29, 4.76]
C5	[4.29, 4.76]	C4	[8.47, 8.89]
C6	[6.17, 6.60]	C5	[5.46, 5.89]
C7	[7.30, 7.75]	C6	[2.46, 2.89]
C8	[8.30, 8.75]	C7	[2.17, 2.61]

Based on the RBO and ROW vectors, the optimal values of the rough weight coefficients of the criteria were calculated. Based on data from Table 6 and Equation (23), a nonlinearly constrained optimization problem was formed, which is represented by specific numbers.

$\min \zeta$

s.t.

$$
\left\{
\begin{array}{l}
\left| \frac{w_B^L}{w_1^U} - 2.28 \right| \leq \zeta; \ \left| \frac{w_B^L}{w_2^U} - 2.79 \right| \leq \zeta; \ \left| \frac{w_B^L}{w_3^U} - 5.59 \right| \leq \zeta; \ \left| \frac{w_B^L}{w_5^U} - 6.6 \right| \leq \zeta; \\[2mm]
\left| \frac{w_B^L}{w_6^U} - 7.75 \right| \leq \zeta; \ \left| \frac{w_B^L}{w_7^U} - 7.75 \right| \leq \zeta; \ \left| \frac{w_B^L}{w_8^U} - 8.75 \right| \leq \zeta; \ \left| \frac{w_B^U}{w_1^L} - 2.02 \right| \leq \zeta; \\[2mm]
\left| \frac{w_B^U}{w_2^L} - 2.3 \right| \leq \zeta; \ \left| \frac{w_B^U}{w_3^L} - 5.17 \right| \leq \zeta; \ \left| \frac{w_B^U}{w_5^L} - 4.29 \right| \leq \zeta; \ \left| \frac{w_B^U}{w_6^L} - 6.17 \right| \leq \zeta; \\[2mm]
\left| \frac{w_B^U}{w_7^L} - 7.3 \right| \leq \zeta; \ \left| \frac{w_B^U}{w_8^L} - 8.3 \right| \leq \zeta; \ \left| \frac{w_1^L}{w_W^U} - 7.7 \right| \leq \zeta; \ \left| \frac{w_2^L}{w_W^U} - 6.46 \right| \leq \zeta; \\[2mm]
\left| \frac{w_3^L}{w_W^U} - 4.29 \right| \leq \zeta; \ \left| \frac{w_4^L}{w_W^U} - 8.47 \right| \leq \zeta; \ \left| \frac{w_5^L}{w_W^U} - 5.46 \right| \leq \zeta; \ \left| \frac{w_6^L}{w_W^U} - 2.46 \right| \leq \zeta; \\[2mm]
\left| \frac{w_7^L}{w_W^U} - 2.17 \right| \leq \zeta; \ \left| \frac{w_1^U}{w_W^L} - 7.97 \right| \leq \zeta; \ \left| \frac{w_2^U}{w_W^L} - 6.87 \right| \leq \zeta; \ \left| \frac{w_3^U}{w_W^L} - 4.76 \right| \leq \zeta; \\[2mm]
\left| \frac{w_4^U}{w_W^L} - 8.89 \right| \leq \zeta; \ \left| \frac{w_5^U}{w_W^L} - 5.89 \right| \leq \zeta; \ \left| \frac{w_6^U}{w_W^L} - 2.89 \right| \leq \zeta; \ \left| \frac{w_7^U}{w_W^L} - 2.61 \right| \leq \zeta; \\[2mm]
\sum_{j=1}^{8} w_j^L \leq 1; \sum_{j=1}^{8} w_j^U \geq 1; \\[2mm]
w_j^L \leq w_j^U, \ \forall j = 1, 2, \ldots, 8 \\[2mm]
w_j^L, w_j^U \geq 0, \ \forall j = 1, 2, \ldots, 8
\end{array}
\right.
$$

By solving the model presented, the optimal values of the rough weight coefficients of the criteria were obtained.

$$
\begin{aligned}
RN(w_1) &= [0.1708, \ 0.1780], \\
RN(w_2) &= [0.1864, \ 0.1900], \\
RN(w_3) &= [0.0942, \ 0.0987], \\
RN(w_4) &= [0.2358, \ 0.2391], \\
RN(w_5) &= [0.1191, \ 0.1202], \\
RN(w_6) &= [0.0632, \ 0.0811], \\
RN(w_7) &= [0.0527, \ 0.0554], \\
RN(w_8) &= [0.0431, \ 0.0473].
\end{aligned}
$$

By analyzing the rough weight coefficients of the optimality criteria, we see that the conditions $\sum_{j=1}^{n} w_j^L \leq 1$ and $\sum_{j=1}^{n} w_j^U \geq 1$ are satisfied, since $\sum_{j=1}^{8} w_j^L = 0.9654 \leq 1$ and $\sum_{j=1}^{8} w_j^U = 1.0098 \geq 1$. In addition, the condition $0 \leq w_j^L \leq w_j^U \leq 1$ is also satisfied; that is, the general condition is satisfied that the values of the weight coefficients of the criteria are found in the interval $w_j \in [0, 1], (j = 1, 2, \ldots, 8)$.

By solving model Equation (23), the value of ζ^* is obtained which is $\zeta^* = 0.945412$. The value of ζ^* is used to determine the consistency ratio (Equation (28)). Since we obtain the value of $\bar{a}_{BW}$ that is a_{BW}^U based on the aggregated decisions of the experts, it is not possible in advance to define the consistency index ζ. Rezaei [50] defined the values of the consistency index (ζ) for crisp BWM. Since this is about RBWM, using Equation (27) for the value $a_{BW}^U = 8.89$ the value was defined for the CI (maxζ) = 5.1406 and the value CR = 0.183911 was obtained. Based on [50], the value obtained for the CR was considered satisfactory.

According to the weights obtained for the criteria, load capacity is the most important criterion for selecting a wagon, while the next most important are maintenance conditions and price of the wagon. Load capacity is an important factor in the field of rail transport, because by selecting suitable wagons from the aspect of capacity, costs are reduced and transport capacity is increased [57]. After obtaining the weight values of the criteria, the expert team carried out the evaluation of the alternatives (Figure 4) based on the defined linguistic scale (Table 2) in the second step of the Rough SAW method. Evaluation of the alternatives is shown in Table 7.

Figure 4. Different types of wagon representing the alternatives.

Table 7. Assessment of the alternatives according to the criteria using the linguistic scale.

	E_1								E_2							
	C_1	C_2	C_3	C_4	C_5	C_6	C_7	C_8	C_1	C_2	C_3	C_4	C_5	C_6	C_7	C_8
A_1	VG	G	VP	P	VP	P	M	P	VG	VG	P	P	VG	P	M	VG
A_2	G	G	VP	G	P	P	M	M	G	G	M	VG	G	M	M	G
A_3	P	M	M	G	G	M	G	M	VP	P	VG	G	M	G	G	M
A_4	G	M	M	P	G	M	VG	M	P	P	VG	P	M	G	G	M
A_5	VP	M	G	VG	G	G	VG	M	VP	P	VG	VG	M	G	VG	M
A_6	P	P	G	G	M	M	P	P	P	M	G	G	M	G	P	M
A_7	P	G	G	G	G	G	P	M	M	P	G	G	M	G	P	M
A_8	P	G	VG	VG	G	G	P	M	P	P	VG	VG	M	VG	P	M

	E_3								E_4							
	C_1	C_2	C_3	C_4	C_5	C_6	C_7	C_8	C_1	C_2	C_3	C_4	C_5	C_6	C_7	C_8
A_1	VG	P	VP	VG	VP	P	P	VP	G	M	VP	P	P	P	P	VP
A_2	G	P	VP	M	P	P	M	P	VG	M	P	M	M	P	VP	VP
A_3	G	M	P	P	G	M	M	P	M	G	P	M	M	P	P	P
A_4	M	M	P	VG	M	M	P	P	G	G	M	P	M	P	P	P
A_5	M	M	M	M	VG	G	M	M	M	P	M	G	G	VP	P	M
A_6	P	P	M	P	P	G	P	P	M	M	M	M	P	P	P	M
A_7	M	G	M	P	G	G	M	M	G	M	G	M	G	P	P	G
A_8	M	G	G	G	G	G	G	M	M	G	G	G	G	P	M	G

	E_5								E_6							
	C_1	C_2	C_3	C_4	C_5	C_6	C_7	C_8	C_1	C_2	C_3	C_4	C_5	C_6	C_7	C_8
A_1	VG	G	VP	VP	P	P	M	G	G	M	VP	P	VP	M	P	M
A_2	G	G	VP	G	M	M	M	VG	G	M	VP	G	VP	M	M	M
A_3	P	M	M	M	VG	G	G	VG	M	M	P	G	VG	M	M	G
A_4	M	M	M	VP	VG	G	VG	VG	G	M	P	P	G	M	P	M
A_5	VP	M	VG	VG	VG	VG	VG	VG	M	M	P	VG	VG	M	M	G
A_6	P	P	G	G	G	G	P	G	M	M	P	G	VP	M	P	M
A_7	P	G	VG	G	VG	VG	P	VG	M	M	M	VG	VG	M	M	VG
A_8	P	M	VG	VG	VG	VG	P	VG	P	M	M	VG	VP	M	G	G

The Gbs-z is a two-axle closed wagon intended for the transport of different cargos that can be packaged or not. It is suitable for the transport of cargos subject to different atmospheric influences. Its capacity is 26 t. The Gas-z is a multi-purpose four-axle wagon designed for the transport of cargos that need to be protected against atmospheric influences. The important characteristic of this wagon is the limited capacity for manipulation in its interior only with a hand forklift. It is possible to adapt the doors to enable the use of other forklifts. Its load capacity is 57.5 t. The Habis is a four-axle closed wagon for the transport of cargos affected by different atmospheric influences. When the side sliding door opens, access to the loading area is gained at full height and up to half its length. This makes it possible to load goods mechanically more easily using forklifts. The load capacity is 50.7 t. The Hbis-z is a two-axle closed wagon suitable for the transport of cargos that are subject to different atmospheric influences. Similar to the previous wagon, access is gained to the loading area by means of a side sliding door. Its load capacity is 25 t. The Habbinss-z is a four-axle wagon with movable aluminum doors, two on each side, which makes it easier to manipulate the goods. It is used for transporting individual units and palletized goods that are subject to atmospheric influences. Its load capacity is 62 t. The Hrrs-z is a four-axle wagon obtained by connecting two wagons from the Gbs-z series. The wagon has two loading spaces with the same dimensions as the wagons in the Gbs-z series. It is built from an aluminum profile and it has two-part sliding side which are built in on each side and which make it possible for the doors to move freely along their guides on the wagon. It has the same function as the wagon from series G, but, because of its large volume, it is very suitable for the transport of bulky cargos. Its load capacity is 52 t. The wagon from series Rils-z is a four-axle flat wagon primarily intended for the transport of goods that must be protected against atmospheric influences. Its load capacity is 53 t. The wagon from the series Shimmns-z je is intended for transporting sheet metal plates that are loaded in a horizontal position and must be protected against atmospheric influences. It has built-in protective tarpaulin on wheeled carriers, by means of which the wagon is closed, sealing the tarpaulin on the front of the wagon. When the tarpaulin is opened, it releases two-thirds of the wagon-length for loading. Its load capacity is 68 t.

After evaluation of the alternatives by the expert team and converting the linguistic values into numerical ones, it was necessary to convert the individual matrices of each of the experts into a group matrix by applying Equations (1)–(6). An example of calculating the elements of the group matrix is presented in Table 8:

$$\tilde{x}_{25} = \{3,7,3,5,5,1\}$$

$$\overline{Lim}(1) = 1, \overline{Lim}(1) = \frac{1}{6}(3+7+3+5+5+1) = 4$$

$$\underline{Lim}(3) = \frac{1}{3}(3+3+1) = 2.33, \overline{Lim}(3) = \frac{1}{5}(3+7+3+5+5) = 4.6$$

$$\underline{Lim}(5) = \frac{1}{5}(3+3+5+5+1) = 3.4, \overline{Lim}(5) = \frac{1}{3}(7+5+5) = 5.67$$

$$\underline{Lim}(7) = \frac{1}{6}(3+7+3+5+5+1) = 4, \overline{Lim}(7) = 7$$

$$RN(x_{25}^1) = RN(x_{25}^3) = [2.33, 4.6]; RN(x_{25}^2) = [4,7]; RN(x_{25}^4) = RN(x_{25}^5) = [3.4, 5.67]; RN(x_{25}^6) = [1,4]$$

$$x_{25}^L = \frac{x_{25}^1 + x_{25}^2 + x_{25}^5 + x_{25}^4 + x_{25}^5 + x_{25}^6}{S} = \frac{2.33 + 4 + 2.33 + 3.4 + 3.4 + 1}{6} = 2.74$$

$$x_{25}^U = \frac{x_{25}^1 + x_{25}^2 + x_{25}^5 + x_{25}^4 + x_{25}^5 + x_{25}^6}{S} = \frac{4.6 + 7 + 4.6 + 5.67 + 5.67 + 4}{3} = 5.26$$

Table 8. Group rough matrix.

	A$_1$	A$_2$	A$_3$	A$_4$	A$_5$	A$_6$	A$_7$	A$_8$
C$_1$	[1.22, 2.11]	[2.39, 2.95]	[4.74, 7.26]	[3.49, 5.22]	[6, 8]	[5.89, 6.78]	[4.53, 6.12]	[5.89, 6.78]
C$_2$	[4.74, 7.26]	[4.78, 6.51]	[4.4, 5.6]	[4.4, 5.6]	[3.89, 4.78]	[3.5, 4.5]	[4.78, 6.51]	[4.78, 6.51]
C$_3$	[8.39, 8.95]	[7.17, 8.77]	[4.06, 6.46]	[3.83, 6.07]	[2.22, 5.11]	[3.49, 5.22]	[2.53, 4.12]	[1.49, 3.22]
C$_4$	[1.89, 2.78]	[5.88, 7.47]	[4.78, 6.51]	[1.89, 2.78]	[7.17, 8.77]	[5.17, 6.77]	[5.12, 7.47]	[7.89, 8.78]
C$_5$	[1.6, 4.67]	[2.74, 5.26]	[6, 8]	[5.49, 7.22]	[6.78, 8.51]	[2.74, 5.26]	[6.53, 8.12]	[4.37, 7.42]
C$_6$	[6.39, 6.95]	[5.5, 6.5]	[3.88, 5.47]	[3.88, 5.47]	[2.58, 5.63]	[3.49, 5.22]	[2.53, 4.88]	[1.99, 4.78]
C$_7$	[3.5, 4.5]	[3.78, 4.89]	[4.78, 6.51]	[4.06, 7.22]	[5.28, 8.08]	[3, 3]	[3.22, 4.11]	[3.64, 5.72]
C$_8$	[2.32, 6.44]	[3.2, 6.8]	[3.99, 6.78]	[3.93, 6.17]	[5.23, 6.83]	[3.88, 5.47]	[5.64, 7.72]	[5.49, 7.22]

After carrying out normalization in the fourth step, using Equations (30)–(33), the normalized matrix shown in Table 9 was obtained, while the weighted normalized matrix was obtained using

Equation (35). Normalization of the group matrix elements for benefit criteria was carried out in the following way:

$$\tilde{r}_{25} = \left[\frac{x_{ij}^L}{x_{ij}^{+U}}, \frac{x_{ij}^U}{x_{ij}^{+L}} \right] = \left[\frac{2.74}{8.51}, \frac{5.26}{6.78} \right] \rightarrow \tilde{r}_{25} = [0.32, 0.78]$$

and for the cost criteria:

$$\tilde{r}_{23} = \left[\frac{x_{ij}^{-L}}{x_{ij}^U}, \frac{x_{ij}^{-U}}{x_{ij}^L} \right] = \left[\frac{1.49}{8.77}, \frac{3.22}{7.17} \right] \rightarrow \tilde{r}_{23} = [0.17, 0.45]$$

Table 9. Normalized matrix.

	C_1	C_2	C_3	C_4	C_5	C_6	C_7	C_8
A_1	[0.58, 1.73]	[0.65, 1.52]	[0.17, 0.38]	[0.22, 0.35]	[0.19, 0.69]	[0.29, 0.75]	[0.43, 0.85]	[0.32, 1.14]
A_2	[0.41, 0.88]	[0.66, 1.36]	[0.17, 0.45]	[0.67, 0.95]	[0.32, 0.78]	[0.31, 0.87]	[0.47, 0.93]	[0.44, 1.21]
A_3	[0.17, 0.45]	[0.61, 1.17]	[0.23, 0.79]	[0.54, 0.83]	[0.71, 1.18]	[0.36, 1.23]	[0.59, 1.23]	[0.55, 1.20]
A_4	[0.23, 0.60]	[0.61, 1.17]	[0.25, 0.84]	[0.22, 0.35]	[0.65, 1.06]	[0.36, 1.23]	[0.50, 1.37]	[0.54, 1.09]
A_5	[0.15, 0.35]	[0.54, 1.00]	[0.29, 1.45]	[0.82, 1.11]	[0.80, 1.26]	[0.35, 1.85]	[0.65, 1.53]	[0.72, 1.21]
A_6	[0.18, 0.36]	[0.48, 0.94]	[0.29, 0.92]	[0.59, 0.86]	[0.32, 0.78]	[0.38, 1.37]	[0.37, 0.57]	[0.54, 0.97]
A_7	[0.20, 0.47]	[0.66, 1.36]	[0.36, 1.27]	[0.58, 0.95]	[0.77, 1.20]	[0.41, 1.89]	[0.40, 0.78]	[0.78, 1.28]
A_8	[0.18, 0.36]	[0.66, 1.36]	[0.46, 2.16]	[0.90, 1.11]	[0.51, 1.09]	[0.42, 2.40]	[0.45, 1.08]	[0.76, 1.28]

After weighting the normalized matrix using Equation (35):

$$v_{23}^L = \left[w_3^L \times r_{23}^L, w_3^U \times r_{23}^U \right] = [0.17 \times 0.094, 0.45 \times 0.099] \rightarrow v_{23}^L = [0,016, 0.044]$$

the values were summed for all the alternative by rows and the final rank of the alternatives is obtained which is shown in Table 10. The ranking is carried out in descending order, whereby the highest value presents the best solution, and the lowest the worst. The table also shows the conversion of a rough number into a crisp number using the average values of the lower and upper limits of the rough number.

Table 10. The results and ranking of the alternatives.

	s_{ij}^L	s_{ij}^U	AV	Rank
A_1	0.364	0.963	0.664	6
A_2	0.469	0.959	0.714	4
A_3	0.454	0.944	0.699	5
A_4	0.377	0.853	0.615	8
A_5	0.529	1.105	0.817	2
A_6	0.392	0.821	0.606	7
A_7	0.500	1.095	0.797	3
A_8	0.553	1.249	0.901	1

Alternative 8 represents the most acceptable solution according to the results obtained.

5. Sensitivity Analysis

To determine the stability of the results, a sensitivity analysis was performed; the first part of which includes a change in the weights of the criteria through 15 different sets. In the first eight sets, the value of each criterion was increased by 14%, and those remaining were reduced by 2% of the value obtained by RBWM, respectively; in the ninth set, all of the criteria were equally important; in the 10th set, the first, second and fourth criteria (as the most important criteria) were reduced by 10%, while the values of other criteria remained unchanged; in the 11th set, the values of the three least influential criteria were increased by 25%; in the 12th set, the values of the two most influential criteria were reduced by 25%, and the values of the two least significant criteria were increased by 25%; in the 13th set, the values of the first four criteria were reduced by 30% each, and the values of the remaining four criteria were increased by 30% each; in the penultimate set of values, the two most important criteria were eliminated; and, in the final set, the values of the three most important criteria were eliminated, while the values of the remaining criteria remained unchanged. Figure 5 shows the ranks of alternatives across all sets.

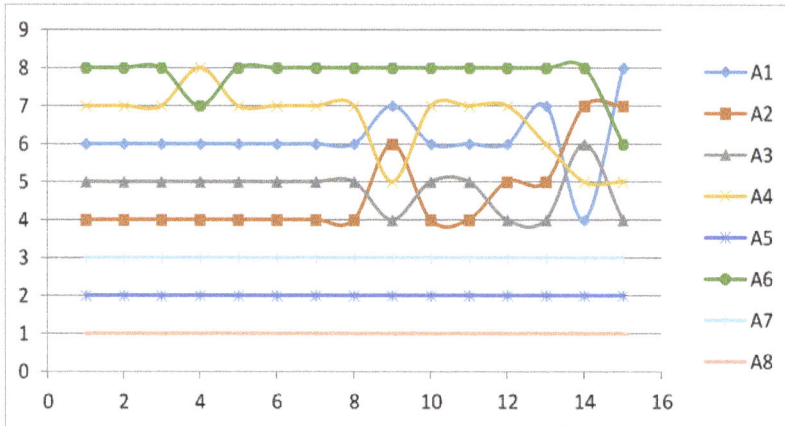

Figure 5. Ranking of the alternatives through the scenarios.

Figure 5 shows the ranking of the alternatives through the sets that were formed and as can be seen alternative eight represents the best solution in all cases. The highest value of A8 is reached in the 11th and fourth sets, although there are completely different values for the weights of the criteria. In the first case, the advantage is given to the least important criteria by a whole quarter of their value, and in the second case the fourth most important criterion has an advantage of 14%. The lowest values for alternative A8 are in scenarios 14 and 15, since in these cases there is a reduction in the total number of criteria, based on which the decisions are made; that is, individual criteria are given a value of zero. In addition, Alternatives 5 and 7 do not change their original ranking and they are ranked in second and third places, respectively. The remaining alternatives generally retain their positions with certain changes. Alternative 2 is in fourth place in 10 sets, and in fifth and seventh position twice. Alternative 4 has a consistent ranking (its rank does not change) in 10 sets, where it is in seventh place, while in the fourth set it is in the last place. It is in fifth place twice, and sixth once. Alternative 6 is in last position for 12 sets, while it is once in seventh and once in sixth position.

The results show that assigning different weights to the criteria through the sets leads to a change in the ranks of individual alternatives, which confirms that the model is sensitive to changes in weight coefficients. However, we can conclude that the changes were not drastic, which also confirms the correlation of the ranks through the scenarios (Table 11).

Table 11. Spearman correlation coefficient (*SCC*) of the ranks through 15 sets.

Set	Set 1	Set 2	Set 3	Set 4	Set 5	Set 6	Set 7	Set 8
SCC	0.869	0.869	0.869	0.702	0.869	0.869	0.869	0.869
Set	Set 9	Set 10	Set 11	Set 12	Set 13	Set 14	Set 15	Average
SCC	0.869	0.869	0.869	0.774	0.845	0.917	0.631	0.837

The *SCC* values in Table 11 were obtained by comparing the initial ranks from the RBWM-SAW model (Table 10) with the ranks obtained through 15 sets. We see in Table 11 that there is a high correlation between the ranks, since, in 80% of the sets (13 sets), the *SCC* is greater than 0.845, while in three sets it is greater than 0.630. The average *SCC* value through all the scenarios is 0.837, which shows an extremely high correlation. Based on recommendations by Ghorabaee et al. [58], all *SCC* values over 0.8 show an extremely high correlation. Since 80% of the *SCC* values are significantly greater than 0.8, we can conclude that there is a very high correlation (closeness) of ranks and that the proposed ranking is confirmed and credible.

In addition to the stability shown by the first part of the sensitivity analysis, the proposed model was compared with other hybrid multi-criteria models. The hybrid models used for comparison of the results are shown in Table 12. The methods used to determine the weights of the criteria were the traditional AHP method [59], BWM [50] and the rough AHP method [49]. The following methods were used to rank the alternatives: TOPSIS [60], rough TOPSIS [44], SAW [61], MABAC [35] and rough MABAC [62]. The combinations of these methods and the ranking of alternatives are shown in Table 12.

Table 12. Ranking the alternatives by combining different methods.

AHP-TOPSIS		AHP-MABAC		AHP-SAW		BWM-TOPSIS		BWM-MABAC		BWM-SAW	
0.415	5	−0.086	7	0.563	6	0.427	5	−0.093	7	0.565	6
0.484	4	0122	3	0.650	4	0.485	4	0.091	4	0.637	4
0.413	6	0.035	5	0.607	5	0.422	6	0.045	5	0.616	5
0.371	8	−0.078	6	0.493	8	0.390	7	−0.051	6	0.519	8
0.486	3	0.118	4	0.719	2	0.485	3	0.123	3	0.724	2
0.393	7	−0.154	8	0.522	7	0.383	8	−0.165	8	0.520	7
0.506	2	0.166	2	0.679	3	0.509	2	0.174	2	0.689	3
0.583	1	0.204	1	0.757	1	0.570	1	0.200	1	0.760	1

RAHP-RTOPSIS		RAHP-RSAW		RAHP-RMABAC		RBWM-RTOPSIS		RBWM-RSAW		RBWM-RMABAC	
0.417	7	1.691	6	−0.633	7	0.434	7	0.664	6	−0.169	6
0.550	2	1.847	4	1.292	3	0.562	3	0.714	4	0.075	4
0.474	5	1.702	5	0.843	5	0.488	5	0.699	5	0.066	5
0.385	8	1.400	8	−1.055	8	0.411	8	0.615	8	−0.196	7
0.544	3	1.998	2	1.047	4	0.555	4	0.817	2	0.158	3
0.456	6	1.506	7	−0.390	6	0.445	6	0.606	7	−0.216	8
0.528	4	1.929	3	2.471	2	0.568	2	0.797	3	0.341	2
0.575	1	2.177	1	2.556	1	0.604	1	0.901	1	0.437	1

When comparing the results, a total of 12 hybrid models were formed, and the results for the models are shown in Table 12 and Figure 6. Figure 6 graphically presents the results of applying the 12 hybrid models.

Based on Figure 6, we notice that A8 represents the best solution in all 12 models, which is an adequate verification of the proposed model. Alternative 3 is in fifth position in 10 out of 12 combinations, and twice in sixth. The second position most often belongs to Alternative 7, which is the same number of times it appears in that position. When it comes to the ranking of the other alternatives there are some changes, but, in most cases, they retain their ranking obtained using the proposed new approach. The ranks obtained were compared with each other and with the initial ranking of the RBWM-SAW model. *SCC* was used to compare the ranks. The results obtained from comparing the 12 hybrid models are presented in Table 13.

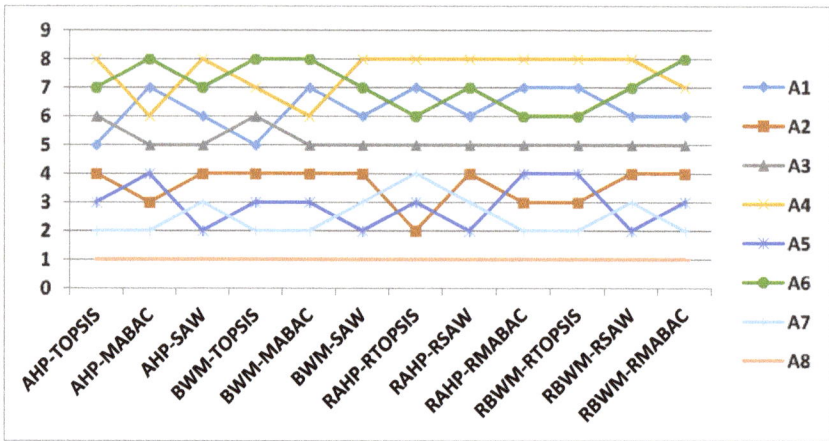

Figure 6. Ranking of the alternatives according to different MCDM methods.

Table 13. Correlation of the results of the hybrid models.

Methods	A	B	C	D	E	F	G	H	I	J	K	L	Average
A	1.000	0.886	0.833	0.976	0.838	0.833	0.976	0.833	0.881	0.881	0.833	0.862	0.861
B	-	1.000	0.857	0.833	0.976	0.857	0.833	0.857	0.905	0.905	0.857	0.952	0.894
C	-	-	1.000	0.810	0.905	1.000	0.905	1.000	0.905	0.905	1.000	0.952	0.938
D	-	-	-	1.000	0.862	0.810	0.952	0.810	0.929	0.929	0.810	0.886	0.865
E	-	-	-	-	1.000	0.905	0.810	0.905	0.881	0.881	0.905	0.976	0.908
F	-	-	-	-	-	1.000	0.905	1.000	0.905	0.905	1.000	0.952	0.952
G	-	-	-	-	-	-	1.000	0.905	0.929	0.929	0.905	0.833	0.917
H	-	-	-	-	-	-	-	1.000	0.905	0.905	1.000	0.952	0.952
I	-	-	-	-	-	-	-	-	1.000	1.000	0.905	0.905	0.953
J	-	-	-	-	-	-	-	-	-	1.000	0.905	0.905	0.937
K	-	-	-	-	-	-	-	-	-	-	1.000	0.952	0.976
L	-	-	-	-	-	-	-	-	-	-	-	1.000	1.000
					Overall average								0.933

A, AHP-TOPSIS; B, AHP-MABAC; C, AHP-SAW; D, BWM-TOPSIS; E, BWM-MABAC; F, BWM-SAW; G, RAHP-RTOPSIS; H, RAHP-RSAW; I, RAHP-RMABAC; J, RBWM-RTOPSIS; K, RBWM-RSAW; L, RBWM-RMABAC

In the next section, a comparison of the ranks from the 12 hybrid models with the initial ranks from Table 12 is carried out. The results are shown in Table 14.

Table 14. SCC comparison of the ranks by scenarios in relation to the initial rank.

Scenario	AHP-TOPSIS	AHP-MABAC	AHP-SAW	BWM-TOPSIS	BWM-MABAC	BWM-SAW	RAHP-RTOPSIS
SCC	0.952	0.952	0.952	0.976	0.976	0.952	0.833
Scenario	RAHP-RSAW	RAHP-RMABAC	RBWM-RTOPSIS	RBWM-RSAW	RBWM-RMABAC	Average	
SCC	0.952	0.905	0.905	0.952	1.000	0.942	

We see in Tables 13 and 14 that there is a high correlation between the ranks of the compared models. Since all the *SCC* values are significantly greater than 0.81, and the average values are 0.933 and 0.942, we can conclude that there is an extremely high correlation (closeness) between the proposed approach and the other models tested. Thus, we can conclude that the proposed RBWM-SAW model gives credible ranks.

6. Conclusions

The paper presents a new model for decision making, which is verified on the selection of wagons for carrying out the internal transport of a paper manufacturing company. It also presents a new scale for evaluating alternatives according to the criteria, depending on their type. In the case of qualitative assessment, which is a very common case in group decision making, from the outset, cost and benefit criteria are treated differently, which can be seen throughout the paper.

One of the contributions of this paper is the new RBWM-RSAW model that enables the objective aggregation of expert decisions with full consideration of their precision and subjectivity that prevail during group decision making. Another significant contribution of this paper is the development of the new RBWM and RSAW models, which contribute to the advancement of the literature that considers the theoretical and practical application of multi-criteria techniques. The proposed models allow the evaluation of alternatives despite the imprecision and lack of quantitative information in the decision-making process. The third contribution of the paper is to improve the methodology for evaluating railway wagons through a new approach for dealing with imprecision.

By using this hybrid model, it is possible to solve the problems of multi-criteria decision making in a simple way and make decisions that have a significant impact on achieving business efficiency, as is the case in this paper. By applying rough numbers in combination with these methods, imprecision in group decision making is taken into account and more objective results are obtained than with crisp approaches. It is very important to mention that the ranks of the alternatives obtained using the rough BWM and rough SAW were also confirmed through the application of the traditional SAW method. When it comes to determining the weight coefficients of the evaluation criteria, the results of the rough BWM were compared with the results given by the AHP, BWM and rough AHP algorithms and they were used to further check the stability of the initial solution. Analysis of the results showed that the ranks of the alternatives for the rough BWM-SAW algorithm fully correlated with the ranks obtained for the other methods.

Further research connected with this study concerns a post-analysis of the internal transport in the given company in order to verify the savings that arise from the proposed method of carrying out the internal transport. In terms of the area of multi-criteria decision making, further research directions relate to the application of rough numbers in combination with other methods and the attempt to develop new methods that would further enrich this widely applied field.

Author Contributions: Each author has participated and contributed sufficiently to take public responsibility for appropriate portions of the content.

Conflicts of Interest: The authors declare no conflict of interest.

References

1. Koskinen, P.; Hilmola, O.P. Supply chain challenges of North-European paper industry. *Ind. Manag. Data Syst.* **2008**, *108*, 208–227. [CrossRef]
2. Pereira, M.; Adelaide, M.; Resgate, L.; Telhada, J. Multicriteria Methodology to Mode of Transport Choosing—The Portuguese Case. In Proceedings of the XI Congreso Galego de Estatística e Investigación de Operacións, A Coruña, Galicia, Spain, 24–26 October 2013; pp. 1–15.
3. Vashist, J.K.; Dey, A.K. Selection Criteria for a Mode of Surface Transport: An Analytic Hierarchy Process Approach. *Amity Glob. Bus. Rev.* **2016**, *11*, 86–95.
4. Bai, C.; Fahimnia, B.; Sarkis, J. Sustainable transport fleet appraisal using a hybrid multi-objective decision making approach. *Ann. Oper. Res.* **2017**, *250*, 309–340. [CrossRef]
5. Barbosa, S.B.; Ferreira, M.G.G.; Nickel, E.M.; Cruz, J.A.; Forcellini, F.A.; Garcia, J.; de Andrade, J.B.S.O. Multi-criteria analysis model to evaluate transport systems: An application in Florianópolis, Brazil. *Transp. Res. Part A Policy Pract.* **2017**, *96*, 1–13. [CrossRef]
6. Tsamboulas, D.A. A tool for prioritizing multinational transport infrastructure investments. *Transp. Policy* **2007**, *14*, 11–26. [CrossRef]

7.   Keshavarz Ghorabaee, M.; Amiri, M.; Kazimieras Zavadskas, E.; Antuchevičienė, J. Assessment of third-party logistics providers using a CRITIC-WASPAS approach with interval type-2 fuzzy sets. *Transport* **2017**, *32*, 66–78. [CrossRef]

8.   Yazdani, M.; Zarate, P.; Coulibaly, A.; Zavadskas, E.K. A group decision making support system in logistics and supply chain management. *Expert Syst. Appl.* **2017**, *88*, 376–392. [CrossRef]

9.   Stević, Ž.; Tanackov, I.; Vasiljević, M.; Vesković, S. Evaluation in logistics using combined AHP and EDAS method. In Proceedings of the XLIII International Symposium on Operational Research, Belgrade, Serbia, 20–23 September 2016; pp. 309–313.

10. Mardani, A.; Zavadskas, E.K.; Khalifah, Z.; Jusoh, A.; Nor, K.M. Multiple criteria decision-making techniques in transportation systems: A systematic review of the state of the art literature. *Transport* **2016**, *31*, 359–385. [CrossRef]

11. Turskis, Z.; Zavadskas, E.K. A new fuzzy additive ratio assessment method (ARAS-F). Case study: The analysis of fuzzy multiple criteria in order to select the logistic centers location. *Transport* **2010**, *25*, 423–432. [CrossRef]

12. Stević, Ž.; Vesković, S.; Vasiljević, M.; Tepić, G. The selection of the logistics center location using AHP method. In Proceedings of the 2nd Logistics International Conference, Belgrade, Serbia, 27–29 May 2015; pp. 86–91.

13. Zhang, Y.; Zhang, Y.; Li, Y.; Liu, S.; Yang, J. A Study of Rural Logistics Center Location Based on Intuitionistic Fuzzy TOPSIS. *Math. Probl. Eng.* **2017**, *2017*, 2323057. [CrossRef]

14. Stević, Ž.; Tanackov, I.; Vasiljević, M.; Novarlić, B.; Stojić, G. An integrated fuzzy AHP and TOPSIS model for supplier evaluation. *Serb. J. Manag.* **2016**, *11*, 15–27. [CrossRef]

15. Tamošaitienė, J.; Zavadskas, E.K.; Šileikaitė, I.; Turskis, Z. A novel hybrid MCDM approach for complicated supply chain management problems in construction. *Procedia Eng.* **2017**, *172*, 1137–1145. [CrossRef]

16. Pohekar, S.D.; Ramachandran, M. Application of multi-criteria decision making to sustainable energy planning—A review. *Renew. Sustain. Energy Rev.* **2004**, *8*, 365–381. [CrossRef]

17. Carlsson, C.; Fullér, R. Fuzzy multiple criteria decision making: Recent developments. *Fuzzy Sets Syst.* **1996**, *78*, 139–153. [CrossRef]

18. Zavadskas, E.K.; Turskis, Z.; Dejus, T.; Viteikiene, M. Sensitivity analysis of a simple additive weight method. *Int. J. Manag. Decis. Mak.* **2007**, *8*, 555–574. [CrossRef]

19. Huang, Y.S.; Chang, W.C.; Li, W.H.; Lin, Z.L. Aggregation of utility-based individual preferences for group decision-making. *Eur. J. Oper. Res.* **2013**, *229*, 462–469. [CrossRef]

20. Shameli, S.A.; Shajari, M.; Hassanabadi, M.; Jabbarifar, M.; Dagenais, M. Fuzzy-Multi-Criteria Decision-Making for Information Security Risk Assessment. *Open Cybern. Syst. J.* **2012**, *6*, 26–37. [CrossRef]

21. Safavian, S.T.S.; Fataei, E.; Ebadi, T.; Mohamadian, A. Site Selection of Sarein's Municipal Solid Waste Landfill Using the GIS Technique and SAW Method. *Int. J. Environ. Sci. Dev.* **2015**, *6*, 934. [CrossRef]

22. Mokhtari, E.; Khamehchian, M.; Montazer, G.; Nikudel, M. Landfill Site Selection Using Simple Additive Weighting (SAW) Method and Artificial Neural Network Method; A Case Study from Lorestan Province, Iran. *Int. J. Geogr. Geol.* **2016**, *5*, 209–223. [CrossRef]

23. Pamučar, D.S.; Božanić, D.; Ranđelović, A. Multi-criteria decision making: An example of sensitivity analysis. *Serb. J. Manag.* **2017**, *12*, 1–27. [CrossRef]

24. Deni, W.; Sudana, O.; Sasmita, A. Analysis and Implementation Fuzzy Multi-Attribute Decision Making SAW Method for Selection of High Achieving Students in Faculty Level. *Int. J. Comput Sci.* **2013**, *10*, 674–680.

25. Gupta, S.; Gupta, A. A Fuzzy Multi Criteria Decision Making Approach for Vendor Evaluation in a Supply Chain. *Intersci. Manag. Rev.* **2012**, *2*, 10–16.

26. Kumar, M.; Jayaswal, P.; Kushwah, K. Exploring Fuzzy SAW Method for Maintenance Strategy Selection Problem of Material Handling Equipment. *Int. J. Curr. Eng. Technol.* **2013**, *3*, 600–605.

27. Đorović, B.; Pamučar, D. Fuzzy mathematical model for design and evaluation of the logistic organisational structure. *Econ. Comput. Econ. Cybern. Stud. Res.* **2012**, *36*, 139–156.

28. Afshari, A.; Mojahed, M.; Yusuff, R.M. Simple Additive Weighting approach to Personnel Selection problem. *Int. J. Innov. Manag. Technol.* **2010**, *1*, 511–515.

29. Azizollah, J.; Mehdi, J.; Abalfazl, Z.; Farzad, Z. Using Fuzzy Delphi Method in Maintenance Strategy Selection Problem. *J. Uncertain Syst.* **2008**, *2*, 289–298.

30. Chen, T.Y. Comparative analysis of SAW and TOPSIS based on interval-valued fuzzy sets: Discussions on score functions and weight constraints. *Expert Syst. Appl.* **2012**, *39*, 1848–1861. [CrossRef]
31. Bureika, G. Study of traction rolling-stock using in lithuanian sector of railway line "Rail Baltica". *Transp. Probl.* **2012**, *7*, 49–56.
32. Jakimavičius, M.; Burinskiene, M. Assessment of Vilnius city development scenarios based on transport system modelling and multicriteria analysis. *J. Civ. Eng. Manag.* **2009**, *15*, 361–368. [CrossRef]
33. Chou, S.Y.; Chang, Y.H.; Shen, C.Y. A fuzzy simple additive weighting system under group decision-making for facility location selection with objective/subjective attributes. *Eur. J. Oper. Res.* **2008**, *189*, 132–145. [CrossRef]
34. Wang, Y.J. A fuzzy multi-criteria decision-making model based on simpleadditive weighting method and relative preference relation. *Appl. Soft Comput.* **2015**, *30*, 412–420. [CrossRef]
35. Pamučar, D.; Ćirović, G. The selection of transport and handling resources in logistics centers using Multi-Attributive Border Approximation area Comparison (MABAC). *Expert Syst. Appl.* **2015**, *42*, 3016–3028. [CrossRef]
36. Manokaran, E.; Senthilvel, S.; Subhashini, S.; Muruganandham, R.; Ravichandran, K. Mathematical Model for Performance Rating in Software industry—A study using Artificial Neural Network. *Int. J. Sci. Eng. Res.* **2012**, *3*, 4–7.
37. Zadeh, L.A. Fuzzy sets. *Inf. Control* **1965**, *8*, 338–353. [CrossRef]
38. Pawlak, Z. *Rough Sets: Theoretical Aspects of Reasoning about Data*; Springer: Berlin, Germany, 1991.
39. Pawlak, Z. Anatomy of conflicts. *Bull. Eur. Assoc. Theor. Comput. Sci.* **1993**, *50*, 234–247.
40. Kuang, H.; Kilgour, D.M.; Hipel, K.W. Grey-based PROMETHEE II with application to evaluation of source water protection strategies. *Inf. Sci.* **2015**, *294*, 376–389. [CrossRef]
41. Arce, M.E.; Saavedra, A.; Míguez, J.L.; Granada, E. The use of grey-based methods in multi-criteria decision analysis for the evaluation of sustainable energy systems. *Rev. Renew. Sustain. Energy Rev.* **2015**, *47*, 924–932. [CrossRef]
42. Kang, B.; Wei, D.; Li, Y.; Deng, Y. A Method of Converting Z-number to Classical Fuzzy Number. *J. Inf. Comput. Sci.* **2012**, *9*, 703–709.
43. Azadeh, A.; Kokabi, R. Z-number DEA: A new possibilistic DEA in the context of Z-numbers. *Adv. Eng. Inf.* **2016**, *30*, 604–617. [CrossRef]
44. Song, W.; Ming, X.; Wu, Z.; Zhu, B. A rough TOPSIS approach for failure mode and effects analysis in uncertain environments. *Q. Reliab. Eng. Int.* **2014**, *30*, 473–486. [CrossRef]
45. Pamučar, D.; Mihajlović, M.; Obradović, R.; Atanasković, P. Novel approach to group multi-criteria decision making based on interval rough numbers: Hybrid DEMATEL-ANP-MAIRCA model. *Expert Syst. Appl.* **2017**, *88*, 58–80. [CrossRef]
46. Pamučar, D.; Gigović, L.; Bajić, Z.; Janošević, M. Location selection for wind farms using GIS multi-criteria hybrid model: An approach based on fuzzy and rough numbers. *Sustainability* **2017**, *9*, 1315. [CrossRef]
47. Pamučar, D.; Petrović, I.; Ćirović, G. Modification of the Best-Worst and MABAC methods: A novel approach based on interval-valued fuzzy-rough numbers. *Expert Syst. Appl.* **2018**, *91*, 89–106. [CrossRef]
48. Zhu, G.N.; Hu, J.; Qi, J.; Gu, C.C.; Peng, J.H. An integrated AHP and VIKOR for design concept evaluation based on rough number. *Adv. Eng. Inf.* **2015**, *29*, 408–418. [CrossRef]
49. Zhai, L.Y.; Khoo, L.P.; Zhong, Z.W. A rough set based QFD approach to the management of imprecise design information in product development. *Adv. Eng. Inf.* **2009**, *23*, 222–228. [CrossRef]
50. Rezaei, J. Best-worst multi-criteria decision-making method. *Omega* **2015**, *53*, 49–57. [CrossRef]
51. Rezaei, J. Best-worst multi-criteria decision-making method: Some properties and a linear model. *Omega* **2016**, *64*, 126–130. [CrossRef]
52. Rezaei, J.; Wang, J.; Tavasszy, L. Linking supplier development to supplier segmentation using best worst method. *Expert Syst. Appl.* **2015**, *42*, 9152–9164. [CrossRef]
53. Gupta, H. Evaluating service quality of airline industry using hybrid best worst method and VIKOR. *J. Air Transp. Manag.* **2017**. [CrossRef]
54. Wan Ahmad, W.N.K.; Rezaei, J.; Sadaghiani, S.; Tavasszy, L.A. Evaluation of the external forces affecting the sustainability of oil and gas supply chain using Best Worst Method. *J. Clean. Prod.* **2017**, *153*, 242–252. [CrossRef]

55. Guo, S.; Zhao, H. Fuzzy best-worst multi-criteria decision-making method and its applications. *Knowl.-Based Syst.* **2017**, *121*, 23–31. [CrossRef]
56. Hafezalkotob, A.; Hafezalkotob, A. A novel approach for combination of individual and group decisions. *Appl. Soft Comput.* **2017**, *59*, 316–325. [CrossRef]
57. Bartuška, L.; Černá, L.; Daniš, J. Costs comparison and the possibilities of increasing the transport capacity with a selection of the appropriate railway wagons. *NAŠE MORE Znanstveno-Stručni Časopis za More i Pomorstvo* **2016**, *63*, 93–97.
58. Keshavarz Ghorabaee, M.; Zavadskas, E.K.; Turskis, Z.; Antucheviciene, J. A new combinative distance-based assessment (CODAS) method for multi-criteria decision-making. *Econ. Comput. Econ. Cybern. Stud. Res.* **2016**, *50*, 25–44.
59. Saaty, T.L. *The Analytic Hierarchy Process*; Mc Graw-Hill: New York, NY, USA, 1980.
60. Hwang, C.L.; Yoon, K. *Multiple Attributes Decision Making Methods and Applications*; Springer: Berlin, Germany, 1981.
61. MacCrimmon, K.R. *Decision Making among Multiple-Attribute Alternatives: A Survey and Consolidated Approach*; RAND Co.: Santa Monica, CA, USA, 1968.
62. Roy, J.; Chatterjee, K.; Bandhopadhyay, A.; Kar, S. Evaluation and selection of Medical Tourism sites: A rough AHP based MABAC approach. *arXiv* **2016**, arXiv:1606.08962.

symmetry

MDPI

Article

Evaluation of Airplane Boarding/Deboarding Strategies: A Surrogate Experimental Test

Shengjie Qiang [1], Bin Jia [2],*, and Qingxia Huang [2]

1   College of Transportation and Logistics, East China Jiaotong University, Nanchang 330013, China; qiangshengjie@163.com
2   School of Traffic and Transportation, Beijing Jiaotong University, Beijing 100044, China; 14114199@bjtu.edu.cn
*   Correspondence: bjia@bjtu.edu.cn; Tel.: +86-010-5168-4240

Received: 15 September 2017; Accepted: 9 October 2017; Published: 11 October 2017

Abstract: Optimally organizing passengers boarding/deboarding an airplane offers a potential way to reduce the airplane turn time. The main contribution of our work is that we evaluate seven boarding strategies and two structured deboarding strategies by using a surrogate experimental test. Instead of boarding a real or mocked airplane, we carried out the experiment by organizing 40 participants to board a school bus with ten rows of four seats, symmetrically distributed on a single, central aisle. Experimental results confirm that the optimized strategies, i.e., Steffen and Steffen-lug, are superior to the traditional ones, i.e., Back-to-front, Window-to-aisle, and Random in time-saving and stability. However, the two structured deboarding strategies failed to reduce the deboarding time, and this result strongly suggests the prerequisites of applying such strategies only when, on average, passengers have a large amount of luggage. Besides, we further carried out a questionnaire survey of participants' preferences on seat layout and discussed how those preferences influence the boarding time.

Keywords: airplane turn time; boarding/deboarding strategies; seat preference; experimental test

---

1. Introduction

Airlines, under increasing competition pressure, are driven to optimize their operations by maximizing their efficiency and profitability. A feasible method is to reduce the airplane turn time, which includes the time to unload an airplane after its arrival and to prepare it for departure again. Turn time processes include boarding/deboarding, refueling, handling of catering, and the off-loading and loading of baggage. Shortening the time required in any of these sections will improve efficiency. Considering the boarding and deboarding processes are the main contribution to an airplane's turn time, improvement in either one or both of these two parts provides the potential to reduce the turn time.

Adopting a fast and friendly boarding/deboarding strategy benefits the airlines, airport operators, and the passengers [1], see Figure 1. Airlines obtain revenue when the airplane is in the air, so they make every effort to minimize the time that their flights stay on the ground. A conservative estimation of a one-minute reduction in the turn time saves $30 for an active airplane staying on the ground [2]. Thus, each minute saved in the turn time of a flight can accumulate to produce considerable annual savings. These savings could inspire the airlines to make better utilization of airplanes, i.e., they may offer additional flights. Reduction of airplane turn time can also benefit the airport operators in three aspects: First, it could mitigate the flight delays caused by inefficient ground service, less redundant ground time could allow for scheduling more flights; Second, it makes a more efficient utilization of the equipment on the ground, such as the boarding bridge; Third, it increases the level of service at the airport by reducing the passengers' waiting time at the departure hall. For passengers, they would enjoy a quick and friendly strategy for less individual boarding/deboarding time and avoid frustrating stoppages.

Figure 1. Optimal boarding and deboarding strategy benefits three principle users.

The currently used boarding strategies perform rather poorly concerning boarding time; flow stoppage occurs in a real boarding process. Such intermittent delays are caused by two kinds of interference: One is aisle interference, which occurs when a passenger is blocked by another passenger in the aisle. The other is seat interference, which happens when a window or middle seat passenger is blocked by other passengers who are already sitting in the same half-row. Efforts have been made to reduce those two interferences in boarding time, and most of them are based on simulation works, either by discrete modeling [3–6] or by continuous modeling [7,8]. Various optimal boarding strategies are proposed to reduce boarding time. In [9], the author presented the most time-saving strategy by applying a Markov Chain Monte Carlo optimization algorithm. In addition, new strategies are also proposed by using linear or nonlinear programming approaches [10,11] based on an assumption that a minimization of the number of interferences leads to a minimal boarding time. In [5,12], the authors emphasized the importance of luggage storage space in stowing the luggage, and they assigned seats to the passengers by considering the number of luggage pieces with which they were traveling.

Though research on the optimal boarding problem has become a bit of a hot topic in the last few years, an obvious shortage is that most of the works are based on simulation analysis and seldom on validation works. Actually, deviation exists between computer simulation and the real process. An example of such deviation is that passengers are basically assumed to be homogeneous in simulations, which is too simple to describe the complex dynamic of the boarding process arising from individual differences. It is, therefore, important to know the real performance of some optimal strategies by experimental test. A primary work was done in [13] as the authors conducted an experimental comparison of airplane boarding in a mock Boeing 757 fuselage, with twelve rows of six seats and a single aisle. The author found a significant reduction in boarding time with the optimized Steffen strategy over the other three traditional strategies, i.e., the block strategy, the window-to-aisle strategy, and random strategy.

In contrast with the experiment using a mock airplane fuselage in [13], here, we designed a surrogate experimental test. Instead of boarding participants onto a real or mock airplane, we had them board a school bus, as there are similarities between passengers boarding these two means of transport. The purpose of our experiment was to evaluate various strategies in boarding time and other qualities such as time stability and time gap. Seven strategies were selected, including Random, Free, Back-to-front, Window-to-aisle, Steffen, Steffen-lug, and CRBF (column rotated in a back to front order, see Section 2.3 for details). In addition, we evaluated two structured deboarding strategies proposed in [6,14], as the authors claimed that deplaning passengers in a structured manner reduces the deboarding time. Moreover, how the passengers' seat preference influences the free boarding time is another interest of our work. By organizing participants to board in a free manner and then carrying out a poll on their seat preferences, we attempted to find the decision making mechanism behind the free boarding process. Based on the experimental results, we could further build or verify the free boarding simulation model, such as utilizing the rules of Boltzmann statistics to describe passengers' preference on seats in a free boarding process [15].

The remaining parts of the paper are organized as follows. In Section 2, a surrogate experimental test is introduced. In Section 3, we present some screenshots to identify the causes of boarding and deboarding time delay. In Section 4, we evaluate various boarding and deboarding strategies by extracting data from the video recordings, and how the seat preference affects the free boarding process is also discussed. In Section 5, conclusions and future works are given.

2. Design of a Surrogate Experimental Test

The equivalent experiment was conducted by organizing passengers to board a school bus instead of a real or mocked airplane. This makes sense because there exists similarities between passenger boarding of these two facilities for four reasons: (1) Seat layout in both facilities are similar, in which seats are symmetrically distributed on both sides of a single aisle, see Figure 2; (2) The aisles are relatively narrow in both facilities which compels the passengers to move in a following manner, that is to say, they do not try to pass others in the boarding and deplaning process; (3) Passengers suffer from aisle and seat interferences in both processes; (4) Passengers will take some time to store (retrieve) luggage in the overhead bins, which is the main cause of boarding (delighting) time delay.

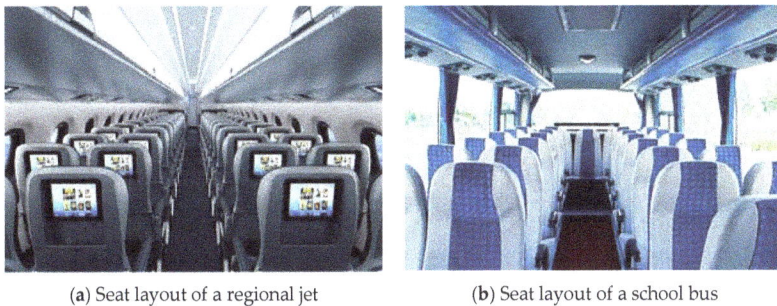

(**a**) Seat layout of a regional jet   (**b**) Seat layout of a school bus

Figure 2. Interior space of a regional jet (**a**) and a school bus (**b**).

2.1. Seat Layout and Dimensions

The maximum passenger capacity of the selected school bus was 45 passengers, and only 40 were selected in the experiment. The layout of available seats and the location of monitors are shown in Figure 3, in which ten rows of four seats are symmetrically distributed on a single, central aisle. The seats are indicated by letters from A to D, and the rows are numbered from 1 in the front to 10 in the rear of the bus.

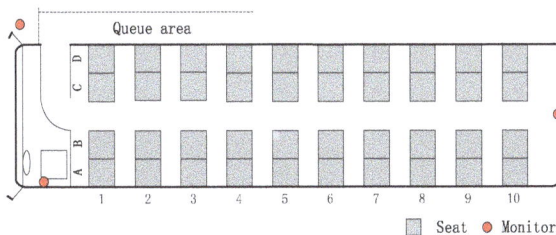

Figure 3. Seat layout of school bus.

The dimensions of interior space of the school bus is shown in Figure 4. The school bus cabin was almost 235 cm in width, 203 cm in height, and contained an affiliated overhead bin with a height of 23 cm. Each seat had a standard 45 cm width and was spaced by 71 cm from the seat in the next row. The aisle was 55 cm wide and only allowed one passenger to pass in a common situation.

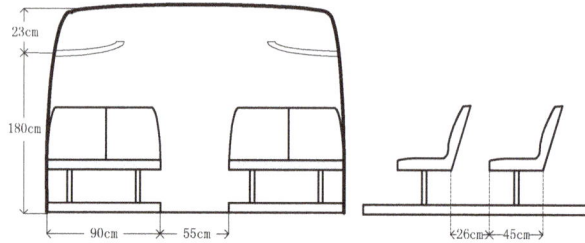

Figure 4. Dimensions of interior space of the school bus.

2.2. Characteristics of Participants

We employed 40 college students with ages ranging from 22 to 30. Male and female students participated in almost equal number. Participants were randomly assigned 0, 1, or 2 pieces of luggage, and the luggage distribution is listed in Table 1. The external dimensions of baggage was about 45 cm (height) × 30 cm (width) × 15 cm (thickness). It needs to be emphasized that the number of luggage carried by an individual remained unchanged in the repeated experiments.

Table 1. Luggage distribution.

Number of luggage pieces	0	1	2
Number of participants	6	28	6

2.3. Boarding and Deboarding Strategies

The purpose of our experiment was to evaluate the boarding strategies, including those that have been used in practice and those that have appeared in scientific literature. Seven boarding strategies and three deboarding strategies were tested and their abbreviations are listed in Table 2. Correspondingly, these strategies are transplanted to fit the bus seat layout, and their configurations are exhibited in Figure 5.

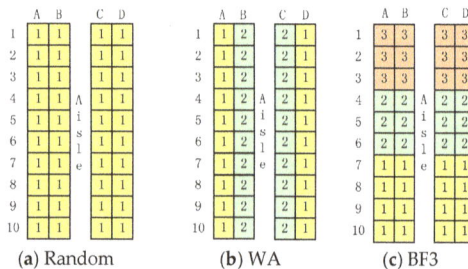

Figure 5. *Cont.*

	A	B		C	D
1	20	40		35	15
2	10	30		25	5
3	19	39		34	14
4	9	29	A	24	4
5	18	38	i	33	13
6	8	28	s	23	3
7	17	37	l	32	12
8	7	27	e	22	2
9	16	36		31	11
10	6	26		21	1

	A	B		C	D
1	20	40		30	10
2	19	39		29	9
3	18	38		28	8
4	17	37	A	27	7
5	16	36	i	26	6
6	15	35	s	25	5
7	14	34	l	24	4
8	13	33	e	23	3
9	12	32		22	2
10	11	31		21	1

(**d**) Steffen   (**e**) CRBF

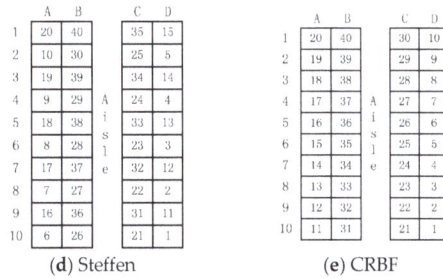

Figure 5. Five boarding strategies. WA = window to aisle manner; BF3 = back to front manner with 3 groups; CRBF = column rotated in a back to front order. (**a**) Random; (**b**) WA; (**c**) BF3; (**d**) Steffen; (**e**) CRBF.

Table 2. Abbreviations of various boarding or deboarding strategies tested.

Abbreviations	Explanation
Random	Board in a random manner
WA	Board in a window to aisle manner
BF	Board in a back to front manner
Steffen	Board according to the Steffen method
CRBF	Board in a manner of column rotated with a back to front order
Steffen-lug	An improved Steffen strategy considering luggage distribution
Free	Select seat freely when boarding
AW	Deboard in an aisle to window manner
FB	Deboard in a front to back manner
Unstructured	Deboard without any instruction

- Random strategy: All participants are allowed to board in one group; each one of them is preassigned to a particular seat and enters the bus in no predefined order (see Figure 5a).
- WA strategy: Participants are divided into two groups according to the type of seats, aisle or window seat. The group with window seats board first, followed by the group with aisle seats; within each group, participants are essentially random (see Figure 5b).
- BF strategy: Participants are divided into several groups along the aisle (three groups in our experiment, indicated by BF3) and board in a back to front order; passengers are essentially random in each group (see Figure 5c).
- Steffen strategy: This method has the passengers lining up in a prescribed order that incorporates, in a specific way, boarding from the back to the front and from the windows to the aisle. Adjacent passengers in line are sitting two rows apart from each other in corresponding seats (e.g., 10D, 8D, 6D, 4D, 2D), so there is a total of eight successive groups (see Figure 5d).
- CRBF strategy: This strategy evolves from the Steffen strategy; one major improvement is that it allows for passengers who are sitting together to be adjacent in line (e.g., 10D, 9D, 8D, 7D, 6D). As seen in Figure 5e, there are a total of four successive groups distinguished by columns. Within each group, passengers are boarded in a back to front order. Here, we name this strategy as CRBF (Column rotated in a back to front order).
- Steffen-lug strategy: This method was first proposed in Reference [5] based on the Steffen strategy. The most prominent characteristic of this method is that the seats of passengers are assigned considering the number of luggage pieces they carry. Passengers are divided into eight successive groups and board the bus in a way that is similar to the Steffen strategy; within each group, passengers with more luggage enter the bus first.

- Free boarding strategy: This strategy is adopted by some budget airlines, such as Southwest in the United States and EasyJet in the UK. Rather than providing assigned seats, they do not offer any numbered tickets. The passengers choose their favorite seats and are free to sit in any available seats when entering the bus.

As an extension of this work, we also tested three deplaning strategies. Unstructured deplaning is the most used strategy adopted by airline companies, in which passengers leave the airplane without any instruction. It has been suggested in [6,14] that structured deplaning strategies may reduce the deplaning time. To what extent these structured deboarding strategies will quicken the deplaning process was also the focus of our attention. Here, we tested two structured deboarding strategies and their rules are summarized as follows.

- AW strategy: Participants are divided into two groups, see Figure 5b. The group with aisle seat deplanes first followed by the group with the window seat, namely in an order of 2→1. Within each group, participants leave freely.
- FB strategy: Participants are divided into three groups (indicated by FB3), see Figure 5c. The groups of participants deboard in a front to back order, namely in an order of 3→2→1. Within each group, participants leave freely.

2.4. Procedure

The test of each of the above boarding or deboarding strategies was repeated two or three times. In each experiment, participants were first required to queue in a line outside of the bus gate. Randomness was guaranteed by varying the sequence order with every test, this largely reduced the probability that participants deliberately stood in the same location of the line or in the same boarding group. Then, for those strategies with preassigned seats, each participant was given a ticket with a unique seat number, for example, number '6D' meant he/she would sit in row 6, column D. The tickets were handed out to the participants according to the boarding strategy being tested. For example, tickets were randomly dispensed to the participants in the random strategy. However, for the Steffen strategy, the first participant in the waiting line was assigned a ticket with seat number '10D', the second participant with a seat number '9D', and the last participant with a seat number "1B". Finally, participants were allowed to board after the ticket assignment work was finished.

Note that the ticket check procedure on the ground was neglected in our experiments. One could suggest that adding the ticket check procedure would make the test more appropriately imitate the real process. In actuality, modern electronic technology nearly makes no delay in reading the ticket at the check desk. Moreover, it has been confirmed by [16] that each strategy has a critical ticket check time. Increasing the delay between successive airplane entries below the critical level will not increase the average boarding time. Usually, the ticket check time is below its critical time, and this makes no apparent difference in those two conditions. Therefore, we chose not to include this step in our experiment so as not to obscure the pure effect of sequencing on boarding time.

3. Time Delay in Boarding and Deboarding

3.1. Seat and Aisle Interferences in Boarding

Similar to the airplane boarding process, the main time delay inside of a school bus was caused by two kinds of interferences. Figure 6 gives a clear picture of how the seat interference occurs. The participant with window seat 6D in the red rectangular box wanted to sit after finishing storing her luggage at $t = 37$ s, but found herself blocked by an already seated participant with an aisle seat 6C in the green rectangular box. In this case, the passenger in seat 6C had to stand and let the participant with seat 6D pass. Both of the participants involved in this seat interference sat down at $t = 44$ s, and the total time to dismiss this interference was 7 s.

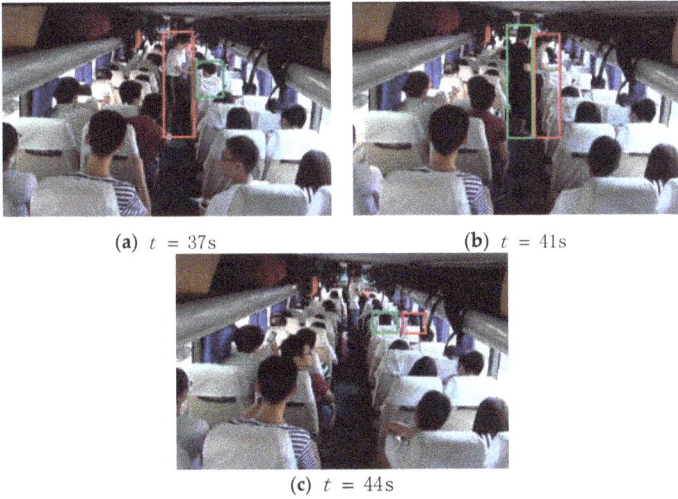

(a) $t = 37\,\text{s}$ (b) $t = 41\,\text{s}$

(c) $t = 44\,\text{s}$

Figure 6. Seat interference when boarding a bus. (a) $t = 37$ s; (b) $t = 41$ s; (c) $t = 44$ s.

Figure 7 gives an example of the formation and dissipation of the aisle interference in boarding the bus. The participant in the red rectangular box (with preassigned seat number 8B) reached his assigned row. It took some time for him to store his carry-on luggage in the overhead bin, during which time he blocked the aisle and prohibited the proceeding of the participant in the green rectangular box (with preassigned seat number 9C). Aisle interference disappeared at $t = 103$ s after the participant in the red rectangular box sat down and the participant in the green rectangular box moved along to the rear of the school bus.

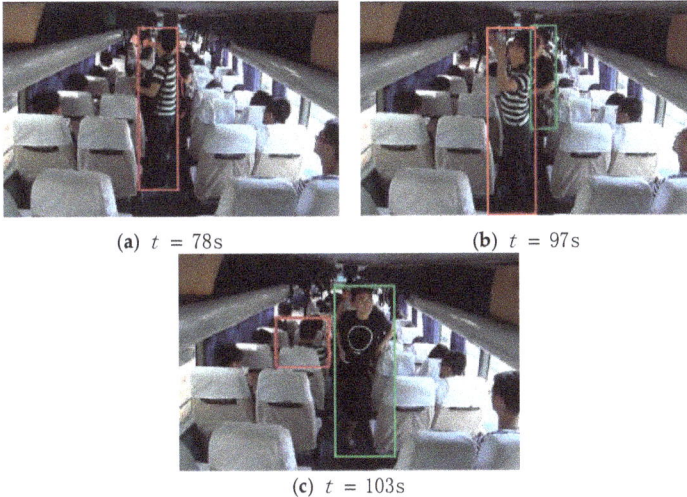

(a) $t = 78\,\text{s}$ (b) $t = 97\,\text{s}$

(c) $t = 103\,\text{s}$

Figure 7. Aisle interference when boarding a bus. (a) $t = 78$ s; (b) $t = 97$ s; (c) $t = 103$ s.

It needs to be stressed that not all seat interferences are responsible for boarding time delays, only those that turn into secondary aisle interferences are important. It makes no effect on the boarding time if the seat interference blocks nobody from advancing further in the aisle during its lifetime, only the individual boarding time of those involved in this seat interference are increased. Similarly, not all aisle interferences are equal, for example, an aisle interference that occurs in the rear of the aisle is somewhat less important compared with one that occurs in the head of the aisle which blocks several passengers.

3.2. Deboarding Time Delays

A deboarding passenger may be prevented from retrieving his luggage because another passenger in the same half row blocks his way (seat interference) or he may be blocked in the aisle and cannot advance to the gate (aisle interference). It is always the aisle interference that causes the time delay, not the seat interference. This is because the passenger in the window seat has the priority to leave first, for the sake of politeness or other reasons. Figure 8 presents the process of participants deboarding the school bus. The participant in the red rectangular box retrieved her luggage at $t = 17$ s, but she could not move ahead, because the aisle was blocked by other participants. She stood still for about 35 s before she could move forward. During this time, a number of participants inserted into the aisle, for example, the participant in the green rectangular box. Once the participant in the red box began to move at $t = 52$ s, the empty space was quickly occupied by other participants, for example, the participant in the yellow rectangular box. It can be easily seen that passengers seated in the back of the school bus will suffer from a lot of aisle interference when deboarding. Even though they can retrieve their luggage quickly after deboarding starts, they will wait for a long time to move.

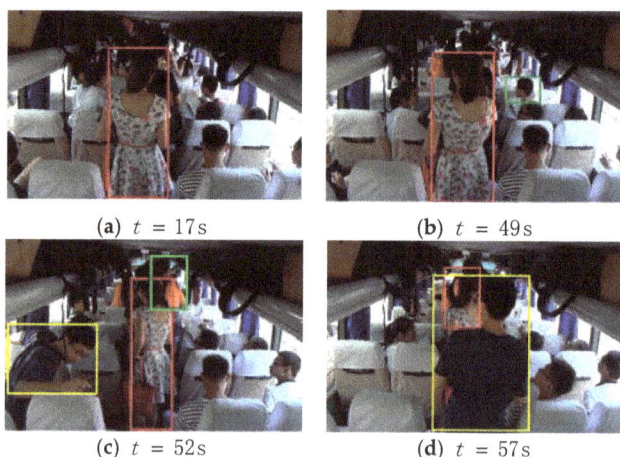

(a) $t = 17$ s   (b) $t = 49$ s

(c) $t = 52$ s   (d) $t = 57$ s

Figure 8. Deboarding in an unstructured manner. (a) $t = 17$ s; (b) $t = 49$ s; (c) $t = 52$ s; (d) $t = 57$ s.

4. Experiment Results

4.1. Boarding and Deboarding Time

The boarding and deboarding time were recorded for these seven strategies, and the results are shown in Table 3. The timing started when the first participant started to board the bus and ended when the last participant settled himself into his seat. The deboarding time was the time interval for all participants to leave the bus after the deboarding started.

According to the results in Table 3, we plot the mean boarding time of each strategy in Figure 9a, as well as their boarding time variance in Figure 9b. Here, the variance of boarding time is the time difference between the maximum and minimum boarding time in the repeated tests.

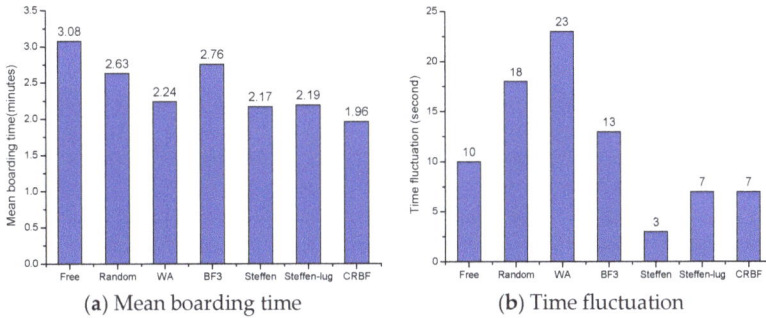

(a) Mean boarding time (b) Time fluctuation

Figure 9. Experiment results. (**a**) Mean boarding time; (**b**) Time fluctuation.

Table 3. Boarding and deboarding time.

Strategies	Test 1		Test 2		Test 3	
	Boarding Time	Deboarding Time	Boarding Time	Deboarding Time	Boarding Time	Deboarding Time
Free	3:10	1:49	3:00	1:40	3:04	1:43
Random	2:29	1:44	2:38	1:39	2:47	1:41
WA	2:26	1:33 *	2:03	1:36 *	-	-
BF3	2:39	1:39 *	2:52	1:37 *	-	-
Steffen	2:11	1:38	2:09	1:36	-	-
Steffen-lug	2:15	1:42	2:08	1:37	-	-
CRBF	1:54	1:39	2:01	1:42	-	-

* Structured deboarding strategies.

Note from Figure 9a, the three strategies that saved the most time are the Steffen, Steffen-lug, and CRBF, which agrees well with the simulation work done by [5]. A common feature of these three strategies is that they define an exact sequence of passengers and every passenger has to board at a given position that is planned in advance. They totally eliminate the seat interferences, and as much as possible aisle interferences. Moreover, they allow passengers to stow their luggage in the aisle simultaneously, for example, three and four participants are spotted storing their luggage simultaneously in Figure 10a,b, respectively. The experimental results, see Figure 9b, prove that a time effective strategy also has time stability. Indeed, boarding the passenger in an exact order makes the above three strategies tolerant of the time fluctuations induced by sequence randomness, or in other words, the stability is achieved.

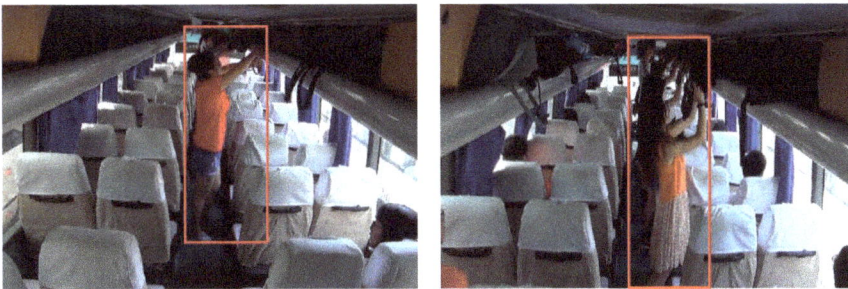

(a)Three participants are dealing with luggage (b) Four participants are dealing with luggage

Figure 10. The number of participants dealing with luggage simultaneously is three in (**a**) and four in (**b**).

It was seen in our experiments and other research, i.e., in references [3–6], that the BF strategy is not necessarily effective in reducing boarding time and is even detrimental with random boarding. It is argued in [3] that the BF strategy was ineffective because it caused local congestion in the airplane. This kind of local congestion in BF3 can be found easily in Figure 11. Taking the first boarding group for example, passengers were constrained to sit in the back of the cabin. When the participant in the red rectangular box was storing his luggage at $t = 19$ s, he blocked his fellow passengers in the same group. Such interference occurs frequently in the back of the aisle, at $t = 41$ s as another example. Due to the length of the aisle along the four rows and the size of the participants, not all of the 16 participants in the first group were able to store their luggage and sit down, some of them needed to queue. What makes this worse is the aisle and seat interference makes the queue move slowly, which triggers a chain of blocks in the aisle. Clearly, a movable bottleneck existed in the aisle and shifted to the head of the aisle during the boarding process.

Though the structured strategies are used in the deboarding process, no apparent efficiency is achieved in reducing the deboarding time, see Figure 12. This result is far from with simulation work done by [6,14], as they claimed that adopting the AW strategy could considerably reduce the deboarding time. Inconsistency comes from two main reasons. First, the effect of system size on the alighting time. The 40 participants in our experiment are much smaller than the approximately 150 passengers in a simulation model, which will largely weaken the difference between those structures. Second, but more importantly, there was nearly no time delay when passengers retrieved their luggage, because the bags carried by participants were small and portable. In such cases, participants left the bus successively, and thus mitigated the advantage of deplaning in order. In actuality, blocks occur frequently in a real deboarding process that is why the simulation model set the amount of time when retrieving luggage. Though they failed to reduce the deboarding time, these two strategies make the deboarding much more orderly. In addition, they appear fairer as they allow the groups to deplane with the first to board being the last to deboard.

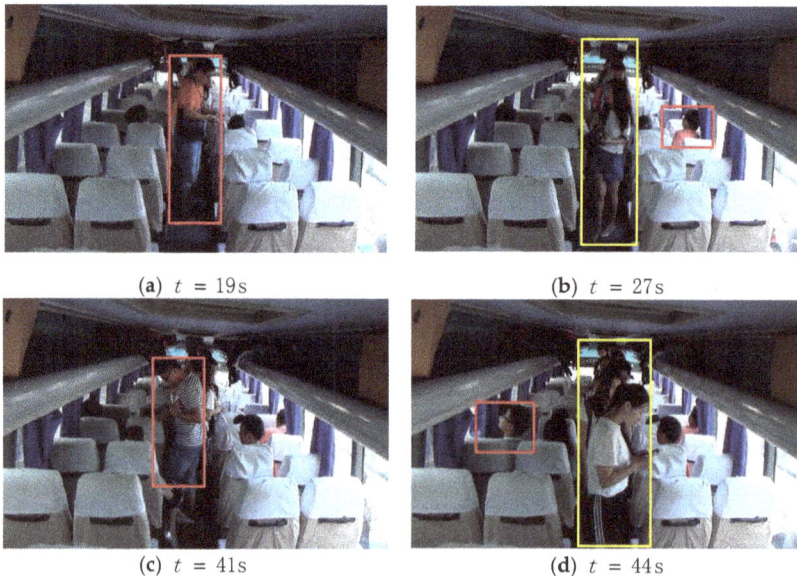

(a) $t = 19$ s

(b) $t = 27$ s

(c) $t = 41$ s

(d) $t = 44$ s

Figure 11. Phenomenon of local congestion in BF3 strategy. (**a**) $t = 19$ s; (**b**) $t = 27$ s; (**c**) $t = 41$ s; (**d**) $t = 44$ s.

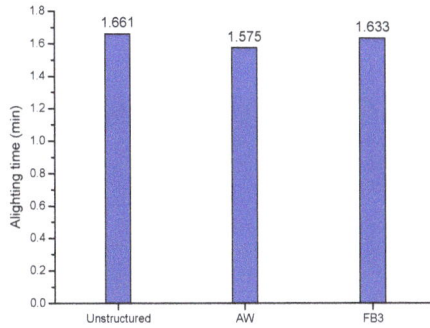

Figure 12. Deboarding time for three strategies.

Here, we want to emphasize the importance of the number of pieces of luggage, as well as their size and weight on the boarding and deboarding time. The stowing and retrieving of luggage are the main causes of time delay. If the airline allows passengers to take luggage, the differences between various boarding strategies are apparent, and those strategy that consider luggage distribution in the cabin show the most time-savings, i.e., Steffen-lug. If less or no luggage is allowed when boarding, there is no need to improve boarding strategy as they have almost equal performance.

4.2. Time Gap

In order to further test the formation and diffusion of the aisle congestion caused by seat and aisle interference, we recorded the time gap of each participant getting through the bus gate in Figure 13. The time gap is the time interval for two successive participants getting through the bus gate. A steady stream of passengers would be preferred since gaps would indicate flow stoppage. One could except that if no congestion occurs in the aisle, the distribution of the time gap should be stable. An abrupt increment of the time gap is caused by the transition of aisle congestion to the gate; the longer the time it takes a participant to go through the gate, the more severe is the congestion in the aisle.

A quick conclusion can be made that the time gap fluctuations were considerably smaller for those strategies boarding quickly, such as Steffen, Steffen-lug, and CRBF. In those strategies, participants got through the bus gate smoothly, less than 3 s per participant. In contrast, some of the participants took a long time to get through the gate, such as in the Free, Random, and BF3 strategies.

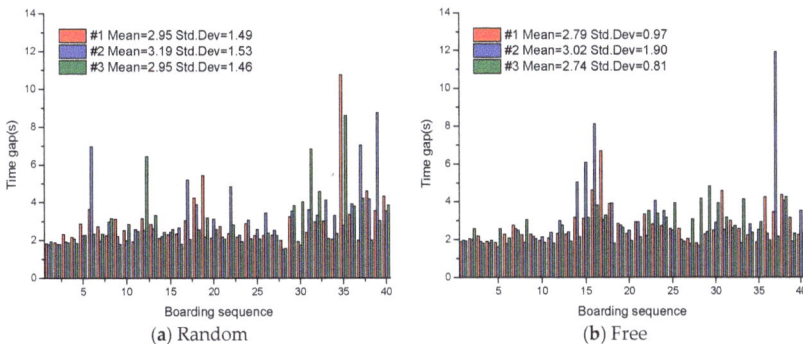

(a) Random

(b) Free

Figure 13. *Cont.*

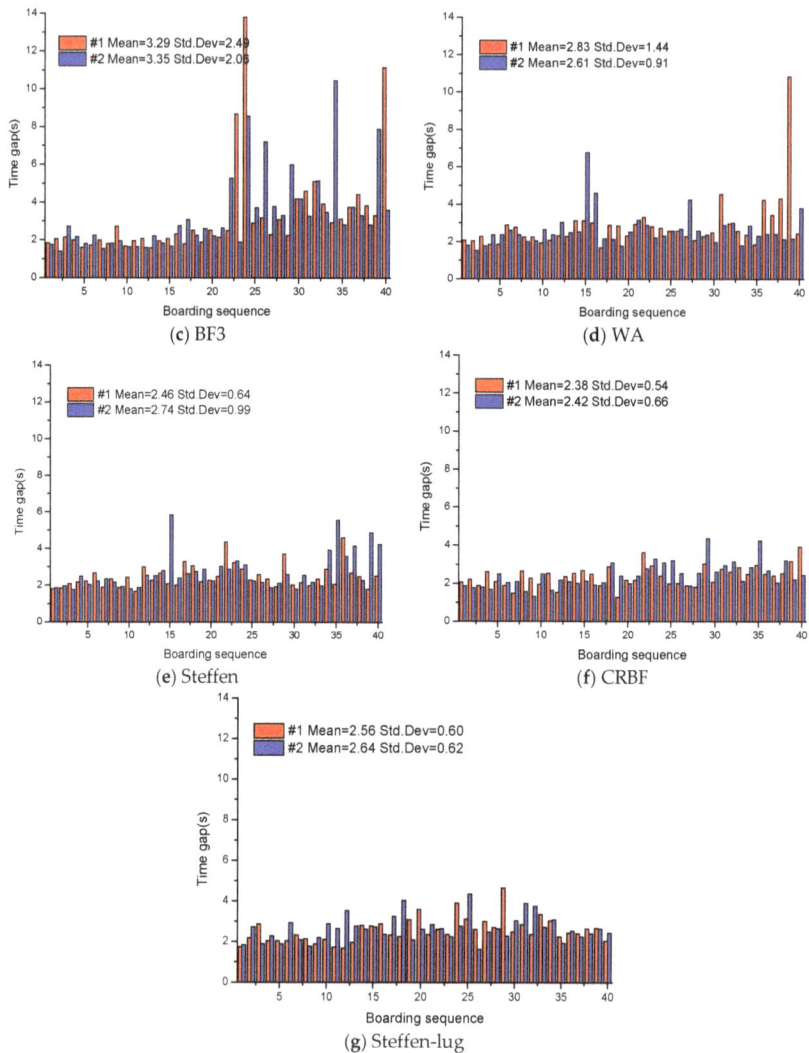

Figure 13. Time gap of getting through the bus gate. (**a**) Random; (**b**) Free; (**c**) BF3; (**d**) WA; (**e**) Steffen; (**f**) CRBF; (**g**) Steffen-lug.

4.3. Seat Preference in Free Boarding

Passengers prefer to select their favorite seats when they travel, especially in a tight and stressful environment like airplane cabins or bus carriages. Their preferences on the seats might be conditioned by a range of factors, such as distance to the entrance, the seat type (aisle or window seat), and the degree of crowding. The window seats are good for sightseeing and sleeping and less affected by seatmates. Passengers in the aisle seat would enjoy an extra benefit of stretching their legs occasionally without disturbing their seatmates. The front seats in a bus are preferred by most of the passenger because they are easy to exit from, and they avoid the noise and vibration of a rear-engine bus. The above differences could be reflected in the free boarding experiments in Figure 14, in which the number indicates the queue order in the waiting line. An intuitive conclusion confirms the preferences among different seat types; in general, participants preferred the head and window seats in boarding.

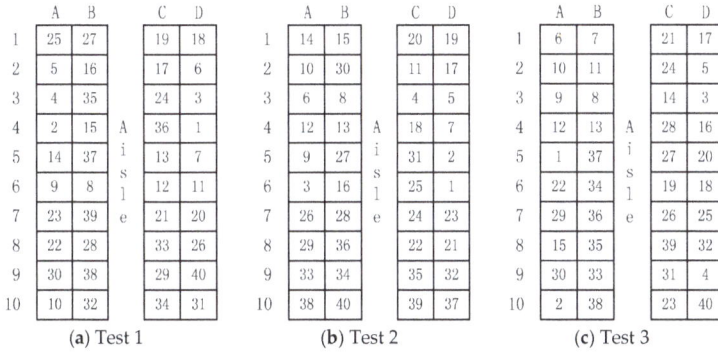

(a) Test 1

	A	B		C	D
1	25	27		19	18
2	5	16		17	6
3	4	35		24	3
4	2	15	A	36	1
5	14	37	i	13	7
6	9	8	s	12	11
7	23	39	l	21	20
8	22	28	e	33	26
9	30	38		29	40
10	10	32		34	31

(b) Test 2

	A	B		C	D
1	14	15		20	19
2	10	30		11	17
3	6	8		4	5
4	12	13	A	18	7
5	9	27	i	31	2
6	3	16	s	25	1
7	26	28	l	24	23
8	29	36	e	22	21
9	33	34		35	32
10	38	40		39	37

(c) Test 3

	A	B		C	D
1	6	7		21	17
2	10	11		24	5
3	9	8		14	3
4	12	13	A	28	16
5	1	37	i	27	20
6	22	34	s	19	18
7	29	36	l	26	25
8	15	35	e	39	32
9	30	33		31	4
10	2	38		23	40

Figure 14. Results of free boarding in three tests. (**a**) Test 1; (**b**) Test 2; (**c**) Test 3.

To further reveal the characteristics of passenger preference on seat layout, we carried out a poll of 40 respondents on their seat preferences. They were required to select their favorite three seats and the most disliked three seats in a bus seat layout as shown in Figure 2. The survey results are shown in Table 4. For simplicity, they are classed into two divisions for their 'position' features. In the first division, cabin seats are divided into head seats (from rows 1 to 3), middle seats (from rows 4 to 7) and rear seats (from rows 8 to 10); in the second division, cabin seats are divided into window seats (columns A and D) and aisle seats (columns B and C).

Table 4. Questionnaire results.

Attitude	Division 1			Division 2	
	Head	Middle	Rear	Window Seats	Aisle Seats
Like	46%	42%	12%	79.3%	12.7%
Dislike	30.7%	3.3%	66%	40.7%	59.3%

Apparently, in Table 4, the respondents prefer the head and middle seats, rather than the seats in the rear of the cabin; compared with the aisle seats, they prefer to choose the window seat. This rightly explains the reason why free boarding is time-consuming in our experiment. Participants boarding the bus at an early time will probably select the seat in the head of the bus, this will cause local congestion in the front aisle, and thereby block the subsequent passengers from boarding the bus. This is confirmed in Figure 13, the boarding interval suddenly increased. Of course, it can be seen from Figure 8 that the time function of the free strategy was less volatile, which implies a stable preference on seats.

Note that the free boarding was the worst strategy in our experimental test. This is somewhat different with its real performance in practice. In the real boarding process, passengers 'rush' to board the airplane to select their favorite seats, however, this activity was mitigated by fatigue as the passengers in our test had to board several times, thus slowing down the pace of boarding. As mentioned above, in a free boarding process, passengers are only allowed to take some handy belongings, which will greatly reduce the storage time.

5. Conclusions

The main contribution of our work is that we evaluate seven boarding and three deboarding strategies by using a surrogate experimental test. This test was conducted by having 40 participants board a school bus. Results confirm the following:

(1) The most time-saving strategies are those defining an exact sequence of the passengers when boarding, i.e., Steffen, CRBF, and Steffen-lug. They are fairly efficient because they eliminate seat interferences and, as much as possible, aisle interferences while allowing multiple passengers to stow their luggage simultaneously. In the light of this standard, the commonly used BF strategy is inefficient, even worse than the Random strategy.

(2) Those time-saving strategies are also time stable, so they benefit both airlines and airport operators to make a reliable schedule.

(3) All the strategies are sensitive to the quantity and quality of luggage taking by the passengers. If the airlines restrict the number pieces of luggage a passenger can take, as well as its weight and size, there will be no apparent difference between various strategies. If not, the strategy considering the luggage distribution is much more efficient, i.e., Steffen-lug, since it largely reduces the aisle interference between two successive boarding groups.

(4) Both the experimental tests and the questionnaire survey reveal that the free boarding process is affected by passengers' preference on the seat. This provides the opportunity to improve the free strategy by redesigning the seat size or layout in the cabin to change passengers' preference on seats, and finally reduce the boarding time.

We should also address here that this paper has the following limitations: First, the number of participants in our experiment (40) is smaller than a real airplane with about 150 passengers; second, some of the new proposed boarding strategies were not tested, such as using online seat assignment [17]. In view of the above limitations, we will in the future redesign an experiment to test various boarding or deboarding strategies by organizing a large number of participants. Moreover, as a foundation of micro modeling the free boarding process, we will further explore the passengers' seat selection behavior.

Acknowledgments: The authors would like to express their thanks to their colleague at the Institution of Transportation System Science and Engineering in Beijing Jiaotong University for their work in preparing and executing this experiment. We also wish to thank the participants who devoted their valuable time to the project. This work was supported by the National Natural Science Foundation of China (Grant Nos. 71390332, 71471012 and 71621001).

Author Contributions: The correspondence author Bin Jia conceived the experiments and provided the financial support; Shengjie Qiang and Qingxia Huang designed the experiments and analyzed the data; Shengjie Qiang wrote the paper; all authors contributed to critically revising the submitted version.

Conflicts of Interest: The authors declare no conflict of interest.

References

1. Jaehn, F.; Neumann, S. Airplane boarding. *Eur. J. Oper. Res.* **2015**, *244*, 339–359. [CrossRef]
2. Nyquist, D.C.; McFadden, K.L. A study of the airline boarding problem. *J. Air Transp. Mag. Manag.* **2008**, *14*, 197–204. [CrossRef]
3. Van Landeghem, H.; Beuselinck, A. Reducing passenger boarding time in airplanes: A simulation based approach. *Eur. J. Oper. Res.* **2002**, *142*, 294–308. [CrossRef]
4. Ferrari, P.; Nagel, K. Robustness of efficient passenger boarding strategies for airplanes. *Transp. Res. Rec. J. Transp. Res. Board* **2005**, *1915*, 44–54. [CrossRef]
5. Qiang, S.J.; Jia, B.; Xie, D.F.; Gao, Z.Y. Reducing airplane boarding time by accounting for passengers' individual properties: A simulation based on cellular automaton. *J. Air Transp. Mag. Manag.* **2014**, *40*, 42–47. [CrossRef]

6. Qiang, S.J.; Jia, B.; Jiang, R.; Huang, Q.X.; Radwan, E.; Gao, Z.Y.; Wang, Y.Q. Symmetrical design of strategy-pairs for enplaning and deplaning an airplane. *J. Air Transp. Mag. Manag.* **2016**, *54*, 52–60. [CrossRef]

7. Tang, T.Q.; Huang, H.J.; Shang, H. A new pedestrian-following model for aircraft boarding and numerical tests. *Nonlinear Dyn.* **2012**, *67*, 437–443. [CrossRef]

8. Tang, T.Q.; Wu, Y.H.; Huang, H.J.; Caccetta, L. An aircraft boarding model accounting for passengers' individual properties. *Transp. Res. C Emerg. Technol.* **2012**, *22*, 1–16. [CrossRef]

9. Steffen, J.H. Optimal boarding method for airline passengers. *J. Air Transp. Mag. Manag.* **2008**, *14*, 146–150. [CrossRef]

10. Bazargan, M. A linear programming approach for aircraft boarding strategy. *Eur. J. Oper. Res.* **2007**, *183*, 394–411. [CrossRef]

11. Soolaki, M.; Mahdavi, I.; Mahdavi-Amiri, N.; Hassanzadeh, R.; Aghajani, A. A new linear programming approach and genetic algorithm for solving airline boarding problem. *Appl. Math. Model.* **2012**, *36*, 4060–4072. [CrossRef]

12. Milne, R.J.; Kelly, A.R. A new method for boarding passengers onto an airplane. *J. Air Transp. Mag. Manag.* **2014**, *34*, 93–100. [CrossRef]

13. Steffen, J.H.; Hotchkiss, J. Experimental test of airplane boarding methods. *J. Air Transp. Mag. Manag.* **2012**, *18*, 64–67. [CrossRef]

14. Wald, A.; Harmon, M.; Klabjan, D. Structured deplaning via simulation and optimization. *J. Air Transp. Mag. Manag.* **2014**, *36*, 101–109. [CrossRef]

15. Steffen, J.H. A statistical mechanics model for free-for-all airplane passenger boarding. *Am. J. Phys.* **2008**, *76*, 1114–1119. [CrossRef]

16. Qiang, S.J.; Jia, B.; Huang, Q.X.; Gao, Z.Y. Mechanism behind phase transitions in airplane boarding process. *Int. J. Mod. Phys. C* **2016**, *27*. [CrossRef]

17. Notomista, G.; Selvaggio, M.; Sbrizzi, F.; Di, M.G.; Grazioso, S.; Botsch, M. A fast airplane boarding strategy using online seat assignment based on passenger classification. *J. Air Transp. Mag. Manag.* **2016**, *53*, 140–149. [CrossRef]

MDPI
St. Alban-Anlage 66
4052 Basel
Switzerland
Tel. +41 61 683 77 34
Fax +41 61 302 89 18
www.mdpi.com

Symmetry Editorial Office
E-mail: symmetry@mdpi.com
www.mdpi.com/journal/symmetry

www.ingramcontent.com/pod-product-compliance
Lightning Source LLC
Chambersburg PA
CBHW051852210326
41597CB00033B/5868